RENORMALIZATION METHODS

Renormalization Methods
A Guide for Beginners

W. D. McComb
School of Physics,
University of Edinburgh

CLARENDON PRESS · OXFORD
2004

OXFORD
UNIVERSITY PRESS

Great Clarendon Street, Oxford OX2 6DP

It furthers the University's objective of excellence in research, scholarship,
and education by publishing worldwide in

Oxford New York

Auckland Bangkok Buenos Aires Cape Town Chennai
Dar es Salaam Delhi Hong Kong Istanbul Karachi Kolkata
Kuala Lumpur Madrid Melbourne Mexico City Mumbai Nairobi
São Paulo Shanghai Taipei Tokyo Toronto

Oxford is a registered trade mark of Oxford University Press
in the UK and in certain other countries

Published in the United States
by Oxford University Press Inc., New York

© Oxford University Press, 2004

The moral rights of the author have been asserted

Database right Oxford University Press (maker)

First published 2004

A catalogue record for this title is available from the British Library

Library of Congress Cataloging in Publication Data

(Data available)

ISBN 0 19 850694 5

10 9 8 7 6 5 4 3 2 1

Typeset by Newgen Imaging Systems (P) Ltd., Chennai, India
Printed in Great Britain
on acid-free paper by
Biddles Ltd, www.biddles.co.uk

For Sophie, Amy and Eugene

PREFACE

The term "renormalization group" (or RG, for short) has passed into the general consciousness of many physicists, mathematicians, engineers, and other scientists. They may not know what it means; but they are nevertheless aware that an esoteric technique from the arcane world of quantum field theory has been used in a decidedly practical and successful way in the study of critical phenomena. Naturally, this awareness tends to stimulate interest in the subject and the questions that I am most frequently asked are: "Is there an easy account of the subject?" and "Is there some application to a simple toy model which I could study?"

The short answer to these questions is that such treatments do exist, but are not readily accessible to those outside a specific school of activity or discipline. The aim of the present book, therefore, is to make the basic ideas of the subject available to a wider audience.

However, it should be pointed out that the subject of renormalization is much wider than just RG (and even that comes in a variety of flavors), and embraces many different ways of taking account of collective interactions in many-body systems. With this generalization, the subject of renormalization methods should be of interest to those working on any problem which involves coupling or interactions between many degrees of freedom. As examples, quite apart from the obvious applications in microscopic physics where these methods were first developed, there are potential engineering applications in nonlinear dynamics of structures; diffusive processes, such as percolation; and fluid turbulence; along with numerous applications in applied physics, soft condensed matter, the life-sciences and even economics.

The main part of the book is divided into three broad sections:

1. What is renormalization?

2. Renormalized perturbation theories (RPT)

3. Renormalization Group (RG).

These are supplemented by a set of appendices with a view to making the book complete in itself for those readers who do not have the necessary background in statistical physics. The purpose of these appendices is to treat topics like ensemble theory in a way which is both accessible and concise, but which also covers all the essential points needed for an understanding of the main text.

Paradoxically, the first section, which comprises Chapters 1–3, is also intended to be complete in itself, without any need for reference to the appendices. Indeed this section could be subtitled "How to learn the meaning of renormalization without reading the rest of the book." It is pitched at an easy level and is aimed at both the complete beginner, who needs to be given some confidence before tackling the more demanding theory to come later; and the busy established researcher, who invariably has little time available for general study. Naturally, I hope that it will act as a "hook" for all readers, and induce them to read on. For those who do, Chapter 1 will provide overviews of, and easy introductions to, later developments.

The second section of the book deals with the topic of renormalized perturbation theory, in which some global technique (such as partial summation) is used to "fix up" a perturbation series which does not converge. At this point we draw a distinction between the terms *renormalization* and *renormalization group*.[1] The term renormalization has been borrowed from quantum field theory and has long been used in many-body physics to describe a technique for taking interactions into account. For instance, in an electron gas, the "bare" electron can be replaced by an effective electron which is said to be "dressed in the interactions with the other electrons." We can interpret the physical situation as either a physical screening of the coulomb potential, due to a cloud of electrons around any one, or as a charge renormalization of each individual electron. Exactly analogous effects are found in the hydrodyamic screening which occurs between macroscopic particles suspended in a fluid, and this concept of renormalization is applicable to virtually all systems which involve any kind of collective interaction. The problem then is to find a way of carrying out such a renormalization, and for some systems there are successful *ad hoc* methods. The only general method is by means of perturbation theory. Renormalization group, by contrast, is a method of establishing scale invariance under a set of transformations (which form a simple group, hence the terminology) and the renormalization of, for example, coupling constants occurs as part of this process. These ideas, and the distinctions between the various concepts, are explained in Chapter 1.

As a topic, RPT is currently unfashionable, when compared with RG methods, but there are three reasons for including it here and indeed for putting it first. These are as follows:

1. It comes first historically, which helps us to establish the importance of these methods in statistical physics before particle theory even existed as a discipline. This has the advantage of both providing a general treatment and also to some extent distancing the topic from quantum field theory. In this way we bear in mind the needs of the reader who does not have a background in quantum theory.

2. Developments of this approach to fluid turbulence have been very much more successful and extensive that is generally realized and this process is now being extended to problems in physics which come under the heading of soft condensed matter.

3. Many of the techniques and ideas in this section can be taken over into RG. This allows us to introduce the basic ideas of RG with a minimum of complication.

We could also add to these points, that although RPT is unable to either estimate or control its own errors (and hence the reasons for its success still remain to be understood), nevertheless it appears to offer a better prospect for the development of analytical theories than does RG.

Realistically, for most readers, the third section comprising Chapters 7–11, will be the main point of the book. Building on the introductory material of Chapters 1–3, we set the scene in Chapter 7 with a brief account of relevant topics in the theory of critical phenomena. Then in Chapter 8 we introduce the well-known treatment of RG in real space as applied to the two-dimensional Ising model. The ideas in this chapter are then used, in conjunction with the work on perturbation theory presented earlier in Chapters 1 and 4, to develop what

---

[1] To many people, the term renormalization is actually synonymous with renormalization group. To some particle theorists, whose education has been even narrower, it is synonymous with a procedure for the elimination of certain divergences in perturbation theory.

is usually called "Momentum space RG." For "non-quantum" readers, it should be noted that there is no real need to worry about the meaning of this, as we shall use the equivalent Fourier wave number space, which for many people will be a more familiar concept. This chapter may be thought of as presenting a simplified treatment of the quantum field theory for the case of a scalar field. In particular, we show how the famous ϵ-expansion can be used to obtain improved numerical values of critical exponents up to order ϵ. In Chapter 10, we show how more formal techniques from quantum field theory can be used to obtain the results of Chapter 9 more elegantly and are inherently more powerful. In the process we cover the techniques needed to take the ϵ-expansion to higher orders. In Chapter 11 we close with an account of the current application of RG to macroscopic systems and in particular to fluid turbulence.

While this book is intended to make topics from theoretical physics more accessible to the general reader who has a scientific background, it is perhaps worth pointing out that Chapters 1–4 and 7–10 are based on lecture courses in critical phenomena and statistical physics which I have given over the years to final-year MPhys and first-year postgraduate students in physics. Although the majority of such students are mathematical physicists the occasional experimental physicist has coped successfully with these courses. My view is that the book may also serve as a text-book at that level and for this reason I have put some exercises at the end of these chapters (with the exception of Chapter 10, where the available pedagogic material is rather specialized to particle theory). Solutions to these exercises, along with further exercises and their solutions, may be found at: **www.oup.com**

ACKNOWLEDGEMENTS

The idea for this book arose out of a set of lectures which I gave as a Guest Professor to an audience of fluid dynamicists and rheologists at the Technical University of Delft in the summer of 1997. The lecture course was given at the invitation of Professor F.T.M. Nieuwstadt and it is a pleasure to thank him for his encouragement, support and kindness throughout.

I also wish to thank my students Adrian Hunter, Iain Gallagher, Craig Johnston and Alistair Young for their help with the preparation of some of the figures and Khurom Kiyani for reading some chapters and pointing out errors and possible improvements.

David McComb

May 2003

Edinburgh

CONTENTS

II RENORMALIZED PERTURBATION THEORIES

NORMALIZATION OF FOURIER INTEGRALS

This is essentially a matter of where we put factors of 2π. Unfortunately the convention can be different in different fields. In this book we abide by the following conventions:

- In Chapters 5, 6, and 11 we base our analysis on eqn (5.1), thus:

$$\mathbf{u}(\mathbf{x}) = \int d^d k \, \mathbf{u}(\mathbf{k}) \exp(i\mathbf{k} \cdot \mathbf{x}).$$

- In Chapters 9 and 10, we base our analysis on eqn (9.5), thus:

$$\phi(\mathbf{x}) = \int \frac{d^d k}{(2\pi)^d} \phi(\mathbf{k}) \exp(i\mathbf{k} \cdot \mathbf{x}).$$

It should be noted that we follow a modern practice of using the same symbol for the field in both real space and wave number space. This does NOT imply that $u(x)$ is the same function as $u(k)$, although purists would be quite entitled to insist on drawing some distinction between the two symbols, such as $u(x)$ and $\hat{u}(k)$. However normally we change over from x-space to k-space at some point and then, for the most part, stay there. Hence there is no real necessity for a plethora of little hats or (worse) a completely different symbol.

NOTATION FOR CHAPTERS 1–4 AND 7–10

Latin symbols

a	lattice constant
b	length re-scaling parameter in RG
B	magnetic field
B'	molecular magnetic field
B_{E}	effective magnetic field
c	speed of light *in vaccuo*
C_{B}	heat capacity at constant magnetic field
d	dimensionality of space or lattice
D	dimensionality of spin, diffusion coefficent
e	electronic charge
E	total energy
$\bar{E}$	mean value of energy
F	Helmholtz free energy
G	Green's function; Gibbs free energy
$G(\mathbf{x}_i, \mathbf{x}_j), G(i, j), G_{ij}, G(r)$	correlation
$G_{\mathrm{c}}(i, j), G_{\mathrm{c}}(r)$	connected correlation
h	Planck's constant; $\hbar = h/2\pi$
H	total energy or Hamiltonian
J	exchange interaction energy for a pair of spins on a lattice
k	wave number in Fourier space; Boltzmann constant
$K = \beta J$	coupling constant
l_{D}	Debye length
L_x	differential operator
m	mass of a particle
M	magnetization

Latin symbols

N	number of particles in a system
p	momentum
P	thermodynamic pressure
$p(E)$	probability of state with energy E
p_i	probability of state $|i\rangle$
S_i	spin vector at lattice site i
T	absolute temperature, kinetic energy
T_c	critical temperature
V	potential, volume of system
$\mathcal{Z}$	partition function

Greek symbols

$\alpha, \beta, \gamma, \delta, \nu, \eta$	critical exponents
β	$= 1/kT$
λ	expansion parameter for perturbation theory
χ_T	isothermal susceptibility
μ	magnetic moment
ω	angular frequency
ϕ	scalar field, order parameter field
ψ	wave function
ξ	correlation length

NOTATION FOR CHAPTERS 5, 6, AND 11

Latin symbols

$D_{\alpha\beta}(\mathbf{k})$	transverse projector
E	total turbulence energy per unit mass of fluid
$E(k)$	energy spectrum
G	exact response function
$G^{(0)}$	Green's function (viscous response)
H	propagator function
k	wavenumber in Fourier space
k_{c}	cut-off wavenumber
k_{d}	Kolmogorov dissipation wavenumber
$M_{\alpha\beta\gamma}$	inertial transfer operator
$Q_{\alpha\beta}$	two-point velocity correlation tensor
$Q_{\alpha\beta\gamma}$	three-point velocity correlation tensor
R_{λ}	Taylor–Reynolds number
$T(k, t)$	energy transfer spectrum
$u_{\beta}(\mathbf{k}, t)$	Fourier component of the velocity field
$u_{\alpha}^{<}(\mathbf{k}, t)$ or $u_{\alpha}^{-}(\mathbf{k}, t)$	low-pass filtered Fourier component for RG
$u_{\alpha}^{>}(\mathbf{k}, t)$ or $u_{\alpha}^{+}(\mathbf{k}, t)$	high-pass filtered Fourier component for RG

Greek symbols

α	Kolmogorov prefactor
$\alpha, \beta, \gamma, \delta \ldots$	tensor indices each taking the values 1, 2, or 3
$\phi(k, t)$	one-dimensional projection of the energy spectrum
λ	book-keeping parameter for iterative perturbation theory
ν_0	kinematic viscosity
$\Pi(k, t)$	transport power

PART I

WHAT IS RENORMALIZATION?

1

THE BEDROCK PROBLEM: WHY WE NEED RENORMALIZATION METHODS

As we have pointed out in the Preface, the term *renormalization* comes from quantum field theory and is usually thought of as being the prerogative of that discipline. As a result, renormalization is seen by many as probably being difficult, esoteric, and highly mathematical. Yet, the basic concept is older than quantum theory of any kind and was first developed in classical physics. Probably the first true examples were the Weiss theory of ferromagnetism in 1907, followed by the Debye–Hückel theory of the screened potential in electrolytes in 1922, both of which are (as we shall see) what are known as *mean-field theories*. But, it is arguable that the earliest example of the use of renormalization, even if only in an *ad hoc* way, was the introduction of the concept of the turbulence diffusivity by Boussinesq in 1877.

In this case, the concept was applied to the diffusion of particles suspended in a fluid. If the fluid is at rest, the particles diffuse from a point source at a rate determined by molecular collisions. If the fluid is then in set into turbulent motion, the rate of diffusion of particles is augmented by the random motion of the turbulent eddies. This is something which can be easily established experimentally by putting a spot of coloring matter in a basin of water (or even possibly in one's bath). Left to itself, the colored spot will remain quite static, a time scale of (at least) hours being necessary to observe molecular diffusion. However, if one stirs the water, the colored spot can be rapidly dispersed. Boussinesq characterized this enhanced dispersion due to turbulence by an increased diffusion coefficient, or turbulence diffusivity, and in this simple example we have all the typical features of a renormalization process:

- An element of randomness (fluid turbulence is chaotic).

- Many length and/or timescales (the turbulent eddies come in a range of sizes).

- A "bare" quantity (here the molecular diffusivity) is replaced by a "renormalized" quantity in the form of an effective turbulence diffusivity.

- The renormalized quantity typically depends on some relevant "scale" (here the eddy size).

Nowadays the term "renormalization" is widely employed in many-body physics, both quantum and classical, and the associated techniques can be applied quite generally to systems with many degrees of freedom. Such systems will typically have many component parts. For instance, a nonideal gas can be modeled by particles which interact in pairs. One set of degrees of freedom is given by the translational velocities of the particles but the interesting degrees of freedom are those associated with the pairwise interactions of the particles. In this way the energy of the system is determined by its configuration. This latter aspect is the key feature of a many-body system, from our present point of view, and the following list of many-body systems is representative but not exhaustive:

1. A gas modeled as N microscopic particles in a box, where the particles interact with each other

2. A magnetic material, modeled as a lattice of miniature magnets, approaching a state of permanent magnetization

3. A turbulent fluid with eddies in a range of sizes

4. Stock-market price variations or currency exchange rates plotted as a function of time

5. Random-coiling long-chain polymers in solution

6. Anharmonic vibrations of complex mechanical structures

7. Queuing and bunching behavior in road traffic.

In all cases, we may expect some degree of *interaction* between the "microscopic" constituents of the system. This is known as *coupling* and is the defining characteristic of many-body systems. We shall constantly use the words interaction/interacting and coupling/couple: in many-body physics they are keywords.

In this book we shall present an introductory account of the development and use of renormalization methods, with the emphasis being put on simple systems and essentially classical problems. This is true even in Chapter 9, where we give an account of what quantum field theorists would consider to be ϕ^4 scalar field theory. In fact this can be regarded as a purely classical field theory or even, if we prefer, as simply an interesting equation to solve! However, it does actually have considerable physical relevance and is not only to be regarded as a toy model. Accordingly, by the end of the book we hope to have achieved some degree of completeness in presenting the general ideas, albeit in a very simple way.

1.1 Some practical matters

1.1.1 Presumed knowledge: what you need to know before starting

Some of the topics treated in this book would normally be encountered by theoretical physicists in their final undergraduate or first postgraduate year.[2] Our aim is to make them more widely available by means of simplification and careful selection. At this point it seems sensible to try to indicate the minimum basic knowledge required to follow the exposition given here.

Realistically, you will need to know at least some classical physics such as mechanics and thermodynamics up to about second-year university level. Elementary quantum mechanics is

---

[2] Other topics (especially the chapter on classical nonlinear systems and turbulence) are unlikely to be encountered at all in a physics degree.

desirable but not essential: we will explain the minimal necessary background at each point, along with giving some guidance on background reading.

As you will see, we can get along quite well by just taking some things as "given." That is, we state a particular equation, briefly explain its nature and importance, and then turn our attention to the question of how one solves it. Thus, for example, someone who has studied mechanical engineering may not have previously met the Klein–Gordon equation, but will be prepared to take it on trust. By the same token, a physicist who wants to read Chapter 5 on turbulence will probably have to take the equations of fluid motion on trust.

Turning now to the necessary mathematical skills, the simplest thing to do here is to make a list, as follows:

1. Differentiation and integration, with respect to several variables.

2. Expansion of a function in series (Binomial expansion, Taylor series, and Fourier series).

3. Solution of ordinary linear differential equations.

4. Simultaneous equations.

5. Determinants, eigenvectors and eigenvalues.

6. Fourier transforms.

Other specific techniques will be explained as and when they are required.

1.1.2 The terminology minefield

One of the difficulties of communicating across disciplines is the unthinking use of esoteric terminology or jargon. It is not easy to avoid this particular pitfall but I should warn physicists that I shall make every effort to avoid using standard physics terminology, or at least will give an explanation.

To take one crucial example, the word "body" is used in physics to mean "something possessing mass"[3] or "an amount of matter." It can refer to something large, like a planet or even a star. It can refer to something small, like an atom or even a sub-atomic particle. Indeed, many of the topics which we shall be discussing in this book belong to the subject of *many-body physics*.

However, outside physics, the colloquial sense of "body" has almost become restricted to "human body," (and for aficionados of crime fiction, if the word is used on its own, to mean a dead body).

Accordingly, we shall try to avoid it and instead use the generic term *particle*. Of course to powder technologists, a particle would be something like a grain of sand whereas to a high-energy physicist it would be something like a meson, so one cannot expect to please everyone, but in general the context should make it clear whether our "particle" is microscopic or macroscopic.

---

[3] Or, more strictly, inertia.

1.1.3   The formalism minefield

Physics employs many different mathematical formalisms, and many people would regard the path integral (and related) formalisms of quantum field theory as being among the most daunting. In turn these formalisms rest securely upon the classical formalisms of Lagrange and Hamilton. In practice, the various topics of physics are taught at different levels, ranging from elementary to the advanced, and at each stage the formalism is appropriate to that level.

It should be emphasized that the formalism employed at the more elementary levels may be simpler, less powerful or even less rigorous. But it is not wrong. Accordingly, in order to present the simplest possible treatment in this book we shall use the simplest possible formalism, even although by some standards this may be seen as unconventional.

For instance, initially we shall use the "energy," rather than the Hamiltonian as being the more familiar concept to most people. Later on in the book, when our approach has been established, we shall blur the distinction and use the symbol H for the energy, so that our exposition can be reconciled with more advanced texts. This will be helpful to those readers who wish to explore these matters further.

1.2   Quasi-particles and renormalization

In order to introduce the idea of renormalization, we begin with a very simple system. We consider the effect on one electron of many others as a specific example. Then we develop the general approach showing how elementary statistical physics has to be modified to include effects of coupling.

1.2.1   A first example of renormalization: an electron in an electrolyte

Referring to Fig. 1.1, the potential at a distance r from one electron in isolation is

$$V(r) = e/r, \qquad (1.1)$$

where e is the electron charge. This is Coulomb's law. In an electrolyte or a plasma, a cloud of charge around any one electron *screens* the Coulomb potential. According to the

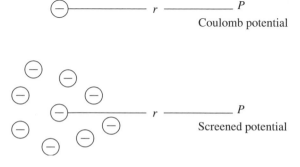

FIG. 1.1 A cloud of electrons screens the potential due to any one electron.

self-consistent theory of Debye and Hückel (see Section 3.2), the actual potential looks like:

$$V(r) = \frac{e \exp\{-r/l_D\}}{r},$$ (1.2)

where l_D is the Debye–Hückel length and depends on the number density of electrons and the absolute temperature. Note that the Debye–Hückel potential is just the Coulomb potential with the replacement

$$e \rightarrow e \times \exp\{-r/l_D\},$$ (1.3)

That is,

"bare charge" $\rightarrow$ "renormalized charge."

Note that the renormalized charge depends on the space coordinate r.

1.2.2 From micro to macro: a quantum system

Once we have decided on a microscopic picture or model we need a method of obtaining information about the macroscopic properties or behavior of a system. The ways in which we can do this make up the large and complicated subject of statistical mechanics. In Appendices A and B we give a brief outline of the basic ideas involved. Here we give an even briefer outline: just enough to understand where the difficulties can lie.

This time we shall adopt a quantum description and, in order to have a definite example, we shall consider a crystalline solid. Our microscopic model of this material will consist of N identical particles (atoms) on a three-dimensional lattice. The distance between lattice sites (the so-called "lattice constant") will be taken to be unity. Then taking the edges of the crystal to define rectangular coordinate axes, we may label each particle by its position on the lattice, starting with Particle 1 at $(0, 0, 0)$ and ending with Particle N at (L, L, L), where $L = N^{1/3} - 1$.

In a quantum mechanical description, each atom will be in a potential well and will be able to exist on some discrete energy level within that potential well. That is, the jth particle will have access to the set of energy levels $\{\epsilon_1, \epsilon_2, \ldots, \epsilon_j, \ldots\}$, where we denote the energy levels of individual particles by the symbol ϵ_j.

At any instant (see Fig. 1.2), each of the N particles will have some energy selected from the above set of permissible levels and we can write the instantaneous energy of the N-particle system as

$$E_i = \epsilon_{i_1} + \epsilon_{i_2} + \cdots + \epsilon_{i_N}.$$ (1.4)

This corresponds to the situation where

- Particle 1 has energy ϵ_{i_1}
- Particle 2 has energy ϵ_{i_2}
  $\vdots$
- Particle $N - 1$ has energy $\epsilon_{i_{N-1}}$
- Particle N has energy ϵ_{i_N}.

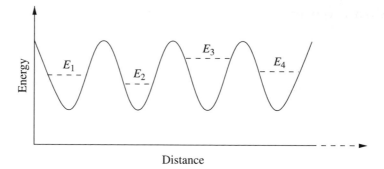

F1G. 1.2 Energy levels available to particles on a regular periodic lattice.

As time goes on, individual particles will make transitions from one level to another and accordingly the total system energy E_i will fluctuate, with each value of i corresponding to a realization defined by the set of numbers $\{i_1, i_2, \ldots, i_N\}$.

From Appendices A and B, we have the probability P of the system having some particular total energy E_i in the form:

$$P(E_i) = \frac{e^{-E_i/kT}}{\mathcal{Z}}, \tag{1.5}$$

where $\mathcal{Z}$ is given by

$$\mathcal{Z} = \sum_i e^{-E_i/kT}. \tag{1.6}$$

Obviously $\mathcal{Z}$ is the normalization of the probability distribution but it is often refered to as the partition function or "the sum over states." In this particular case, it can be written as

$$\mathcal{Z} = \sum_{i_1, i_2, \ldots, i_N} \exp[-(\epsilon_{i_1} + \epsilon_{i_2} + \cdots + \epsilon_{i_N})/kT]. \tag{1.7}$$

Now the vital point here is that this expression factorizes. That is we can write it as the product of N identical terms. Thus, using the property of the exponential

$$e^{a+b+c\ldots} = e^a \times e^b \times e^c, \ldots,$$

we can write

$$\mathcal{Z} = \sum_{i_1} \exp[-\epsilon_{i_1}/kT] \times \sum_{i_2} \exp[-\epsilon_{i_2}/kT] \times \cdots \times \sum_{i_N} \exp[-\epsilon_{i_N}/kT]$$

$$= \left\{ \sum_j \exp[-\epsilon_j/kT] \right\}^N, \tag{1.8}$$

where j stands for any of the $\{i_1, i_2, \ldots, i_N\}$.

We can further write this result in a particularly neat way by defining the single-particle partition function $\mathcal{Z}_1$ as

$$\mathcal{Z}_1 = \sum_j \exp[-E_j/kT].$$

Hence the N-particle partition function can be expressed in terms of the single-particle partition function, thus:

$$\mathcal{Z} = \mathcal{Z}_1^N.$$

This seems reasonable: if the N particles are independent of each other then we expect the probability distributions to factorize in this way. Then from Appendix B we have the bridge equation

$$F = -kT \ln \mathcal{Z} = -NkT \ln \mathcal{Z}_1, \tag{1.9}$$

the last step following from the substitution of (1.8) for $\mathcal{Z}$. At this point, the problem in statistical physics has been solved. Once we know the free energy F, then all thermodynamical properties can be obtained by differentiation using standard relationships.

The essential conclusion to take from this section is that the independence of the individual particles allowed us to work out the partition function (and hence the thermodynamical properties of the crystalline solid) in terms of the energy levels of the individual particles. In the next section we shall revert to a classical picture of a gas as an assembly of minute hard spheres and then in the following section we shall see what happens when particles are coupled together in pairs.

1.2.3 A simple model of a gas

The microscopic theory of matter essentially began in the late 19th century, with the kinetic theory of gases. In this theory, a gas is imagined as being made up of microscopic particles, which are widely separated and which are assumed to be spherical, hard, and elastic. As the actual properties of these "particles" (or molecules) were unknown, it made good sense to choose properties which would allow one to readily apply classical mechanics to the problem. However, just to be on the safe side, we refer to such a gas as an *ideal gas*.

In passing we should note that we have essentially defined the two characteristic properties of a good physical model. That is,

1. *The model should be physically plausible.* In the present case, the system, being a gas, is highly compressible and indeed can be turned into a liquid and then a solid by compression. Accordingly it is plausible to reverse this process in one's mind and imagine a gas to be small fragments of the incompressible solid which have been spread out in space. We call these fragments "molecules."

2. *The model should be soluble.* That is why we assume that the constituent particles are hard elastic spheres. This is one of the easiest cases to consider in classical mechanics. Indeed, for a dilute gas, the actual size of the molecules is unimportant and we can model them as mass "points."

The general idea is illustrated in Fig. 1.3.

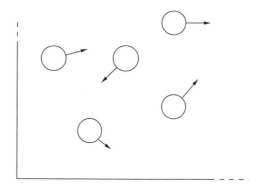

F<small>IG</small>. 1.3 Dilute gas modeled as microscopic hard spheres in random motion.

Now suppose that our ideal gas has an amount of energy E. We know that for a given volume V and number of molecules N the energy E depends on the temperature. However, we would like to obtain a relationship between the energy of the gas as a whole and the energy of its constituent molecules. This could not be easier.

Label the molecules (this is a "classical physics" picture) 1, 2, 3, etc., up to N. The first molecule has speed v_1, the second one has speed v_2, and so on. They all have the same mass m so the total energy of the ideal gas is the sum of all the kinetic energies of the molecules, thus:

$$E = \tfrac{1}{2}mv_1^2 + \tfrac{1}{2}mv_2^2 + \cdots + \tfrac{1}{2}mv_N^2 = \tfrac{1}{2}m\sum_{j=1}^{N} v_j^2. \qquad (1.10)$$

As we know from Appendices A and B, in statistical physics what we really want to know is the Helmholtz free energy F, as we can then use the standard techniques of thermodynamics to obtain virtually any other property of the gas. We discuss this in the next section. For the moment, we note that we are going from macroscopic to microscopic. In our model, the gas is made of N identical little balls. Hence from the mass M of the gas we can infer the mass m of the molecule:

$$m = \frac{M}{N}. \qquad (1.11)$$

Of course the energies of the molecules do not have to be all the same but we can work out the average value as

$$\tfrac{1}{2}m\bar{v}^2 = \frac{E}{N}. \qquad (1.12)$$

That is, if the average speed of a molecule is $\bar{v}$, then the average kinetic energy per molecule is just the energy of the gas divided by the number of molecules.

We can take the quantized results over into the classical limit (which will be valid for a dilute gas) in which the discrete fluctuations of E_i about the mean $\bar{E}$ can be replaced by a continuously fluctuating variable $E(t)$ as a function of time. Then, from (1.5) and (1.10), we can evaluate the probability of $E(t)$ as

$$p(E) \sim \exp[-E/kT] \sim \exp\left[-\tfrac{1}{2}m\sum_{j=1}^{N}v_j^2/kT\right]. \tag{1.13}$$

For a representative molecule, the probability that its velocity lies between $\mathbf{v}$ and $\mathbf{v} + d\mathbf{v}$ can be written as

$$p(v)dv \sim \exp\left[-\tfrac{1}{2}\frac{mv^2}{kT}\right]dv. \tag{1.14}$$

1.2.4 A more realistic model of a gas

If a gas has a very low density then the molecules will be widely separated on the average and their individual size may be unimportant in comparison to their mean separation. However, if we wish to consider dense gases then the finite size of molecules becomes important. Also, we need to consider whether any energy is associated with the interaction between particles. So in this section we generalize the energy to include both *kinetic energy T* and *potential energy V*, thus:

$$E = T + V. \tag{1.15}$$

In the previous section, we had

$$E = \sum_{j=1}^{N}\frac{1}{2}mv_j^2 = \sum_{j=1}^{N}T_j, \tag{1.16}$$

as the total energy of a gas of N molecules, where we now write this in terms of $T_j = \tfrac{1}{2}mv_j^2$, the kinetic energy of an individual molecule.

In order to introduce some interparticle interactions suppose that our gas is made of electrons. Then the potential energy of interaction between electron 1 and electron 2 is given by

$$V_{12} = \frac{e^2}{|\mathbf{x}_1 - \mathbf{x}_2|} = \frac{e^2}{|\mathbf{x}_2 - \mathbf{x}_1|} = V_{21}, \tag{1.17}$$

where the situation is as shown in Fig. 1.4 and e is the electronic charge. This is the well known Coulomb electrostatic potential and e is measured in appropriate units such that the prefactor is unity.

We can generalize this result to any electron labeled by n interacting with any other electron labeled by l, thus:

$$V_{nl} = \frac{e^2}{|\mathbf{x}_n - \mathbf{x}_l|}. \tag{1.18}$$

Now we wish to add up these interactions for all possible pairs of particles in order to find the total potential energy of the system. It can be helpful to think of V_{ne} as being a matrix

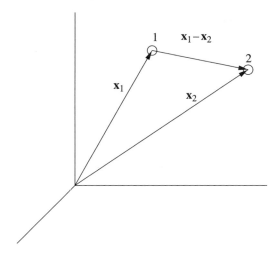

F IG. 1.4 Interaction between a pair of particles labeled 1 and 2.

element and all the possible interactions can be written as a matrix, thus:

$$\begin{pmatrix} 0 & V_{12} & V_{13} & \cdots \\ V_{21} & 0 & V_{23} & \cdots \\ V_{31} & V_{32} & 0 & \\ \vdots & \vdots & & \end{pmatrix}$$

Evidently when we add up the pair potentials we do not wish to count terms on the diagonal (which are zero), nor do we wish to count V_{21}, as well as V_{12}. These are the same pair interaction and should not be counted twice. An easy way to do the double sum is to add up EITHER all the terms above the diagonal OR all the terms beneath the diagonal. This can be achieved by writing the total potential energy of the gas as

$$V = \sum_{n<l=1}^{N} V_{nl} = \sum_{n<l=1}^{N} \frac{e^2}{|\mathbf{x}_n - \mathbf{x}_l|}, \tag{1.19}$$

for the particular case of the Coulomb potential. This notation implies a double sum over n and l. For $n = 1$, the inequality implies that we start the sum over l at $l = 2$. For $n = 2$, the sum over l starts at $l = 3$. Writing out some terms will quickly establish that we are summing the upper off-diagonal elements of V_{nl}.

In general, for any pairwise potential written as V_{nl} we have the total energy of the gas as

$$E = \sum_{j=1}^{N} T_j + \sum_{n<l=1}^{N} V_{nl}. \tag{1.20}$$

We can now see what the "many body" problem is. In the previous section we could find the probability distribution for the velocity by simply factorizing the expression for the probability of the system having an energy E. That is, we could write the exponential as the product of N exponentials. Obviously if we now write down the equivalent of (1.13), we have:

$$p(E) \sim e^{-E/kT},$$ (1.21)

and so,

$$p(E) \sim \exp\left[\sum_{j=1}^{N} T_j/kT + \sum_{n<l=1}^{N} V_{nl}/kT\right]$$

$$\sim \exp\left[\sum_{j=1}^{N} T_j/kT\right] \times \exp\left[\sum_{n<l=1}^{N} V_{nl}/kT\right].$$ (1.22)

Evidently the difficulty lies in the fact that the second factor (unlike the first) does not factorize any further. All particles in the system are coupled together in pairs.

If we pursue our idea of representing individual pair interactions as off-diagonal matrix elements, then we can regard the kinetic energy terms as the diagonal elements of a matrix and the potential energy matrix can be replaced by the energy matrix, thus:

$$\begin{pmatrix} T_{11} & V_{12} & V_{13} \cdots \\ V_{21} & T_{22} & V_{23} \cdots \\ V_{31} & V_{32} & T_{33} \\ \vdots & \vdots & \end{pmatrix}$$

In this way it is usual to speak of the first term on the right hand side of (1.21) as being *diagonal* whereas the second term is *non-diagonal*.

1.2.5 Example of a coupled system: lattice vibrations

We can give a nice example of these diagonal and non-diagonal forms by considering the lattice vibrations of solids. Let us picture a crystalline solid as being made up of atoms on a regular lattice, with each atom being free to vibrate about its mean position. The "toy model" is obtained by assuming that the individual atoms can oscillate independently of each other and this is shown schematically in Fig. 1.5. In a more realistic model, the atomic oscillations must be assumed to be coupled together as is shown in Fig. 1.6.

If the solid is heated up, then the atoms oscillate more energetically. By considering this energy of oscillation we can use the methods of statistical physics to obtain macroscopic properties such as the specific heat of the solid. We shall just look at the first steps of obtaining the partition function.

We begin by considering a solid made up of N atoms to be an array of N simple harmonic oscillations. This is the toy model. At the same time, we shall introduce the momentum of each oscillating particle as $p = mv$, which means that we can write the kinetic kinergy as

$$E = mv^2/2 = p^2/2m.$$ (1.23)

The latter is the more usual form in this type of theory.

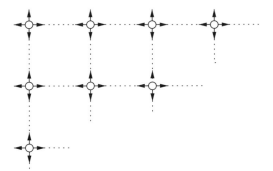

FIG. 1.5 A lattice of independent oscillators.

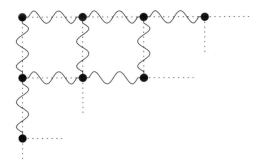

FIG. 1.6 A lattice of coupled oscillators.

Next, we change from using the position coordinate x (relative to some origin) to the coordinate q which measures the distance that an atom is from its mean position at any point in its oscillation. Then the usual expression for the energy of a harmonic oscillator may be written for the jth atom as

$$E_j(p, q) = \frac{p_j^2}{2m} + \frac{mw^2q_j^2}{2},\qquad(1.24)$$

where w is the angular frequency of the oscillation. This is the classical picture and we may obtain the total energy by simply adding up all the E_j to obtain the total energy of the system of N atoms. We note that in this example, both the kinetic energy and the potential energy are in diagonal form.

The non-diagonal problem arises when we consider the more realistic model where the oscillators are coupled. This is shown schematically in Fig. 1.6. In this case we may expect that the total energy of the oscillating system can be written as:

$$E(p, q) = \sum_j \frac{p_j^2}{2m} + \sum_{jn} A_{jn}q_jq_n,\qquad(1.25)$$

where A_{jn} is a matrix which depends on the nature of the interaction between atoms and which also takes account of inertia effects such as particle mass.

We can see the similarity between this result and the energy as given by Eqn (1.20) for the nonideal gas. However in this case a simple trick allows us to handle the awkward non-diagonal term. A standard technique can be used to diagonalize the interaction term by transforming to the *normal coordinates*.

This may be illustrated by considering the easiest nontrivial example involving coupled oscillators. See, for example [22]. Let us first consider the transverse oscillations of a mass fixed to a string under tension. If we take the displacement from equilibrium to be q, then the application of Newton's second law allows us to obtain this as a function of time and the natural frequency of oscillation is easily worked out in terms of the mass m and the string tension T.

We now extend this analysis to two masses attached to a stretched string, as illustrated in Fig. 1.7. We represent the transverse displacements from the equilibrium position by q_1 and q_2, respectively and again invoke Newton's second law to obtain equations of motion for q_1 and q_2. However, the two equations of motion are coupled together. This is obvious because the restoring force (due to string tension) experienced by one mass is affected by the position of the other mass. We shall not go into details, but essentially the two coupled equations of motion can be written as a matrix equation, in terms of a 2×2 matrix which is not diagonal.

This matrix can be diagonalized leading to two independent equations of motion for new coordinates

$$Q_A \sim q_1 + q_2$$

and

$$Q_B \sim q_1 - q_2,$$

where Q_A and Q_B correspond to the normal modes of vibration, as illustrated in Fig. 1.7. In mode A, both masses move together with frequency ω_A (the lower frequency) and in mode B they move in opposite directions with frequency ω_B (the higher frequency). Here Q_A and Q_B are known as the *normal coordinates*, while ω_A and ω_B are known as the *normal frequencies*.

This is as much as we need to know about this topic for our present purposes. Applying the same general idea to the problem of lattice vibrations, we may diagonalize the component

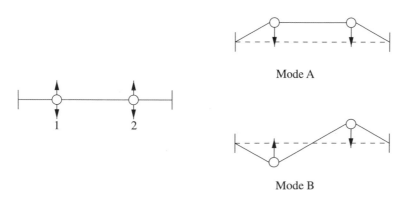

Mode A

1   2

Mode B

FIG. 1.7 Transverse oscillations of two equal masses attached to a stretched string showing the normal modes.

matrix by transforming to normal coordinates P_j (momentum) and Q_j (displacement). Then the energy as given by (1.25) can be written as

$$E_j(P, Q) = \frac{P_j^2}{2m} + \frac{mw_j^2 Q_j^2}{2},\tag{1.26}$$

where w_j are the angular frequencies of the normal modes of vibration. Note that (1.26) in terms of P_j and Q_j now takes the simple uncoupled form of (1.24). However, it should be emphasized that this example is a special case and that in general we can only hope to diagonalize the interaction terms approximately.

1.2.6 The general approach based on quasi-particles

A very powerful *approximate* method is to diagonalize E by replacing the interaction by the *overall effect* of all the other particles on the nth particle. The result can be an approximation to (1.20) in the form:

$$E = \sum_{j=1}^{N} E_j'.\tag{1.27}$$

Here each of the N particles is replaced by a *quasi-particle* and E_j' is the *effective* energy for the jth quasi-particle. Each quasi-particle has a portion of the interaction energy added on to its single-particle form. In order to describe this process, we borrow the term "renormalization" from quantum field theory. A *renormalization process* is one where we make the replacement:

$$\text{"bare" quantity} \quad + \quad \text{interactions} \quad \rightarrow \quad \text{"dressed" quantity}.$$

For example, we could consider the case of conduction electrons in a metal. In this case we have:

$$\text{effect of the lattice potential} \quad \rightarrow \quad \text{"quasi electron" with an effective mass}.$$

Or, a case which we have touched on in Section 1.2.1 and which we shall discuss in some detail later, that of electrons in a classical plasma. Here we have:

$$\text{effect of all the other electrons} \quad \rightarrow \quad \text{quasi-electron with effective charge}.$$

A general systematic *self-consistent* approach along these lines is usually known as a *mean-field theory*. In Chapter 3 we shall illustrate this approach with two examples: the Weiss theory of ferromagnetism and the Debye–Hückel theory of electrolytes.

1.2.7 Control parameters: how to weaken the coupling

Our "trick" of using normal modes to diagonalize the equations of motion for lattice vibrations is only possible when the oscillators are harmonic or linear. Any nonlinearity would couple the normal modes together and hence frustrate our attempt to diagonalize the coefficient matrix. In practice this limits the method to the case where the amplitude of oscillation is small.

This conclusion could suggest a general line of approach: we should look for circumstances where the coupling term is small. If, for example, we consider a nonideal gas, then however strong the intermolecular potential is at short distances we can expect that it will be weak at sufficiently long range. Thus, if the molecules are widely separated on average, the overall effect of the potential will be small. Hence, a sufficiently dilute gas should, in effect, be an ideal gas.

This suggests that the density is a relevant parameter which in some sense controls the strength of the coupling. However, this would not work so well with a gas of charged particles because the Coulomb potential is very long range and the potential term could only be neglected in the limit of zero density (we are pretending at this point that we do not yet know about the screening effects mentioned in Section 1.2.1).

However, a more general approach can be found by noticing that in the partition function energy always appears in the exponent divided by kT. Hence if we increase the temperature, we can weaken the effect of the interaction energy.[4]

In this context, both density and temperature are known as *control parameters*. There are others and, as we shall see later, a particularly intriguing control parameter is the number of spatial dimensions d of the system being studied, with $d = 3.999\ldots$ being especially interesting.

1.2.8 Perturbation theory

If we can weaken the coupling in some way, then this holds out the possibility of a practical approach, in which the coupling is small enough for us to regard it as a *perturbation* of the soluble, uncoupled problem. This approach, which originated in the study of planetary motions, is known as *perturbation theory*. There, the motion of (say) a planet round the sun might be partially determined by the presence of another planet. In principle this would be an example of the insoluble three-body problem. But, if the effect of the other planet was small, it could be regarded as:

three-body problem   =   two-body problem plus a perturbation.

There are several variants of perturbation theory and in the next section we shall develop the basic tools by studying its application to the solution of differential equations.

1.3   Solving the relevant differential equations by perturbation theory

We shall work up to the idea of perturbation theory by first considering the well-known method of Frobenius or the method of solution by assuming a series. In the process we encounter the idea of a *recurrence relation* which will be of great interest later on, when we encounter the *renormalization group*.

---

[4] The kinetic energy increases with increasing temperature and hence its significance is not affected by this argument.

1.3.1   Solution in series

This is a general method, but we shall illustrate it by applying it to a simple example where we already know the solution.

1.3.1.1   Example: A simple differential equation.   Consider the equation:

$$\frac{d^2 y}{dx^2} = -\omega^2 y,$$

and pretend that we do not recognize it, nor know its solution. So we try a solution of the form:

$$y(x) = x^k (a_0 + a_1 x + a_2 x^2 + \cdots) = \sum_{n=0}^{\infty} a_n x^{k+n}, \quad a_0 \neq 0.$$

Differentiate the series for y twice with respect to x to obtain, first:

$$y' = \sum_{n=0}^{\infty} a_n (k + n) x^{k+n-1};$$

and second:

$$y'' = \sum_{n=0}^{\infty} a_n (k + n)(k + n - 1) x^{k+n-2},$$

and substitute for y'' and y into the differential equation to obtain:

$$\sum_{n=0}^{\infty} a_n (k + n)(k + n - 1) x^{k+n-2} + \omega^2 \sum_{n=0}^{\infty} a_n x^{k+n} = 0.$$

Note that this equation:

- Holds for all values of x.
- Hence it is only true in general if the coefficient of each power of x vanishes.

1.3.1.2   The recurrence relation.   Let us re-write the differential equation as:

$$\sum_{j=-2}^{} C_j x^{k+j} = 0.$$

This is true for all x, if and only if, $C_j = 0$, for all values of j. We obtain the coefficients C_j by:

(1) setting $n = j + 2$ in the first series;

(2) setting $n = j$ in the second series; thus:

$$C_j = A_{j+2}(k + j + 1) + \omega^2 A_j = 0,$$

or, re-arranging,

$$A_{j+2} = -A_j \frac{\omega^2}{(k + j + 2)(k + j + 1)}. \tag{1.28}$$

This result is called a *two-term recurrence* relationship, because it connects two coefficients A_{j+2} and A_j.

Now as we started from a second-order differential equation we have two arbitrary constants which we shall take to be A_0, and A_1, where $A_0 \neq 0$. If we:

- Start from A_0, then from (1.28) we obtain the even coefficients.
  Thus $A_0 \to A_2, A_4, \ldots$.

- Start from A_1, then from (1.28) we obtain the odd coefficents.
  Thus $A_1 \to A_3, A_5, \ldots$.

- *Choose $A_1 = 0$, and we get only *even* coefficients.

1.3.1.3 Indicial equation: $k = 0$ or 1. Consider a special case: take $j = -2$. Then the coefficient $C_j = C_{-2}$ is given by:

$$C_{-2} = A_0(k - 2 + 2)(k - 2 + 1) = 0;$$

or

$$A_0 \times (k) \times (k - 1) = 0.$$

Thus we conclude that $k = 0$ or $k = 1$, because $A_0 \neq 0$. Hence we obtain two solutions corresponding to $k = 0$ and $k = 1$. These are as follows:

- *Solution for $k = 0$:*
  Solve for $A_2, A_4, \ldots$, in terms of A_0:

$$y(x) = A_0 \left[1 - \frac{(\omega x)^2}{2!} + \frac{(\omega x)^4}{4!} - \frac{(\omega x)^6}{6!} + \cdots \right] \equiv A_0 \cos \omega x.$$

- *Solution for $k = 1$:*
  Similarly

$$y(x) = A_0 \omega \left[\omega x - \frac{(\omega x)^3}{3!} + \frac{(\omega x)^5}{5!} - \frac{(\omega x)^7}{7!} + \cdots \right] = A_0 \omega \sin \omega x.$$

Thus, by ignoring the fact that we recognized the equation of simple harmonic motion, and that the usual method of solution is to assume exponential forms, we have illustrated the most general approach to solving differential equations: solution in series. In fact this is a very powerful method and can be used to solve much more difficult equations. However, as we delve into many-body problems, we shall see that matters are never as simple as this.

1.3.2   Green's functions

A great labor-saving device in mathematical physics was invented early in the nineteenth century by a Nottingham miller called Green and is now normally referred to as the *Green's function*. In practice it means that if we know the solution of a given differential equation in one set of circumstances then we know it in all circumstances. Its use in a symbolic form is absolutely crucial to modern perturbation theory. We begin by considering a simple and familiar differential equation.

1.3.2.1   Example: driven, damped harmonic oscillator.   Consider a one-dimensional harmonic system subject to a force, thus:

$$m\ddot{x} + 2\gamma\dot{x} + kx = f(t), \tag{1.29}$$

where m is the mass of particle, γ is the damping, k is the spring constant, $f(t)$ is the driving force, and the parameters m, γ, k are fixed.

1.3.2.2   Problem: given $f(t)$ find $x(t)$.   For given initial or boundary conditions, we wish to find the displacement as a function of time. Obviously we can invent as many functions $f(t)$ as we please. The driving force could be of the form sine, cos, square wave, periodic or random. In fact there is an *infinite* number of possibilities.

1.3.2.3   Homogeneous case.   If we write the differential equation without an applied force then we have:

$$m\ddot{x} + 2\gamma\dot{x} + kx = 0. \tag{1.30}$$

In these circumstances we know that there is only one solution for given initial or boundary conditions. This is known as the homogeneous case.

1.3.2.4   Inhomogeneous case.   If we add a driving force or input term on the right hand side, then we have:

$$m\ddot{x} + 2\gamma\dot{x} + kx = f(t), \tag{1.31}$$

and this is known as the inhomogeneous case. There are as many solutions as there are functions $f(t)$. That is, there is an infinite number of solutions!

1.3.2.5   Intermediate case: introducing the Green's function.   It is of interest to consider a case which is intermediate between homogeneous and inhomogeneous. This is when the equation is homogeneous except for one particular time $t = t'$. To achieve this we use a delta function input:

$$m\ddot{x} + 2\gamma\dot{x} + kx = \delta(t - t'), \tag{1.32}$$

with solution

$$x(t - t') = G(t - t'), \tag{1.33}$$

satisfying the initial or boundary conditions. Then for *any* $f(t)$ as "input" the general solution is

$$x(t) = \int_{-\infty}^{\infty} G(t - t') f(t') \, dt',$$

(1.34)

where $G(t - t')$ is known as the Green's function.

1.3.2.6 General procedure. The method of using Green's functions can be extended to any differential equation which can be written in terms of a differential operator L (say). For simplicity we shall look at a problem in one dimension: extension to any number of dimensions is easy. For variety we change the independent variable to space x, as opposed to time t. We wish to solve an equation of general form:

$$L_x \phi(x) = f(x),$$

(1.35)

where L_x is some given differential operator and $f(x)$ is a given input function. We require a form for $\phi(x)$, subject to initial or boundary conditions, so first we solve for the Green's function

$$L_x G(x - x') = \delta(x - x'),$$

(1.36)

subject to the *same* initial or boundary conditions. Then we may write the required solution as:

$$\phi(x) = \int_{-\infty}^{\infty} G(x - x') f(x') \, dx'.$$

(1.37)

1.3.2.7 Proof. Operate on both sides of equation (1.37) with L_x, thus:

$$
\begin{aligned}
L_x \phi(x) &= L_x \int_{-\infty}^{\infty} G(x - x') f(x') \, dx' \\
&= \int_{-\infty}^{\infty} \left[L_x G(x - x') \right] f(x') \, dx' \\
&= \int_{-\infty}^{\infty} \left[\delta(x - x') \right] f(x') \, dx' \\
&= f(x), \quad \text{by the sifting property of delta functions.}
\end{aligned}
$$

(1.38)

We should note that the Green's function solution takes the form of a convolution,[5]

$$\phi(x) = \int G(x - x') f(x') \, dx'. \tag{1.39}$$

or, in words,

$$\text{"output"} = \text{"system response"} * \text{"input."}$$

In other words, we interpret the *Green's function* as the *response function* of the system.

1.3.3 Example: Green's function in electrostatics

The electrostatic potential $\phi(\mathbf{x})$, due to a charge distribution $\rho(\mathbf{x})$ satisfies the Poisson equation:

$$\nabla^2 \phi(\mathbf{x}) = -\frac{1}{\epsilon_0} \rho(\mathbf{x}), \tag{1.40}$$

where $1/\epsilon_0$ is a constant of proportionality which occurs in Coulomb's law. We have tacitly taken it to be unity in Sections 1.2.1 and 1.2.4, but now we shall be more formal. We can guess the form of the solution. The volume element $d\mathbf{x}'$ centred at position $\mathbf{x}'$ contains the differential charge $dq = \rho(\mathbf{x}')d\mathbf{x}'$. The potential at position $\mathbf{x}$ due to this little bit of charge is

$$d\phi(\mathbf{x}) = \frac{dq}{4\pi \epsilon_0 |\mathbf{x} - \mathbf{x}'|} = \frac{\rho(\mathbf{x}')d\mathbf{x}'}{4\pi \epsilon_0 |\mathbf{x} - \mathbf{x}'|}. \tag{1.41}$$

By superposition (i.e. we add up the effects of all the elementary changes), we get

$$\phi(\mathbf{x}) = \int \frac{\rho(\mathbf{x}')d\mathbf{x}'}{4\pi \epsilon_0 |\mathbf{x} - \mathbf{x}'|}. \tag{1.42}$$

Equation (1.42), when compared to equation (1.39), suggests that the Green's function for Poisson's equation is

$$G(\mathbf{x} - \mathbf{x}') = \frac{1}{4\pi \epsilon_0 |\mathbf{x} - \mathbf{x}'|}. \tag{1.43}$$

This is a sensible result, since this Green's function is supposed to satisfy eqn (1.40) with a delta function charge distribution at $\mathbf{x}'$. That is, it is supposed to be the potential due to a unit point charge, which eqn (1.43) in fact is.

1.3.4 Example: Green's function for the diffusion equation

To keep things simple we stick to one space dimension: the analysis is easily extended to three. Suppose we carry out a simple experiment in which we take a pail of water and put a drop of ink at (roughly) the center of the pail. As time goes on, the ink will spread out under

---

[5] Note that a convolution is often written symbolically in terms of an asterisk. That is, $a * b$ denotes "a convolved with b."

the influence of molecular motions and the drop will become a diffuse blob. This diffusion process will continue until the ink is uniformly distributed throughout the water.

Considering the diffusion process in only one dimension labeled by x, and letting the concentration of ink molecules at time t be $n(x, t)$, we may write the diffusion equation as

$$\frac{\partial n(x, t)}{\partial t} = D\frac{\partial^2 n(x, t)}{\partial x^2} \tag{1.44}$$

where D is known as the *diffusion coefficient* or *diffusivity*, and has dimensions $L^2 T^{-1}$. We take the initial condition to be

$$n(x, 0) = n_0\delta(x), \tag{1.45}$$

where n_0 is a constant and $\delta(x)$ is the Dirac delta function, corresponding to a point source at the origin at $t = 0$.

In physics we usually assume that the physical system under consideration is so large when compared with any region of interest that it can be safely assumed to be infinite. That is, we assume that $-\infty \leq x \leq \infty$ and that the solution obeys the boundary conditions

$$n(x, t) \to 0 \quad \text{and} \quad \frac{\partial n}{\partial x} \to 0 \quad \text{as } x \to \pm\infty.$$

This allows us to introduce the Fourier transform

$$n(x, t) = \frac{1}{\sqrt{2\pi}} \int_{-\infty}^{\infty} n(k, t)e^{-ikx}\, dk, \tag{1.46}$$

where $n(k, t)$ is the Fourier transform of $n(x, t)$ and is given by the relation

$$n(k, t) = \frac{1}{\sqrt{2\pi}} \int_{-\infty}^{\infty} n(x, t)e^{ikx}\, dx. \tag{1.47}$$

It should be noted that the definitions given here for the Fourier transforms correspond to one of several possible conventions. Reference should be made to the note *Normalization of Fourier integrals* at the beginning of this book for the conventions adopted in other chapters. Also note that if $n(x, t)$ is a field in real (x) space then in modern physics it is usual to interpret $n(k, t)$ as a field in wave number (k) space. It is also modern practice to use the same symbol (here, n) for both fields, and let the argument x or k tell us which is intended.[6]

Now substitute (1.46) for $n(x, t)$ into (1.44) to obtain

$$\frac{1}{\sqrt{2\pi}} \int_{-\infty}^{\infty} \frac{\partial n(k, t)}{\partial t}e^{-ikx}\, dx = \frac{D}{\sqrt{2\pi}} \int_{-\infty}^{\infty} n(k, t)(-ik)^2 e^{-ikx}\, dx. \tag{1.48}$$

This relationship must hold for arbitrary values of e^{-ikx} and hence it implies:

$$\frac{\partial n(k, t)}{\partial t} = -Dk^2 n(k, t). \tag{1.49}$$

---

[6] Of course this does not imply that $n(x, t)$ is the same function as $n(k, t)$.

This is just an ordinary first-order differential equation, with immediate solution:

$$n(k, t) = Ae^{-Dk^2 t}, \tag{1.50}$$

where A is constant with respect to time and is determined by the initial condition, here chosen to be (1.45). Fourier transforming both sides of (1.45), gives

$$n(k, 0) = \frac{n_0}{\sqrt{2\pi}} \tag{1.51}$$

and hence $A = n_0/\sqrt{2\pi}$ and (1.50) becomes

$$n(k, t) = \frac{n_0}{\sqrt{2\pi}} e^{-Dk^2 t}. \tag{1.52}$$

Next generalize this procedure to the case where the initial condition is given by

$$n(x, 0) = f(x), \tag{1.53}$$

where $f(x)$ is some prescribed function of x. Then by Fourier transform we have

$$n(k, 0) = f(k), \tag{1.54}$$

where $f(k)$ is just the Fourier transform of $f(x)$. For this more general case, $A = f(k)$ and (1.50) becomes

$$n(k, t) = f(k)e^{-Dk^2 t} = e^{-Dk^2 t} \times f(k). \tag{1.55}$$

Fourier transforming back to x-space, and using the convolution theorem, we have

$$n(x, t) = \int_{-\infty}^{\infty} G(x - x') f(x') \, dx'. \tag{1.56}$$

It follows from the comparison of (1.55) and (1.56) that the Green's function in k-space is

$$G(k, t) = e^{-Dk^2 t}, \tag{1.57}$$

while in x-space we have

$$G(x, t) = \frac{1}{\sqrt{2\pi}} \int_{-\infty}^{\infty} \exp(-Dk^2 t) \exp(ikx) \, dk,$$

$$= \frac{1}{\sqrt{2\pi}} \int_{-\infty}^{\infty} \exp\left[-Dt\left(k^2 - \frac{ix}{Dt}k\right)\right] dk. \tag{1.58}$$

The basic trick now is to complete the square in the exponent: this is one of the most important techniques in theoretical physics.[7] We note that the exponent can be written as:

$$-Dt\left(k^2 - \frac{ix}{Dt}k\right) = -Dt\left\{\left(k - \frac{ix}{2Dt}\right)^2 - \frac{x^2}{4D^2 t^2}\right\}. \tag{1.59}$$

---

[7] It will be needed in Chapters 7 and 9.

Thus we can re-write (1.58) as:

$$G(x,t) = \frac{1}{\sqrt{2\pi}} \exp\left(-\frac{x^2}{4Dt}\right) \int_{-\infty}^{\infty} \exp\left(-Dtk'^2\right) dk', \tag{1.60}$$

where we have made the change of variable

$$k' = k - ix/2Dt. \tag{1.61}$$

Lastly, we make use of the fact that the integral in (1.60) is now a standard form[8]

$$\int_{-\infty}^{\infty} \exp\left(-Dtk'^2\right) dk' = \sqrt{\frac{\pi}{Dt}}, \tag{1.62}$$

and hence (1.60) gives us

$$G(x,t) = \sqrt{\frac{1}{2\pi Dt}} \exp\left(-\frac{x^2}{4Dt}\right). \tag{1.63}$$

This is a Gaussian function and gives the distribution of the diffusing ink after time t. The height of the Gaussian decreases with time as $1/\sqrt{2\pi Dt}$, while the width at half-height increases with time as $\sqrt{4Dt}$.

1.3.5 Simple perturbation theory: small λ

Consider an equation
$$L\phi(x) = f(x), \tag{1.64}$$

where L is some operator in the space containing the variable x. Typically in many-body or nonlinear problems the governing equation of interest can be written in such a form and typically the resulting equation cannot be solved.

However, suppose that the operator L can be written as

$$L = L_0 + \lambda L_I, \tag{1.65}$$

where λ is a small quantity and the subscript "I" stands for interactions and need not be confused with subscript 1. The operator L_0 is such that we can solve the equation

$$L_0\phi_0(x) = f(x), \tag{1.66}$$

where the resulting solution is NOT the solution of (1.64). That is,

$$\phi_0(x) \neq \phi(x).$$

---

[8] Strictly speaking we should now treat this as an integral involving a complex variable. However, a rigorous treatment shows that changes to the integration path and the limits due to the imaginary part do not affect the result.

In general, as (1.66) is soluble, we should be able to obtain a Green's function $G_0(x - x')$ such that

$$L_0 G_0(x - x') = \delta(x - x') \tag{1.67}$$

and

$$\phi_0(x) = \int dx' G_0(x - x') f(x'). \tag{1.68}$$

We now have everything we need to solve (1.64) by perturbation theory but it will be helpful to re-write (1.68) in a more symbolic way as

$$\phi_0(x) = G_0(x) f(x). \tag{1.69}$$

Strictly we should use an asterisk to denote the operation of taking a convolution, but for later simplicity we write it as a product.

Now we assume that the perturbed, or exact, solution $\phi(x)$ will take the form

$$\phi(x) = \phi_0(x) + \lambda \phi_1(x) + \lambda^2 \phi_2(x) + \cdots \tag{1.70}$$

to all orders in the expansion parameter λ. Obviously, if λ is small, we hope to be able to truncate this expansion at quite a low order.

In order to calculate the coefficients $\phi_0, \phi_1, \phi_2, \ldots$, of the λ-expansion, we first substitute (1.70) for $\phi(x)$ along with (1.65) in (1.64), thus:

$$L_0 \phi_0(x) + \lambda L_0 \phi_1(x) + \lambda^2 L_0 \phi_2(x) + \lambda L_I \phi_0(x) + \lambda^2 L_I \phi_1(x) + \mathcal{O}(\lambda^3) = f(x), \tag{1.71}$$

and then equate coefficients of each power of λ:

$$\lambda^0: L_0 \phi_0(x) = f(x); \tag{1.72}$$

$$\lambda^1: L_0 \phi_1(x) = -L_I \phi_0(x); \tag{1.73}$$

$$\lambda^2: L_0 \phi_2(x) = -L_I \phi_1(x); \tag{1.74}$$

and so on. We next calculate the coefficients iteratively, beginning with ϕ_0 which we already know, thus:

$$\phi_0(x) = G_0(x) f(x); \tag{1.75}$$

then substituting from (1.75) for $\phi_0(x)$ in (1.73),

$$\phi_1(x) = -G_0(x) L_I \phi_0(x),$$
$$= -G_0(x) L_I G_0(x) f(x); \tag{1.76}$$

then substituting from (1.76) for $\phi_1(x)$ in (1.74),

$$\phi_2(x) = -G_0(x) L_I \phi_1(x),$$
$$= G_0(x) L_I G_0(x) L_I G_0(x) f(x); \tag{1.77}$$

and so on. One may carry on indefinitely in this way and collecting together coefficients from (1.75)–(1.77) and substituting into (1.70), one sees the general solution developing as:

$$\phi(x) = G_0(x)[1 - \lambda L_I G_0(x) + \lambda^2 L_I G_0(x) L_I G_0(x) + \cdots] f(x). \qquad (1.78)$$

In most cases λ is *not* small, and later we will pay quite a lot of attention to how we should tackle such problems: essentially that is what renormalization methods are all about. Yet, even at this stage it is worth drawing attention to the regular, repetitive structure which is developing in the perturbation solution. This holds out the possibility of actually summing the series, or at least summing certain subsets of terms. (Our presentation here is very much simplified and, in practice, terms proliferate with the power of λ. In such circumstances it can be helpful to represent the terms by pictures, the best known of these being the *Feynman diagrams*.)

Lastly, it can sometimes be useful to introduce an exact or renormalized Green's function $G(x)$ by generalizing (1.69) to the form

$$\phi(x) = G(x) f(x) \qquad (1.79)$$

in which case (1.78) gives us an expression for $G(x)$, thus:

$$G(x) = G_0(x) - \lambda G_0(x) L_I G_0(x) + \lambda^2 G_0(x) L_I G_0(x) L_I G_0(x) + \cdots, \qquad (1.80)$$

For some problems, λ will be a small parameter, and in practice one can truncate the series at (typically) first or second order. In this book we are concerned with cases where λ is *not* small, but for the sake of completeness we should just mention the issue of convergence. Normally, even when λ is small, the perturbation expansion is not convergent. However, in practice one may still obtain a good approximation to $G(x)$ by taking a finite number of terms and neglecting the remainder. This is often referred to as *asymptotic convergence* and sometimes it can be established that the series is summable in some restricted sense. If, for instance the Borel sum exists, then the series is said to be "summable (B)" or "Bosel summable." Further discussion will be found in [31].

1.3.6 Example: a slightly anharmonic oscillator

A very simple example of the use of perturbation theory is obtained by considering the effect of a small nonlinear term on a linear oscillator. To begin with, we consider the usual form (as previously met in Section 1.3.1)

$$\frac{d^2 X_0}{dt^2} + \omega_0^2 X_0 = 0, \qquad (1.81)$$

with solution

$$X_0 = A \cos(\omega_0 t + \epsilon), \qquad (1.82)$$

where X_0 is the displacement at any time of a particle (say) oscillating with angular frequency ω_0 and amplitude A. The initial phase is ϵ and we take $\epsilon = 0$ for simplicity. We have introduced the subscript zero in anticipation of the fact that X_0 will be our zero-order solution.

Now, if we add a quadratic nonlinearity of the form $-\lambda X^2$, eqn (1.81) becomes

$$\frac{d^2 X}{dt^2} + \omega_0^2 X - \lambda X^2 = 0. \tag{1.83}$$

Note that as we are considering the (unrealistic) undamped oscillator, there is no need to add a forcing term to the right hand side of eqn (1.83). Instead, we re-arrange it as

$$\frac{d^2 X}{dt^2} + \omega_0^2 X = \lambda X^2, \tag{1.84}$$

and assume the trial solution

$$X(t) = X_0(t) + \lambda X_1(t) + \lambda^2 X_2(t) + \cdots, \tag{1.85}$$

so that we may employ the perturbation methods of the previous section. It is, of course, an easy matter to make the identifications

$$L_0 \equiv \frac{d^2}{dt^2} + \omega_0^2, \tag{1.86}$$

and

$$\lambda L_I = -\lambda X. \tag{1.87}$$

As the unperturbed problem is so simple and familiar, there is no need to employ the Green's function formalism (we shall reserve that for the next section) and we substitute (1.85) into (1.83) and equate coefficients of powers of λ as follows:

$$\lambda^0 : \frac{d^2 X_0}{dt^2} + \omega_0^2 X_0 = 0; \tag{1.88}$$

$$\lambda^1 : \frac{d^2 X_1}{dt^2} + \omega_0^2 X_1 = X_0^2; \tag{1.89}$$

$$\lambda^2 : \frac{d^2 X_2}{dt^2} + \omega_0^2 X_2 = 2X_0 X_1; \tag{1.90}$$

and so on.

The structure of these equations is very simple and enables us to see what is going on without too much in the way of mathematics. Equation (1.88) is just eqn (1.81) and has solution (1.82). Evidently eqn (1.89) is just a linear oscillator driven by a force X_0^2 and hence, as $X_0^2 \sim \cos^2(\omega_0 t)$, will oscillate at a frequency $2\omega_0$, as well as its natural frequency ω_0.

Similarly eqn (1.90) describes a system oscillating at ω_0, $2\omega_0$, and $3\omega_0$, and we can carry this on to any order in λ, with more and more harmonics being mixed into the solution. This phenomenon is traditionally known as *nonlinear mixing* and will turn up again later on when we discuss turbulence theory.

1.3.7 When λ is not small

In many problems we are faced with the situation that the natural expansion or coupling parameter is not less than unity and in fact may be very much larger. In these circumstances we can still follow the procedures of simple perturbation theory, and calculate the coefficients in the expansion of the solution iteratively. Of course this means that we cannot truncate the series at low order. Nevertheless it can still be worth carrying out the general procedure in the hope that one can find a way to extract useful information from the infinite series.

In such an approach, it is usual to interpret the expansion parameter λ as either a *book-keeping parameter* or a *control parameter*. The procedure adopted is still much the same although it can have a slightly *ad hoc* air to it.

Suppose we are again faced with an equation of the form (1.64) and we can again find a soluble equation like (1.66), then we define the interaction operator by

$$L_I = L - L_0. \tag{1.91}$$

Now we do not assume that L_I is in any sense small and when we write

$$L = L_0 + L_I \equiv L_0 + \lambda L_I \tag{1.92}$$

then it is clear that $\lambda = 1$. Accordingly, we can follow the methods of perturbation theory, as if for small λ, and simply set $\lambda = 1$ at the end of the calculation. Hence the solution is still given by (1.70), but now we put $\lambda = 1$ and there is no basis for truncating the expansion. The parameter λ has merely been used to keep track of terms and for this reason is known as a *book-keeping parameter*.

A variant on this approach, we shall also use later on, is to think of λ as a variable *control parameter*, such that

$$0 \leq \lambda \leq 1,$$

where

- $\lambda = 0$ corresponds to the soluble case;
- $\lambda = 1$ corresponds to the exact case.

A good general treatment of perturbation theory in classical physics problems can be found in [25], while [9] gives a good account of perturbation methods in fluid mechanics.

1.4 Quantum field theory: a first look

Quantum mechanics is the particle mechanics (analogous to Newtonian mechanics) which describes the motion of microscopic particles such as molecules, atoms or subatomic particles. In making the transition from quantum mechanics to quantum field theory we follow a route which is quite analogous to the classical case and take the limit of $N \to \infty$, where N is the number of degrees of freedom of the system.

Let us begin with the classical case and consider the specific example of the transverse oscillations of a particle with mass m attached to a (massless) string which is subject to tension. This is an example of a system with one degree of freedom and the equation of motion is just the equation of simple harmonic motion. If we add another particle, then two particles of mass m attached to the string lead to two degrees of freedom and are described by two *coupled* equations of motion (see the discussion in Section 1.2.5). Evidently we can carry on in this way for any N identical particles of mass m equispaced with the nth particle a distance x_n along a string of length L. Then as $N \to \infty$ we have

$$x_n = na \to x: \text{a continuous variable;}$$

$$Nm/L \to \mu: \text{continuous mass density per unit length.}$$

If the amplitude of vibration is denoted by $f(x, t)$, then the vibrations satisfy a wave equation

$$\frac{\partial^2 f}{\partial x^2} = \frac{1}{c^2} \frac{\partial^2 f}{\partial t^2}, \tag{1.93}$$

where the wave speed c is given by

$$c = \sqrt{F/\mu}, \tag{1.94}$$

and F is the tension in the string.

This is a quite representative example of a N-particle system going over into a continuous field as $N \to \infty$. After we have briefly discussed quantum mechanics we shall consider a simple quantum field.

1.4.1   What is quantum mechanics?

In quantum mechanics, the basic postulate is that the knowledge which we have of the position and momentum of a particle is given by its wave function $\psi(\mathbf{x}, t)$. This wave function may be interpreted as a probability amplitude such that:

probability of finding the particle between x and $dx + dx = \psi(x, t)\psi^*(x, t) \, dx$,

where the asterisk indicates "complex conjugate." The other main postulates are that the momentum and kinetic energy of the particle are obtained by operations on the wave function, respectively:

$$\mathbf{p} = -i\hbar\nabla\psi(\mathbf{x}, t) \tag{1.95}$$

and

$$E = i\hbar\frac{\partial}{\partial t}\psi(\mathbf{x}, t), \tag{1.96}$$

where $\hbar = h/2\pi$ and h stands for Planck's constant. Then for the case without interactions (i.e. a free particle) we can write the classical relationship between kinetic energy E and momentum p,

$$E = \frac{p^2}{2m}, \tag{1.97}$$

for a particle of mass m, in the quantum formalism, by operating on both sides of the identity

$$\psi = \psi, \tag{1.98}$$

to obtain

$$i\frac{\partial \psi}{\partial t} = \frac{\hbar}{2m}\nabla^2\psi, \tag{1.99}$$

which is the famous Shrödinger equation for a free particle. To take account of interactions, we may add a potential $V(\mathbf{x})$ (again treated as an operator) to obtain the general form:

$$i\frac{\partial \psi}{\partial t} = \frac{\hbar}{2m}\nabla^2\psi + V\psi. \tag{1.100}$$

This is the most usual form of the Shrödinger equation and describes the motion of a single particle under the influence of a potential V.

If we make the transition, as described in Section 1.4 for the classical case, from many single-particles to a continuous field, then the process of replacing a wave function for N particles by a continuous field is often referred to as *second quantization*.

1.4.2 A simple field theory: the Klein–Gordon equation

The simplest field equation in quantum field theory is the Klein–Gordon equation which represents a neutral spinless particle of mass m by a Hermitian scalar field ϕ. This time we start from the relativistic[9] statement connecting the energy and the momentum, viz:

$$E^2 = c^2 p^2 + m^2 c^4, \tag{1.101}$$

where c is the speed of light *in vacuo*.

Following the same procedure as in the preceding section, we can re-write this equation as:

$$\nabla^2\phi - \frac{1}{c^2}\frac{\partial^2\phi}{\partial t^2} = m^2 c^4 \phi. \tag{1.102}$$

In so-called natural units (where $\hbar = 1, c = 1$) this can be written as

$$\nabla^2\phi - \frac{\partial^2\phi}{\partial t^2} - m^2\phi = 0. \tag{1.103}$$

Furthermore, in 4-space it is usual to introduce the notation[10]

$$\partial^2 \equiv -\nabla^2 + \frac{\partial^2}{\partial t^2} \tag{1.104}$$

---

[9] In the sense of Einstein's Special Relativity.
[10] In relativity one works in four-dimensional space–time known as Minkowski space.

and so the Klein–Gordon equation can be written in the compact form:

$$(\partial^2 + m^2)\,\phi(x) = 0. \tag{1.105}$$

If we wish to include interactions then we may add a term involving a potential to the right hand side, thus

$$(\partial^2 + m^2)\,\phi(x) = -\frac{\partial V(\phi)}{\partial \phi}. \tag{1.106}$$

We shall come back to this equation in greater detail later but here we shall take a first look at a perturbative treatment of the Klein–Gordon equation with interactions.

1.4.3   The Klein–Gordon equation with interactions

Let us consider an interaction potential of the form:

$$V(\phi) = \frac{\lambda\phi^4}{4!}. \tag{1.107}$$

We shall enlarge on this choice later, when we discuss Ginsburg–Landau theory and ϕ^4 scalar field theory. For the moment we merely take it on trust that this is an interesting choice. Then, from (1.106), the Klein–Gordon equation becomes:

$$(\partial^2 + m^2)\phi = -\lambda\phi^3. \tag{1.108}$$

This equation can be solved by the perturbation methods discussed in Section 1.3.5, with the identification:

$$L_0 \equiv \partial^2 + m^2, \tag{1.109}$$

$$L_I \equiv \phi^2. \tag{1.110}$$

We can now take over eqn (1.80), but it will be necessary to restore the two-point structure to the Green's functions by writing $x = x_1 - x_2$ on the left hand side and making corresponding amendments to the right hand side. For the moment we should just bear in mind that terms on the right hand side involve convolution integrals (not shown explicitly) over intermediate variables which we shall denote as y and z such that $x_1 \le y, z \le x_2$. With this in mind, eqn (1.80) can now be re-written as

$$G(x_1 - x_2) = G_0(x_1 - x_2) - \lambda G_0(x_1 - y)L_I G_0(y - x_2)$$
$$+ \lambda^2 G_0\,(x_1 - y)\,L_I G_0(y - z)L_I G_0\,(z - x_2) - \cdots\,, \tag{1.111}$$

where this is still in a very symbolic form. However, it should be noted that the labeling variables x_1 and x_2 must balance on both sides of the equation.

The basic difficulty with this approach in quantum field theory has always been the existence of divergent integrals. In order to see how these arise, we can push the analysis forward a little more by noting that the zero-order Green's function can be written as

$$G_0(x_1 - x_2) = \langle \phi(x_1)\phi(x_2) \rangle_0, \tag{1.112}$$

where $\langle \ldots \rangle_0$ denotes the ground-state average (i.e., the average without interactions).[11] Thus L_I, as defined by (1.110), in eqn (1.111) stands for some integral over a zero-order Green's function. This means that when we later work the series out in detail, we shall find terms at first order (for example) which are of the form:

$$\text{First-order terms} = -\frac{1}{2}\lambda \int d^d y\, G_0(x_1 - y)G_0(y - y)G_0(y - x_2)$$

$$- \frac{1}{8}\lambda \left[\int d^d y\, G_0^2(y - y) \right] G_0(x_1 - x_2). \tag{1.113}$$

The divergences arise because of the single-point Green's functions $G_0(y - y) = G_0(0)$ and we shall examine this problem in the next section.

1.4.4 Infrared and ultraviolet divergences

As we have seen, in our expansions, $G_0(0)$ crops up. From Fourier transformation of (1.109) in d dimensions we have:

$$G_0(0) = \int_0^\Lambda G(k)\, dk = \int_0^\Lambda \frac{d^d k}{k^2 + m^2} = \int_0^\Lambda \frac{k^{d-1}\, dk}{k^2 + m^2}. \tag{1.114}$$

Note that we are assuming that the theory has been formulated on a lattice and so the wave number is bounded by an ultraviolet cut-off Λ, such that

$$|\mathbf{k}| \leq \Lambda \sim \pi/a,$$

where a is the lattice constant.

Also note that if the notation for the integral:

$$\int d^d k,$$

is unfamiliar, then the form with $d = 3$ should be quite familiar. So this is just a generalization to d dimensions.

1. For large Λ/m^2, the integral goes like Λ^{d-2}. In quantum field theory we expect the formulation to go over into the continuum limit $\Lambda \to \infty$, and so the integral diverges for $d > 2$. We refer to this as an *ultraviolet divergence*.

2. On the other hand, if $d \leq 2$ and $m = 0$ (zero mass), the integral diverges as $k \to 0$. This is known as an *infrared divergence*.

---

[11] Don't worry! This will be explained later on, in Chapter 10.

1.4.4.1   Example: the photon propagator.   The infra-red divergence can be cured by adding a fictitious mass, which we shall call m_{ph}, to the photon and so we have:

$$G_0(0) = \int_0^\Lambda \frac{k^{d-1}}{k^2 + m_{ph}^2}.$$

Then, at the end of the calculation, we take the limit $m_{ph}^2 \to 0$. This is an example of *regularization*.

We can also multiply by a function which depends on the cut-offs: known as a *convergence factor*.

1.4.5   Renormalized perturbation theory

We are now in a position to appreciate that there are three problems afflicting the use of perturbation theory. These are:

1. Coefficients in the expansion depend on integrals which are divergent.

2. If λ is not very small, then one cannot truncate the expansion at low order.

3. In practice, when one goes beyond the purely symbolic treatment presented here, the terms at each order of λ can be very complicated mathematically and this complication (and indeed the number of terms) increases rapidly with the power of λ.

Methods have been found of tackling all these problems, although with varying degrees of success; and to a greater or lesser extent they can be *ad hoc* in character. All such methods can be lumped together collectively under the heading of *renormalization*. Once a perturbation expansion has been fixed up in this way, we can refer to it generally as a *renormalized perturbation expansion*.

1.5   What is the renormalization group?

One of the main topics in this book is the theoretical technique known as *renormalization group*, or *RG* for short. The idea began in quantum field theory in the 1950s, where it was introduced as an *ad hoc* method of eliminating divergences. It later found its fulfillment in the 1970s, as a technique for coarse-graining the statistical description of systems with many length and time scales; and, in particular, as a method of providing a deep insight into critical phenomena. It has had great influence on the subject, leading to the development of the new topic of statistical field theory, and has undoubtedly been one of the most successful developments in many-body theory in the last half-century.

In Chapters 8–10, we shall discuss aspects of RG in the reverse of historical order, beginning with critical phenomena and ending up with the original field-theoretic version. In this section we shall introduce the general idea in a rather simple way, by discussing the para-ferromagnetic transition as a specific example. In the following chapter we will discuss the use of the technique to calculate the partition functions for the one- and two-dimensional Ising models. At that stage, RG may not appear to be very different from exact methods of solving for the partition function of the one-dimensional Ising model, as discussed in Appendix C. Nevertheless, these examples will illustrate some of the ideas that we shall need for a more general treatment in Chapter 8.

1.5.1 Magnetic models and "spin"

Traditionally the microscopic model of a magnet has consisted of an array of little (i.e. microscopic) magnets or dipoles arranged in a regular way on a lattice. The set of models will have the following characteristics:

- The lattice can have any number of dimensions. That is, letting d stand for lattice dimension, we could consider $d = 1, d = 2$, and so on, up to $d = \infty$.

- The microscopic magnet at each lattice site can be represented by a vector which may be interpreted as its magnetic moment. This vector can have any number of dimensions. That is, letting D stand for dimension of spin vector (not necessarily the same as lattice dimension d), we could consider $D = 1$, $D = 2$ and so on, up to $D = \infty$.

- It is usual to relate the microscopic magnetic moment to the concept—classical or quantum—of *spin*. That is we can imagine the magnetic moment to be due to a charged particle rotating about an axis.

- The *spin* variable will be denoted by **S** and labeled by its lattice site. For site i there will be a spin vector $\mathbf{S}_i$ and this can take any value (classical picture) or prescribed discrete values (quantum picture).

- In practice the distinction between classical and quantum may be blurred by model choice. For instance, it is convenient to have a Boolean model where the spin can only point "up" or "down." This case is known as the *Ising model* and here the spin vector dimension is $D = 1$ irrespective of the dimension d of the lattice.

In practice, a choice of model amounts to a choice of values for the lattice dimension d and the spin vector dimension D, and we shall discuss such choices in more detail in Chapter 7. However, in all models we assume that the application of an external magnetic field will tend to align the dipoles or spins such that the system acquires a net magnetic moment. At the same time, thermal fluctuations will tend to randomize the spin orientations such that there is no net magnetism. The case of particular interest to us here, is where the *interaction* between spins tends to align the spins leading to spontaneous magnetism even when there is no external magnetic field.

1.5.2 RG: the general idea

Throughout statistical physics we are concerned with the idea of reducing the amount of information necessary to describe a system. For instance, in going from a microstate to a macrostate description, we go from typically $\sim 10^{23}$ numbers (i.e. individual positions and momentum of all molecules) down to five or six (e.g. pressure, density, temperature, etc), assuming that the system is in equilibrium. The *"bridge"* between the two states is the coarse-graining operation in which the *exact distribution* is replaced by the *most probable distribution* of (say) positions and momenta of constituent particles.

It is well known in statistical physics that this coarse-graining can be done in a more progressive fashion, by the introduction of reduced distribution functions and by making the transition to the continuum limit. Here we shall consider an even more progressive method, in which we progressively "coarse grain" a microscopic description of a system by means of a series of transformations.

For each transformation, there are two steps:

1. Coarse-grain our description of the microscopic system.

2. Rescale basic variables such as lengths to try to restore the original picture.

The coarse-graining involves the earlier idea of *block spins* (or, more generally, block variables, if we are not working with spin systems). There are various ways of forming block variables: we shall mention three. In each case we begin by dividing our lattice up into blocks. For example, a square lattice in $d = 2$, could be divided into blocks of 4 spins, as shown in Fig. 1.8.

How do we find the spin value for each site in the new coarse-grained lattice? As we have said, we shall describe three methods, thus:

1. Take one spin from each block (e.g., the top right hand corner). Assign that value to the corresponding site in the "block lattice." This is called *decimation*.

2. Take the commonest value of spin in the block and assign it to the corresponding site in the "block lattice." This is known as *majority rule*. It is a good method for the Ising model. If there are equal numbers of up and down spins in any block, then in order to decide, we simply toss a coin!

3. Average over the values of the spin vector in the block: and assign the mean value of the spin to the corresponding site in the "block lattice." This is a good technique for models with spin vector $D = 2$ or higher.

If we carry out such a procedure for the square lattice in two dimensions (or $d = 2$, as sketched in Fig. 1.8) then the new lattice has to be shrunk by a factor $b = 2$ in each direction to make it similar to the original one. Thus the final lattice has $b^d = 2^2 = 4$ fewer sites than the original one. That is, we have reduced the number of degrees of freedom in the problem by a factor of 4. If there is *scale invariance* then the new lattice may be approximately similar to the old one in its properties. This can be identified mathematically if a sequence of such

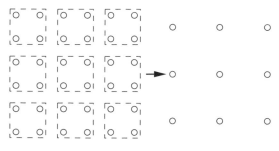

FIG. 1.8 "Blocking lattice spins."

transformations reaches a *fixed point*. Successive transformations satisfy the group closure property: hence the term *Renormalization Group*.

Lastly, we should note that we shall consider two kinds of RG:

1. *Real-space RG*: this is applied to discrete systems on a lattice in physical or real space.

2. *k-space RG*: this is essentially field-theoretic RG and is applied to continuous systems, with the analysis being done in momentum (or Fourier) space.

1.5.3   The problem of many scales

The "strong-coupling" aspect, and the consequent difficulty in doing perturbation theory, puts the study of critical phenomena in the same category as fluid dynamics. This was recognized in the 1960s with the introduction of ideas like *scaling* and *similarity*.

In the next section we shall consider a simple problem in fluid dynamics and see how the existence of only two characteristic scales can simplify our description of such a physical system. However, first we should realize that this is not the case for magnetism.

Supposing we consider a macroscopic piece of magnetic material, then we can immediately identify two length scales:

*   The macroscopic scale is the overall length in any direction of the magnet: denote this by L.

*   The microscopic scale is the distance between lattice sites: denote this by a. This is also known as the lattice constant.

However, magnetization arises as temperature-dependent fluctuations in which groups of spins become correlated (or lined up) over a distance $\xi(T)$, known as the correlation length. At very high temperatures there will be no spin alignment. As the temperature is reduced, correlated fluctuations become longer and larger, until at some temperature T_c (say), they can extend over the whole specimen. At that point (the *critical* or *Curie* temperature) the specimen can have a spontaneous magnetization.

Evidently ξ can take all possible values between the interatomic spacing and the overall size of the magnet, or:

$$a \leq \xi \leq L.$$

In general we shall not write the temperature dependence of ξ explicitly but its existence must be borne in mind.

1.5.4   Problems with a few characteristic scales

To help us to see what underlies the ideas of "scaling" and "similarity" in critical phenomena, we shall begin with a very simple example from fluid dynamics. In order to consider a problem with characteristic scales, we examine the case of flow down a pipe under the influence of an imposed pressure gradient. The situation is as illustrated schematically in Fig. 1.9. For low values of the Reynolds number, we have the well-known parabolic velocity profile of

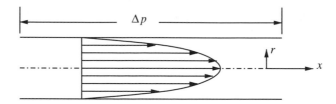

FIG. 1.9 Flow of a viscous fluid along a pipe.

laminar flow. This can be written as:

$$U(r) = \frac{\Delta p}{4\mu l}(R^2 - r^2), \tag{1.115}$$

where $U(r)$ is velocity along the pipe as a function of radial distance r, Δp is the imposed pressure drop over a length of pipe l, and R is the radius of the pipe. We can rewrite this as:

$$U(r) = U(0)\left(1 - \frac{r^2}{R^2}\right), \tag{1.116}$$

where the group in front of the bracket is now denoted by $U(0)$ and may be interpreted as the maximum value of the velocity (which is found at the centreline of the pipe).

In other words, we can now see that the laminar velocity profile scales on the maximum velocity $U(0)$ and the radius R. Evidently, different pipes, fluid viscosities and imposed pressure gradients will produce different values of $U(0)$. But the resulting *profile*:

$$\frac{U(r)}{U(0)} = \left(1 - \frac{r^2}{R^2}\right),$$

is the same for all pipes and fluids and hence is described as *universal*.

However, in critical phenomena (as in fluid turbulence), we have no characteristic scales: fluctuations exist on all scales as $T \to T_c$ and all scales are equally important. How does scaling come about in this case? We shall find the first part of the answer in the concept of geometrical similarity.

1.5.5 Geometrical similarity, fractals, and self-similarity

A square is always a square, however large or small. A 1 cm × 1 cm square can be transformed into a 1 in × 1 in square by multiplying the length of each side by 2.54 . . ... In order to transform the area we multiply it by $(2.54 \ldots)^2$.

In general, any two square areas can be connected by the transformation

$$A(\lambda L) = \lambda^2 A(L),$$

where L is the characteristic length (i.e. the length of one side) or the *length-scale* of the square and λ is the *scaling transformation* (i.e. ratio of the two length scales). Clearly, if we are given one square on one length-scale, we can generate all squares on all length-scales.

This general mathematical idea is expressed in the idea of *homogeneous functions*. A function $f(x)$ is homogeneous of degree n if for all λ it has the property

$$f(\lambda x) = \lambda^n f(x). \qquad (1.117)$$

Evidently this is just a straightforward generalization of the above equation for a simple square. We can also generalize other statements made above. For instance, we may note that a homogeneous function has the property that if we know the function at one point x_0 then we know the function everywhere:

$$f(x) = f(\lambda x_0) = \lambda^n f(x_0), \qquad (1.118)$$

where $x = \lambda x_0$. These arguments can be extended to any number of independent variables.

One way in which we can visualize the kind of self-similar structure which may arise in complicated interacting systems is by means of *fractals*. These are geometrical models which can be generated recursively.

One simple example is the *Koch curve* which can be generated from an equilateral triangle as follows. Trisect each side of the triangle and erect a small equilateral triangle on the central section, The process is illustrated in Fig. 1.10. Then trisect each of the short lines, including two sides of the small equilateral triangle and again erect an equilateral triangle on the central line segment of each trisected line. Carrying on in this way we can obtain a jagged curve with length-scales ranging from the size of the original triangle down to as small as we please.

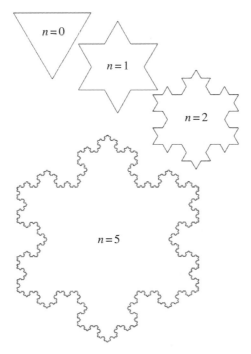

FIG. 1.10 A Koch curve as an example of a fractal curve.

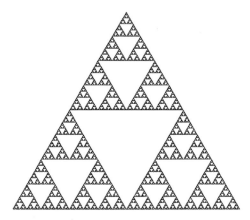

FIG. 1.11 The Sierpinski gasket: another fractal created from an equilateral triangle.

A portion of the resulting curve may, when considered in isolation, look quite random. But clearly this curve is not random as it can be produced on any length-scale by simple recursive scaling; that is, by purely geometrical means.

There are many fractals of interest in the physical sciences but we shall restrict ourselves to just one more: the *Sierpinski gasket*. This is illustrated in Fig. 1.11 and is again based on the equilateral triangle. The basic operation is to inscribe an inverted triangle into an equilateral triangle. This generates three small equilateral triangles and we now inscribe a smaller, inverted triangle into each of these. Evidently this procedure can be carried on down to any length scale, however small, although in practice its pictorial representation is limited by the resolution limits of our apparatus for drawing figures.

1.5.6 Fixed points for a ferromagnet

As we shall see later, when discussing magnetic systems in physics we normally study the Ising model, and indeed this model—in one form or another—will be our primary concern when discussing RG. However, in order to introduce the basic concept of RG, and to illustrate the physical significance of a fixed point, we shall discuss a more realistic (or Heisenberg-type) model, in which lattice spins are envisaged as little arrows which can point in any direction. For sake of simplicity, and without loss of generality, we shall restrict our attention to a lattice in two dimensions.

Consider three cases: $T = 0$, $T \to \gg T_c$, and $T \sim T_c^+$.

Case 1: $T = 0$: see Fig. 1.12
All spins are aligned: therefore RG transformation must give the same results, however much we "coarse grain" our picture.

Case 2: $T \to \infty$: see Fig. 1.13
All spins are randomly oriented: therefore RG transformation must give the same results, at every scale.

Case 3: At $T \geq T_c$: see Fig. 1.14

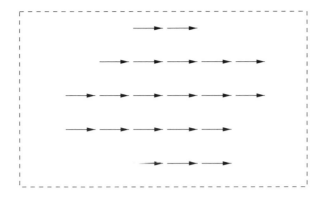

FIG. 1.12 All lattice spins aligned at zero temperature.

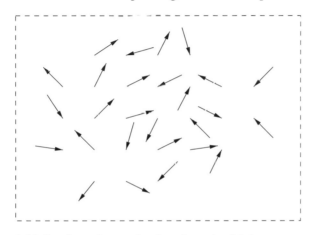

FIG. 1.13 Lattice spins randomly oriented at high temperatures.

The spins are a random sea, with islands of "correlation," each of which only exists for a brief time as a fluctuation. For a finite correlation length ξ, a finite number of RG coarse-graining transformations will "hide" ordering effects. As $\xi(T) \to \infty$ (for $T \to T_c^+$), then no finite number of RG scale changes will hide the correlations. As a result we can identify:

$$\text{fixed point} \equiv \text{critical point.}$$

To sum this up, for a system of this kind we can see that the iteration of RG transformations could reveal the presence of three different kinds of fixed point, viz:

1. Low-temperature, perfect order.

2. High-temperature, complete disorder.

3. The critical fixed point.

From the point of view of describing critical phenomena, the low-temperature and high-temperature fixed points are trivial. In Chapter 8 we shall find out how to identify and

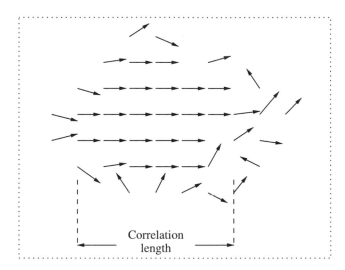

FIG. 1.14 Correlated groups of spins as the critical temperature is approached from above.

distinguish the nontrivial or critical fixed point. In the next section we shall learn a little more about fixed points by considering some relevant aspects of discrete dynamical systems.

1.5.7   Form invariance and scale invariance: RG and the partition function

We conclude this section by pointing out that RG can be discussed more formally in terms of how to calculate the partition function Z. Consider a partition function $Z(a)$ defined on a cubic lattice of spacing a, with a spin S_i at each lattice site $i = 1, 2, \ldots, N$. Decompose this lattice into blocks of length ba and perform a spin averaging within each block as discussed in Section 1.5.2. This averaging creates a new average spin S'_α with a new Hamiltonian where α labels blocks.

This procedure also creates a new partition function $Z'(ba)$ that is defined on a new lattice with spacing ba and spin S'_α at each lattice site. In order to complete the RG transformation we then rescale the new partition function $Z'(ba)$ by simply reducing the final lattice spacing from $ba \to a$.

As we shall see, in general, the energy (or, more strictly, the Hamiltonian) is changed by the RG transformation and hence in general $Z'(a)$ is not the same as $Z(a)$ with which we started. However, it should be emphasized that our goal with RG is the same as our objective in statistical physics: we wish to find the free energy F and hence, by differentiation, all the thermodynamic quantities of interest.

For this reason, we must impose the following constraints:

1. The free energy F of the system must be invariant under RG transformation.

2. By the bridge equation, it follows that the partition function Z should also be invariant under RGT.

In practice, changes to the energy or Hamiltonian under RGT mean that invariance of the partition function can only be maintained as an approximation. However, near the critical point there is the possibility that $Z'(a) = Z(a)$, as under these circumstances there is no dependence on scale.

1.6 Discrete dynamical systems: recursion relations and fixed points

In physics we are accustomed to think of dynamics as being the subject dealing with the motion of bodies under the influence of forces. We expect such motion to take the form of trajectories, in which positions vary continuously with time, and which are in accord with Newton's laws (or their relativistic generalization). Normally we expect to obtain such trajectories by solving differential equations.

However, nowadays there is also a wide class of problems coming under the heading of *discrete dynamical systems*, in which system properties vary in a series of discrete steps. In mathematics, it is said that such systems are described by the *iteration of a simple map*. This is very similar to the idea of a recursion relation and in Section 1.3.1 we saw that a differential equation could be reduced to a recursion relation.

We shall introduce this subject by means of a simple example, which illustrates the idea of a *fixed point*. Then we shall discuss the idea of a fixed point a little more formally, and conclude by discussing the behavior of a specific system with two fixed points.

1.6.1 Example: repaying a loan at a fixed rate of interest

Suppose you borrow £$X(0)$, at 1% interest per month, and repay the loan at £20 per month. After 1 month, you owe:

$$X(1) = X(0) + 0.01X(0) - 20 = 1.01X(0) - 20;$$

and after $n + 1$ months:
$$X(n + 1) = 1.01X(n) - 20. \qquad (1.119)$$

The amount you can borrow is limited by both the interest rate and the repayment rate. For example, if we arbitrarily choose three different initial amounts £$X(0)$ =£1000, £3000, and £2000, in that order, then eqn (1.119) gives us:

(a) $X(0) = £1000$, $X(1) = £990$, $X(2) = £979.90, \ldots$
(b) $X(0) = £3000$, $X(1) = £3010$, $X(2) = £3020.10, \ldots$
(c) $X(0) = £2000$, $X(1) = £2000$, $X(3) = £2000, \ldots$

In case (a), the amount you owe decreases with time, whereas for case (b) it increases with time. However, case (c) is a constant solution and the value £2000 is a *fixed point* of the system. The fixed point is also known as the *critical point* or, *the equilibrium value* or, the *constant solution*.

1.6.2   Definition of a fixed point

Consider a first-order dynamical system

$$X(n + 1) = f(X(n)). \tag{1.120}$$

A number c is a fixed point of the system if

$$X(n) = c \quad \text{for all } n,$$

when $X(0) = c$. That is, when
$$X(n) = c,$$

is a constant solution, the value c is a fixed point. We can put this in the form of a theorem: the number c is a fixed point of

$$X(n + 1) = f(X(n)) \quad \text{if and only if} \quad c = f(c). \tag{1.121}$$

For completeness, we should note that a dynamical system may have many fixed points. For example, if there is an $X^2(n)$ term, then there are two fixed points; an $X^3(n)$ term, then three fixed points; and so on. In general, the more nonlinear a system is, the more fixed points it will have.

1.6.3   Example: a dynamical system with two fixed points

Consider the dynamical system

$$X(n + 1) = [X(n) + 4]X(n) + 2. \tag{1.122}$$

It can be shown that this has two fixed points, $c = -1$ and -2. That is, for an initial value $X(0) = -2$, we find that $X(n) = -2$ for all n, and for $X(0) = -1$, we find that $X(n) = -1$ for all n. Now take different initial values $X(0) = -1.01, -0.99, -2.4$. We sketch the result in Fig. 1.15, from which it may be concluded that:

- $c = -1$ is a repelling fixed point (unstable equilibrium)

- $c = -2$ is an attractive fixed point (stable equilibrium).

It should be noted that $X(0) = -0.99$ is said to be not within the "basin of attraction" of $c = -2$.

Behavior like this will be encountered again when we discuss the general application of RG in Chapter 8.

1.7   Revision of statistical mechanics

In Appendices A and B we present a brief outline of the methods of equilibrium statistical mechanics, as restricted to the canonical ensemble. Here we give a list of useful formulae. We restrict our attention to systems in thermal equilibrium.

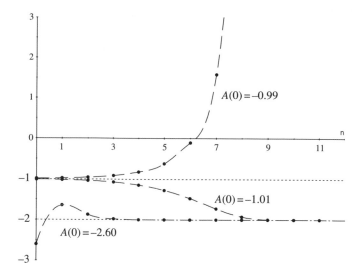

FIG. 1.15 Illustration of recursion relations leading to fixed points.

- *Assembly*: N particles (atoms, molecules, spins...) occupying a macroscopic volume V.

- *Macrostate*: We can specify our assembly in terms of (usually) a few numbers such as N particles in volume V with total energy E, at pressure P and temperature T:

$$\text{macrostate} \equiv (E, V, N, P, T). \tag{1.123}$$

- *Microstate*: We can also specify the state of our assembly at a microscopic level. This requires first a model of the microscopic structure. Then at least $6N$ numbers to specify (for a classical assembly) the velocities and positions of the constituent particles. Evidently such a description would fluctuate rapidly with time.

- *Constraints*: Any variable at the macroscopic level which is fixed is a constraint. For instance, N particles in a closed impermeable container of volume V. Both N and V are fixed and are hence constraints.

    However, from the point of view of the variational procedure used in statistical mechanics (See Appendices A and B for details) the most important constraints are those on the mean values of variables which are free to fluctuate. For instance, for an assembly in a heat-bath, the energy is free to fluctuate about a fixed mean value $\bar{E}$.

$$\text{Set of constraints} \equiv \{V, N, \bar{E}, \ldots\}.$$

- *Ensembles*

    - If an assembly is isolated (i.e. thermally) it has fixed energy E.

    - If an assembly is in energy contact with a reservoir or heat bath, it fluctuates through its energy eigenstates E_i such that its mean value $\bar{E}$ is fixed. The overbar denotes a time average.

Instead of studying one assembly as a function of time, we can look at a lot of identical assemblies at one instant of time. They will all be in different microstates as there is no physical connection between them. We can obtain their mean value $\langle E \rangle$ by summing over the ensemble of assemblies.

- *Ergodic principle*: Amounts to an assumption that $\bar{E} = \langle E \rangle$.
- *Probability distribution*: For thermal equilibrium, this will take the form

$$p_i = \frac{1}{Z} \exp\{-\beta E_i\}, \tag{1.124}$$

where E_i is the energy eigenvalue of the state $|i\rangle$ of the assembly, p_i is the probability of the assembly being in state $|i\rangle$ and $\beta \equiv 1/k_B T$. The *normalization* Z^{-1} is chosen such that

$$\sum_i p_i = 1.$$

- *Partition function*: "sum over states"

$$Z = \sum_i \exp\{-\beta E_i\}.$$

Z depends on the temperature and also on the constraints: $\{V, N, \bar{E}, \ldots\}$.

- *Bridge equation*: The Helmzholtz free energy F can be written as

$$F = -kT \ln Z \equiv -\frac{1}{\beta} \ln Z. \tag{1.125}$$

Given, Z therefore we get F and hence through thermodynamics many quantities of physical interest.

A useful form is:

$$Z = e^{-\beta F}. \tag{1.126}$$

- *Expectation values*: we can work out the average value $\langle X \rangle$ of any quantity X_i associated with the system:

$$\langle X \rangle = \sum_i p_i X_i = \frac{1}{Z} \sum_i X_i \exp\{-\beta E_i\}, \tag{1.127}$$

or

$$\langle X \rangle \equiv \frac{\sum_i X_i \exp\{-\beta E_i\}}{\sum_i \exp\{-\beta E_i\}} \tag{1.128}$$

- *Example*: the mean energy U

$$U \equiv \langle E \rangle = \frac{1}{Z} \sum_i E_i \exp\{-\beta E_i\}. \tag{1.129}$$

From the definition of Z,

$$U = -\left(\frac{\partial \log Z}{\partial \beta} \right)_V. \tag{1.130}$$

- *Heat capacity* at constant V

$$C_V \equiv \left(\frac{\partial U}{\partial T}\right)_V = k_B \beta^2 \left(\frac{\partial^2 \log Z}{\partial \beta^2}\right)_V. \tag{1.131}$$

- *Fluctuations*: for example, energy. If the assembly is not in energy isolation, its total energy can fluctuate about the mean. It is quite easy to show that:

$$\langle (E - \langle E \rangle)^2 \rangle = \left(\frac{\partial^2 \log Z}{\partial \beta^2}\right)_V = \frac{C_V}{k_B \beta^2}. \tag{1.132}$$

- *Expectation values using the density operator ρ*

A full quantum treatment replaces the probability density p_i by the density operator ρ. The expectation value of an observable X is now given by:

$$\langle X \rangle = \mathrm{tr}\langle \rho X \rangle. \tag{1.133}$$

Note 1. In some representation (energy, occupation number etc) the operator is equivalent to its matrix elements.
Note 2. Trace of an operator is independent of the representation used.

Example: For the canonical ensemble (in the energy representation)

$$\langle X \rangle = \frac{\mathrm{tr}(X e^{-\beta H})}{\mathrm{tr}(e^{-\beta H})}. \tag{1.134}$$

Note that H is the Hamiltonian, which is the operator in quantum mechanics which has the energy as an eigenvalue and $Z = \mathrm{tr}(e^{-\beta H})$.

Further reading

Good introductions to thermodynamics and statistical physics may be found in [6] and [16].

1.8 Exercises

Solutions to these exercises, along with further exercises and their solutions, will be found at: *www.oup.com*

1. Show that the number a is a fixed point of the dynamical system

$$X(n + 1) = f((X(n)),$$

if a satisfies the equation:

$$a = f(a).$$

Hence show:

(a) That $a = 3$ is the fixed point of the dynamical system

$$X(n+1) = 2X(n) - 3.$$

(b) That the dynamical system

$$X(n+1) = rX(a) + b$$

only has a fixed point at infinity if $r = 1$.

2. Obtain and verify the two fixed points of the dynamical system

$$X(n+1) = [X(n) + 4]X(n) + 2.$$

Calculate $X(n)$ to four-figure accuracy for initial values $X(0) = -1.01, -0.99$, and -2.4 taking values of n up to $n = 7$ and comment on the results.

3. The *logistic equation* of population growth takes the form

$$X(n+1) = (1+r)X(n) - bX^2(n)$$

where r is the growth rate (r = births – deaths) and $b = r/L$, where L is the maximum population which the environment can support. Obtain the two fixed-point values of the system and briefly comment on their physical significance.

4. For any system which undergoes a single-phase transition, two of the fixed points may immediately be identified as the low-temperature and high-temperature fixed points. Discuss the physical significance of these points and explain why they are attractive.

   How would your conclusions be affected if we chose to apply these arguments to the one-dimensional Ising model as a special case?

5. A system has energy levels $E_i = 0, \epsilon_1, \epsilon_2, \epsilon_3, \ldots$ with degeneracies $g_i = 1, 2, 2, 1, \ldots$. The system is in equilibrium with a thermal reservoir at temperature T, such that $e^{-\beta\epsilon_j} \ll 1$ for $j > 4$. Work out the partition function, the mean energy and the mean squared energy of the system.

   N.B. An energy level with a degeneracy of two implies two states with the same energy. The partition function is a sum over *states*, not *levels*.

6. When a particle with spin $\frac{1}{2}$ is placed in a magnetic field B, its energy level is split into $\pm\mu B$ and it has a magnetic moment $\pm\mu$ (respectively) along the direction of the magnetic field. Suppose that an assembly of N such particles on a lattice is placed in a magnetic field B and is kept at a temperature T. Find the internal energy, the entropy, the specific heat, and the total magnetic moment of this assembly.

2

EASY APPLICATIONS OF RENORMALIZATION
GROUP TO SIMPLE MODELS

In this chapter we illustrate the concept of renormalization group (RG) by applying the method to some very simple models.

We start with a one-dimensional magnet, in which the individual, microscopic magnets either point up or down but not in any other direction. In physics this is known as the one-dimensional Ising model. The problem it poses is something of a "cheat," because not only can it be solved exactly but the approximate RG method is readily converted to an exact solution! Nevertheless, it is a nice introduction to how one may calculate the partition function recursively.

For the benefit of those who are not familiar with Hamiltonian systems and partition functions, we next apply the method to a two-dimensional percolation problem. This is not only a more realistic case but also merely requires simple probabilities, rather than partition functions.

Lastly we consider the method applied to a two-dimensional version of our magnetic model (properly known as the two-dimensional Ising model). This is a more realistic case than the one-dimensional magnet and also has the salutary effect of demonstrating how complications can arise when we attempt to carry out the RG algorithm.

2.1 A one-dimensional magnet

In Section 1.2.3, we discussed the microscopic model of a gas which we referred to as an ideal gas and in Section 1.5.1 we gave a general discussion of magnetic models and their relationship to the quantum concept of "spin." Now we have our first encounter with a simple model for a ferromagnet. We shall keep the treatment very simple and return to a more comprehensive look at the subject of magnetic models in Chapter 7.

The idea of a one-dimensional magnet (or one-dimensional Ising model) is illustrated in Fig. 2.1. We can envisage a long line of N little magnets, each of which can have its north pole pointing up or down but not in any other direction. To be a shade more technical, we shall refer to these microscopic magnets as "dipoles." In later chapters, we shall employ more conventional physics usage and refer to them as spins.

Now, we can envisage two extreme situations:

1. All the dipoles are aligned in the same direction (either up or down). Then our magnet has a net magnetization.

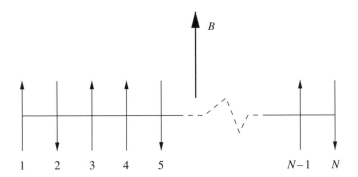

FIG. 2.1 A one-dimensional linear chain of N dipoles or spins. By convention, the arrow head corresponds to the "north pole."

2. The dipoles have a random distribution between up and down such that there is no overall magnetization.

From statistical physics we know that the probability of any actual state (including the above two extreme cases) is determined by the energy associated with the configuration and by the absolute temperature of the system. If we replace the discrete energy states E_i of eqns (1.5) and (1.6) by a continuously varying energy which we denote by H, then these equations may be written as:

$$p(H) = \frac{e^{-H/kT}}{\mathcal{Z}} \tag{2.1}$$

and

$$\mathcal{Z} = \sum_{\text{states}} e^{-H/kT}, \tag{2.2}$$

where $\mathcal{Z}$ is, as usual, the partition function.[12]

Then we follow the basic programme of statistical physics:

1. Evaluate the partition function $\mathcal{Z}$.

2. From $\mathcal{Z}$ we evaluate the free energy F, using the "bridge equation."

3. From F, we calculate the macroscopic variables such as specific heat or the magnetization.

These are our objectives and in the next section we shall put together a simple model with the required many-body properties to allow us to relate the configurational energy H to the microscopic properties of the system.

---

[12] It may of course be objected that if we change to a continuous variable, then the "sum over states" should be an integral over states. We shall simply blur such distinctions at this stage, as we are concerned with establishing the general idea.

2.1.1 The model

We have briefly discussed many-body models for a gas in Section 1.2.4 and now we consider the analogous problem for our line of dipoles as illustrated in Fig. 2.1. Let us label the individual dipoles (or lattice sites) by a number i where i ranges from 1 to N. Then, let us consider the effect of imposing a uniform magnetic field, which we denote by B in a direction of right angles to the line of spins.[13] Then at any particular site, there are two possibilities for the energy H_B of that dipole due to interaction with the applied field, thus:

$$H_B = -B \quad \text{spin up,} \tag{2.3}$$

$$H_B = B \qquad \text{spin down.} \tag{2.4}$$

We can introduce interactions between dipoles by supposing that the energy H_I, of an interacting pair is given by

$$H_I = -J \quad \text{both spins up (or down);} \tag{2.5}$$

$$H_I = J \qquad \text{spins in opposite directions.} \tag{2.6}$$

In order to simplify matters we make the further assumption:

Interactions are limited to nearest-neighbor pairs.

Note that this is still a full many-body problem, because:

- If dipole 1 interacts directly with dipole 2
- And dipole 2 interacts directly with dipole 3
- Then dipole 1 and dipole 3 interact *indirectly*.

We can summarize all this by writing the total energy H as

$$H = H_B + H_I = -B \sum_{i=1}^{N} S_i - J \sum_{\langle i,j \rangle}^{N} S_i S_j, \tag{2.7}$$

where S_i is a site variable such that:
$$S_i = \pm 1, \tag{2.8}$$

and $\langle i, j \rangle$ means that the double sum is restricted to *nearest neighbor pairs*. Note that the double sum counts each dipole pair twice as (for instance) $i = 1$, $j = 2$ and $i = 2$, $j = 1$. Hence, when we carry out the summation, we must be careful to count each pair once only. In eqn (2.13) we shall introduce a notation which avoids this ambiguity.

---

[13] It is usual in physics to represent a magnetic field by the letter H. But in theoretical treatments it is absolutely sanctified by tradition to reserve H for the energy or Hamiltonian. Some books distinguish between the two by using different letter fonts but we prefer to follow the example of others and use B for magnetic fields.

Substituting (2.7) into (2.2) we have the partition function as

$$\mathcal{Z} = \sum_{\text{states}} \exp\left[B \sum_{i=1}^{N} S_i + J \sum_{\langle i,j \rangle} S_i S_j\right] \Big/ kT. \tag{2.9}$$

As we are interested in spontaneous magnetization in the absence of an applied field we simply set $B = 0$ to obtain the partition function as

$$\mathcal{Z} = \sum_{\{S\}} \exp\left\{\frac{J}{kT} \sum_{\langle i,j \rangle} S_i S_j\right\}, \tag{2.10}$$

where the first summation stands for "sum over all possible configurations of the set of variables $\{S_i\}$ for $i = 1, 2, \ldots, N$."

Evidently the next stage is to work out the partition sum but before doing that we should take one more step and write:

$$K = J/kT, \tag{2.11}$$

Here K is known as the *coupling constant* and in future we shall often write our partition function in terms of it, as

$$\mathcal{Z} = \sum_{\{S\}} \exp\left\{K \sum_{\langle i,j \rangle} S_i S_j\right\} = \sum_{\{S\}} \prod_{\langle i,j \rangle} \exp\{K S_i S_j\}, \tag{2.12}$$

where we have again used the property of the exponential:

$$\exp\{a + b + c \cdots\} = e^a \times e^b \times e^c \times \cdots.$$

2.1.2   Coarse-graining transformation

Consider eqn (2.11) for the partition function, thus:

$$Z = \sum_{\{S\}} \prod_{\langle i,j \rangle} \exp\{K S_i S_j\} = \sum_{\{S\}} \prod_{i} \exp\{K S_i S_{i+1}\} \tag{2.13}$$

where the second step is just a different way of representing the "nearest neighbor" characteristic of the model. Re-writing in this way allows us to further write this as:

$$Z = \sum_{\{S\}} \prod_{i=2,4,6,\ldots} \exp\{K S_i (S_{i-1} + S_{i+1})\}. \tag{2.14}$$

That is, the full sum is now expressed in terms of the *sum over even spins only*.

Now do a partial sum over all the even spin states: $S_2 \pm 1$, $S_4 = \pm 1, \ldots$. In other words we put $S_i = \pm 1$ for all $i (= 2, 4, 6, \ldots)$. The result is a new partition function, corresponding to the lattice sites labeled $i = 1, 3, 5, \ldots,$. We call this new partition function $\mathcal{Z}'$ and write it as

$$Z' = \sum_{\{\cdots S_1, S_3, S_5 \cdots\}} \prod_{i=2,4,6,\ldots} [\exp\{+K(S_{i-1} + S_{i+1})\} + \exp\{-K(S_{i-1} + S_{i+1})\}]. \quad (2.15)$$

For convenience, we re-label spins on the new lattice as $i = 1, 2, 3, \ldots, N/2$. Hence

$$Z' = \sum_{\{S\}} \prod_i [\exp\{K(S_i + S_{i+1})\} + \exp\{-K(S_i + S_{i+1})\}]. \quad (2.16)$$

Now if we insist on invariance of Z under the renormalization group transformation, as discussed in Section 1.7, we require Z' to take form:

$$Z' = \sum_{\{S\}} \prod_i f(K) \exp\{K'S_i S_{i+1}\}. \quad (2.17)$$

In other words, this is just like we started out with, but now K has been replaced by K' and there are $N/2$ rather than N lattice sites. Equating the two forms of the partition function, $\mathcal{Z} = \mathcal{Z}'$, yields:

$$\exp\{K(S_i + S_{i+1})\} + \exp\{-K(S_i + S_{i+1})\} = f(K) \exp\{K'S_i S_{i+1}\}. \quad (2.18)$$

We now wish to find a relationship between the new coupling constant K' and the original one K. We also want to identify the form taken by the function $f(K)$.

2.1.3 Renormalization of the coupling constant

Let us consider two specific cases:

Case 1: $S_i = S_{i+1} = \pm 1$
Equation (2.18) becomes:

$$\exp\{2K\} + \exp\{-2K\} = f(K) \exp\{K'\}. \quad (2.19)$$

Case 2: $S_i = -S_{i+1} = \pm 1$
Equation (2.18) becomes:

$$2 = f(K)e^{-K'} \quad (2.20)$$

Equations (2.19) and (2.20) can be written as:

$$f(K)e^{K'} = 2\cosh 2K, \quad (2.21)$$

and

$$f(K)e^{-K'} = 2, \quad (2.22)$$

respectively, where we have changed to exponential notation.

Now to obtain a solution, we divide (2.22) into (2.21) with the result:

$$e^{2K'} = \cosh 2K. \tag{2.23}$$

Or, taking logs of both sides, and dividing across by 2:

$$K' = \tfrac{1}{2} \ln \cosh 2K. \tag{2.24}$$

Next, re-arrange (2.24) as:

$$e^{K'} = \cosh^{1/2}(2K), \tag{2.25}$$

and substitution of this into (2.21) yields:

$$f(K) = 2\cosh^{1/2}(2K). \tag{2.26}$$

We can write the relationship (2.18) between the old and the new partition functions as:

$$\mathcal{Z}(N, K) = f(K)^{N/2} \mathcal{Z}(N/2, K'), \tag{2.27}$$

which should be taken in conjunction with (2.24) and (2.26).

    In the next two subsections, we show how eqn (2.27) may be used to calculate the partition function recursively.

2.1.4   The free energy F

As always in statistical physics, we ultimately wish to find the free energy from the bridge equation, $F = -kT \ln \mathcal{Z}$. Because the free energy is extensive,[14] we must have

$$F = Nq \quad \text{(say)}, \tag{2.28}$$

where q does not depend on N but will depend on K. Take logarithms of both sides of eqn (2.27) and put

$$\ln \mathcal{Z}(N, K) = Nq(K), \tag{2.29}$$

to get

$$Nq(K) = \frac{N}{2} \ln f(K) + \frac{N}{2} q(K'). \tag{2.30}$$

Then re-arrange and substitute for $f(K)$ from (2.26):

$$q(K') = 2q(K) - \ln\left[2\cosh^{1/2}(2K)\right]. \tag{2.31}$$

This recursion relation goes in the direction of reducing K (with the relationship between K' and K given by eqn (2.24)).

---

[14] That is, it depends on the size of the system, in this case, on N.

We can invert these recursion relations as given by (2.23) and (2.31) to go the other way, thus:

$$K = \tfrac{1}{2}\cosh^{-1}(e^{2K'});\qquad(2.32)$$

$$q(K) = \tfrac{1}{2}\ln 2 + \tfrac{1}{2}K' + \tfrac{1}{2}q(K').\qquad(2.33)$$

In the following section we use these equations to calculate the partition function recursively.

2.1.5 Numerical evaluation of the recursion relation

For sake of simplicity, we divide the calculation into steps.

Step 1 We begin at high temperatures. Here K' is small so the interaction between dipoles is negligible and the partition function can be easily calculated as follows. In (2.9), we are now going to consider the high-temperature case where we can neglect the coupling constant. As a temporary device, let us assume that there is an applied field B. Then (2.9) can be written as:

$$\mathcal{Z} = \sum_{\{S\}} \prod_{i}^{N} \exp\{BS_i/kT\},\qquad(2.34)$$

where we again use the "exponential property." Then summing over possible spin states $S_i = \pm 1$, we have

$$\mathcal{Z} = \prod_{i}^{N} \left[\exp\{B/kT\} + \exp\{-B/kT\}\right] = 2^N \cosh^N\{B/kT\}.\qquad(2.35)$$

For sufficiently high temperatures, we can neglect the exponent for any given value of B and hence, recalling that $\cosh(0) = 1$, we have the high-temperature partition function as:

$$\mathcal{Z} = 2^N,\qquad(2.36)$$

so for free spins, $\mathcal{Z} \sim 2^N$.
Let us guess that an appropriate value of the coupling constant is $K' \sim 0.01$, therefore from the definition of q in eqn (2.28) we have:

$$q(K' = 0.01) = \frac{1}{N}\ln \mathcal{Z} = \frac{1}{N}\ln 2^N = \ln 2.\qquad(2.37)$$

Step 2 Substitute for K' in (2.32) to obtain:

$$K = 0.100334.\qquad(2.38)$$

$K=0$ $K=$ infinity

High temperature   Low temperature

FIG. 2.2 Coupling constant "flow" under RG transformations: one-dimensional Ising model.

Step 3 Calculate $q(K)$ from (2.33):

$$q(K) = \tfrac{1}{2}\ln 2 + \tfrac{1}{2} \times 0.01 + \tfrac{1}{2}\ln 2. \tag{2.39}$$

Carrying on in this way, we get the same numerical result (to six significant figures) as the exact result for the one-dimensional Ising model:

$$\mathcal{Z} = \left[2\cosh\left(\frac{J}{kT}\right)\right]^N. \tag{2.40}$$

In general, this procedure yields only the trivial fixed points at $K = 0$ or ∞. It is more difficult in higher dimensions, but for $d = 2$ we get the nontrivial fixed point $K = K_c$. We look briefly at this case in Section 2.3.

However, before we turn to the more realistic case of two-dimensional percolation, it is interesting to consider what is known as *parameter flow*. If we keep on iterating eqns (2.32) and (2.33), then the value of the coupling constant increases at each step, and in this trivial problem we can carry on until K becomes infinite (corresponding to $T = 0$). Each iteration of (2.32) and (2.33) is a renormalization group transformation or RGT, and the coupling constant increases with each such RGT.

The process is illustrated in Fig. 2.2. In this sense the coupling constant is said to *flow* from (in this case), zero to infinity. Or, in general, we say that coupling constants flow under RGT.

We should also note that the operation of the RGT is equivalent to a change of scale and that the same effect can be achieved by varying the temperature. For this reason the temperature is referred to in this context as a *scaling field*.

We shall return to these ideas in greater detail in Chapter 8.

2.2   Two-dimensional percolation

Percolation is something which is familiar to all of us in everyday life, from hot water percolating through coffee grounds to our bath water percolating through a sponge. The phenomenon is of great practical significance in industrial applications such as filtration, chemical processes and catalysis, while in the environment there is interest in the extraction of water or oil from porous rock.

Although porous rocks or bath sponges are probably rather irregular in respect of pore size and distribution, we may nevertheless construct an interesting mathematical model of percolation processes by considering a regular cubic lattice. As usual, for convenience, we look at a two-dimensional version, which is the square lattice. The situation is illustrated in

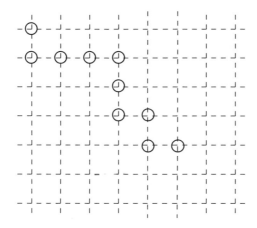

FIG. 2.3 Site percolation. Open circles denote lattice sites which are occupied. Two or more adjacent occupied sites constitute a cluster. Here we show a cluster of 10 occupied sites.

Fig. 2.3. The lattice sites are the intersections of the dotted lines and we show that a site is occupied by enclosing it by a circle. Evidently this is a Boolean model (just like the Ising model of magnetism) in that sites are either occupied or not.

If two or more neighbouring sites are occupied, we refer to that as a *cluster*. It should be noted that "nearest neighbor" means "along a dotted line." Along a diagonal is "next nearest neighbor." Figure 2.3 shows a cluster of ten occupied sites and the situation of interest is when a cluster is large enough to span the lattice from one side to the other.

This is a very simple model which is known as *site percolation*. In the next section we shall consider an equally simple model which is called *bond percolation*. It is worth noting that both these models are relevant to other physical processes ranging from the spread of forest fires to predator–prey biological systems.

2.2.1 Bond percolation in two dimensions: the problem

Figure 2.4 shows the "same" cluster as in Fig. 2.3, but this time the cluster is defined by ten bonds. If the probability of there being a bond between any two sites is p, then our basic problem may be stated as: how does the size of a cluster depend on p?

Evidently if p is small, then clusters will tend to be small, and the converse, must also be true. If follows that, for sufficiently large values of p, we may expect a single cluster which is large enough to extend from one side of the lattice to the other. Hence, as we increase p from zero, there must be some critical value

$$p = p_c \tag{2.41}$$

for which a single cluster extends from one side of the lattice to the other. Obviously for a one-dimensional lattice $p_c = 1.0$, as this amounts to a certainty that every pair of sites will be bridged and this is necessary for the cluster to span the entire lattice.

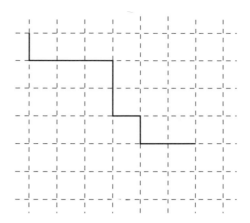

FIG. 2.4 Bond percolation. Bonds are denoted by solid lines joining lattice sites. The cluster
shown here is the analogue of the cluster shown in Fig. 2.3 with bonds between sites
replacing occupied sites.

It can also be shown (see Appendix C) that for a square lattice in two-dimensions the
critical probability is

$$p_c = 0.5. \tag{2.42}$$

This gives us a good test for the RG procedure: we would like to calculate the critical
probability for a two-dimensional square lattice. First we consider some aspects of a statistical
formulation of the problem.

2.2.2   The correlation length

Percolation (as defined here) is a purely geometrical problem and, unlike for many of the
systems we consider, we do not have a Hamiltonian nor need we consider thermal effects.

Nevertheless, it is tempting to draw an analogy between percolation and, for instance,
magnetism. In percolation, as we increase the probability of bond formation, the number
and size of the clusters will grow until at $p = p_c$ the lattice is spanned by at least one
cluster. Similarly, if in the magnetic problem we increase the strength of the coupling K (or,
equivalently, reduce the temperature), then clusters of spontaneously aligned lattice spins
will grow in number and size until the entire lattice is aligned.

In both problems, we must resort to statistics and introduce probabilities into our description of the system. Let $P(r)$ be the probability that two randomly chosen sites, distance r
apart, are connected by a single cluster. Then for $p < p_c$, all clusters will be subcritical and
we can write:

$$P(r) \rightarrow 0 \quad \text{as } r \rightarrow \infty \quad \text{for } p < p_c. \tag{2.43}$$

It is usual to introduce a correlation length ξ by taking the asymptotic (i.e. $r \rightarrow \infty$) form of
the probability in terms of an exponential, thus:

$$P(r) \sim e^{-r/q} \quad \text{as } r \rightarrow \infty, \tag{2.44}$$

where $\xi(p)$ is the *correlation length*.

As $p \rightarrow p_c$, we know that the correlation length must become infinity and it is conventional to write it as

$$\xi \sim (p - p_c)^{-\nu} \quad \text{as} \quad p \rightarrow p_c, \tag{2.45}$$

for positive ν. The exponent ν is an example of a *critical exponent* and we shall discuss these in more detail in Chapter 3, and again in Chapter 7.

2.2.3   Coarse-graining transformation

In the previous section on the one-dimensional magnet we discussed how we could do the "sum over states" for alternate sites. In the percolation problem, we just have the actual sites to work with and coarse-graining means removing sites from the lattice. In one dimension, this means knocking out every other site and in two dimensions knocking out every other site in two directions.

We can explain our coarse-graining transformation by reference to Figs 2.5 and 2.6, and the RG algorithm as follows:

- *Coarse-graining transformation*

    (a) Remove every alternate site from the lattice. Figure 2.5 shows the initial lattice. Open circles denote lattice sites to be removed, while solid circles indicate sites which will be retained and which will be part of the new lattice. Note that the number of sites is reduced $N \rightarrow N/4$ (and in d dimensions $N \rightarrow N/2^d$).

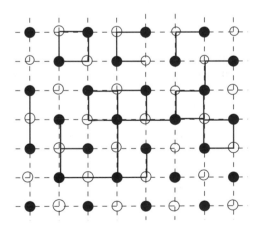

FIG. 2.5 Bond percolation: the original lattice prior to the RG transformation. Open circles denote sites to be removed while solid circles denote sites to be retained. Solid lines are bonds.

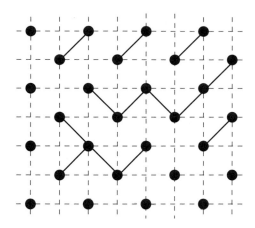

FIG. 2.6 Bond percolation: the new lattice is defined by the solid circles and has been rotated
through $\pi/4$ relative to the original lattice.

    (b) Draw a new bond between any two sites in the new lattice if there were *at least two*
    bonds joining these particular sites on the original lattice.[15] The result is shown in
    Fig. 2.6.

- *Rescaling*. Rescale the new lattice by reducing all lengths by a factor of $b = \sqrt{2}$. This
  is because the new lattice has been rotated through an angle of $\pi/4$ relative to the initial
  lattice and the new bonds lie on the old diagonals.

This completes a possible RG algorithm for bond percolation and our next step is to work
out the correspondingly recursion relation.

2.2.4 The recursion relation

If we consider a unit cell on the old lattice, then it will contribute a bond on its diagonal to
the new lattice, providing it has four, three, or two of its sides as bonds. The three possible
configurations are shown in Fig. 2.7.

    We can work out the probabilities of these configurations as follows:

- Probability of a side being a bond on the old lattice $= p$
- Probability of a side not being a bond on the old lattice $= 1 - p$.

Hence:

- Probability of a configuration (a) $= p^4$;
- Probability of a configuration (b) $= p^3(1 - p)$;
- Probability of a configuration (c) $= p^2(1 - p)^2$.

---

[15] This ensures that any sites on the new lattice which were connected on the old lattice remain connected on the
new lattice. Thus this particular coarse-graining procedure preserves lattice connectivity.

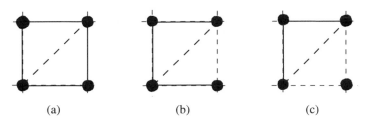

(a)   (b)   (c)

F IG. 2.7 Bond percolation: three types of unit cell on the old lattice which can contribute a
diagonal bond to the near lattice. Note that there are three other versions of (b), depending
on which site of the square is not a bond, and, similarly, one other version of (c).

Now, the probability p' that there is a diagonal bond on the new lattice is just the sum of the
probabilities of these configurations occurring on the old lattice. Noting that (b) can occur, in
four ways (increasing its probability by a factor 4) and (c) can occur in two ways (increasing
its probability by a factor 2), we have:

$$p' = p^4 + 4p^3(1 - p) + 2p^2(1 - p)^2 = 2p^2 - p^4. \tag{2.46}$$

This result is analogous to eqn (2.24) for the coupling constants in the one-dimensional
magnet. Now we have to find the fixed points.

2.2.5   Calculation of the fixed point: the critical probability

The fixed points are defined by the condition

$$p' = p = p^* \tag{2.47}$$

Our last step is to solve (2.46) for its fixed points and identify the nontrivial or critical fixed
point.
    Applying the condition (2.47), eqn (2.46) may be written as

$$p^4 - 2p^2 + p = 0. \tag{2.48}$$

Now take out common factors, p and $p - 1$, to rewrite it as:

$$p(p - 1)(p^2 + p - 1) = 0. \tag{2.49}$$

Then the fixed points are given by

$$p^* = 0 \quad \text{and} \quad p^* = 1; \tag{2.50}$$

along with the roots of the quadratic equation:

$$p^2 + p - 1 = 0, \tag{2.51}$$

thus:

$$p = \frac{-1 \pm \sqrt{5}}{2}. \tag{2.52}$$

Discarding the negative root (probabilities must be positive!) we have the third possibility as

$$p^* = (\sqrt{5} - 1)/2 \simeq 0.62. \tag{2.53}$$

Evidently $p^* = 0$ corresponds to there being no bonds on the lattice while $p^* = 1.0$ corresponds to *all* lattice sites being connected by bonds. In other words they are analagous to the trivial high- and low-temperature fixed points in a magnetic system. Thus we may conclude that the critical fixed point is given by

$$p_c = p^* = 0.62. \tag{2.54}$$

This is not the same as the exact result $p_c = 0.5$, but is not so far away from it either.

2.2.6 General remarks on RG applied to percolation

Other simple calculations can be done. For $d = 1$, in both bond and site percolation, RG is exact, as it is in the Ising model of ferromagnetism. For site percolation on a triangular lattice, the two-dimensional case is exact. This is because the coarse-graining of the triangular lattice leads to another triangular lattice.

We can improve the result for bond percolation on a square lattice by changing the RG algorithm. We do this by giving priority to "left–right" bonds and ignoring effects due to vertical bonds which would change the lattice topology.

This, of course, gives a one-dimensional character to the calculation but the renormalized lattice is also "square" and is re-scaled by the factor $b = 2$. This procedure gives much more accurate results for bond percolation on a square lattice. Further discussion of these points can be found in the books [5, 7, 30].

2.3 A two-dimensional magnet

The magnetic model which we now study is a generalization of the one-dimensional chain of dipoles studied in Section 2.1. In this case the dipoles can still only point up or down but now they are arranged in a two-dimensional square array. The arrangement is illustrated schematically in Fig. 2.8. This system is just the two-dimensional Ising model and its exact solution due to Onsager is one of the triumphs of statistical physics. For further discussion of this, see the book [29].

FIG. 2.8 A two-dimensional array of dipoles: the plus signs denote dipole (or spin) vectors pointing out of the page and the plain circles denote vectors pointing into the page.

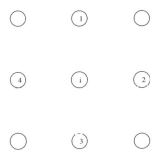

FIG. 2.9 Illustrating the nearest neighbors of site i, as summed in eqn (2.56).

2.3.1 The partition function

Although we have increased the number of dimensions from $d = 1$ to $d = 2$, the expression for the partition function remains the same and we may take over eqn (2.12) as

$$ \mathcal{Z} = \sum_{\{S\}} \exp \left\{ K \sum_{\langle ij \rangle} S_i S_j \right\}. \tag{2.55} $$

In order to see the structure of this summation, we consider a particular value of the site index i. Then, for that i, the particular term will be

$$ \mathcal{Z} = \sum \cdots e^{K S_i (S_1 + S_2 + S_3 + S_4)} \cdots \tag{2.56} $$

where we have labeled the four neighboring sites $j = 1, 2, 3, 4$, as shown in Fig. 2.9.

2.3.2 Coarse-graining transformation

We carry out the same procedure as in one dimension and sum over the alternate spins. A typical term from the sum in eqn (2.55) may be written as:

$$\mathcal{Z} = \sum \cdots [e^{K(S_1+S_2+S_3+S_4)} + e^{-K(S_1+S_2+S_3+S_4)}] \cdots . \tag{2.57}$$

We continue with the same procedure as in $d = 1$ but find that new couplings are generated, that is, not just nearest neighbors. In this way it can be shown that

$$\mathcal{Z} = f(K)^{N/2} \mathcal{Z} \left[K_1 \sum_{nn} S_i S_j + K_2 \sum_{nnn} S_i S_j + K_3 \sum \sum S_i S_j S_r S_t \right], \tag{2.58}$$

where:

$$\sum_{nn} \equiv \text{sum over nearest neighbors}; \tag{2.59}$$

$$\sum_{nnn} \equiv \text{sum over next nearest neighbors}; \tag{2.60}$$

$$\sum \sum \equiv \text{sum over 4 spins round a square.} \tag{2.61}$$

The new coupling constants are given by:

$$K_1 = \tfrac{1}{4} \ln \cosh(4K); \tag{2.62}$$

$$K_2 = \tfrac{1}{8} \ln \cosh(4K); \tag{2.63}$$

and

$$K_3 = \tfrac{1}{8} \ln \cosh(4K) - \tfrac{1}{2} \ln \cosh(2K). \tag{2.64}$$

An approximation is needed to carry this on. For form invariance we really need

$$\mathcal{Z}(N, K) = f^{N/2} \mathcal{Z}(N/2, K'), \tag{2.65}$$

if it is to work like the $d = 1$ case. One way of achieving this is to set

$$K_2 = K_3 = 0. \tag{2.66}$$

Unfortunately there is then no nontrivial fixed point.
   A better approximation is to set:

$$K = K_1 + K_2; \quad K_3 = 0. \tag{2.67}$$

In this case, we get a nontrivial fixed point with a prediction for T_c within 20% of the exact result. The flow of the coupling constant under RGT is illustrated in Fig. 2.10. This is more

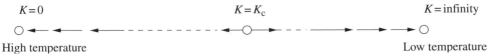

FIG. 2.10 Coupling constant "flow" under RG transformations in the two-dimensinal Ising model. Note that in this case (unlike in Fig. 2.2) there is a nontrivial fixed point.

interesting than the one-dimensional case and we see that under RGT the system will move away from the critical point $K = K_c$ and flow to one of the trivial fixed points at $K = 0$ or $K = \infty$. This is in fact a very important topological property of this type of flow in nontrivial systems and we shall enlarge on this aspect in Chapter 8.

It is not easy to improve upon this. If K_3 is included, then a poorer result is obtained. However, it is possible to obtain nontrivial results by restricting the interactions at each RGT to nearest neighbor and next-nearest neighbor and retaining only second-order in the expansion of $\ln \cosh(4K)$ in (2.62)–(2.64). Denoting these new couplings by K and L, respectively, one can obtain RG equations in the form:

$$K' = 2K^2 + L; \quad L' = K^2, \tag{2.68}$$

and we shall discuss the associated fixed points in Section 8.4.2.

2.4 Exercises

Solutions to these exercises, along with further exercises and their solutions, will be found at: *www.oup.com*

1. A linear chain Ising model has a Hamiltonian given by

$$H = - \sum_{i=1}^{N-1} J_i S_i S_{i+1}.$$

Show that the partition function takes the form:

$$Z_N = 2^N \prod_{i=1}^{N-1} \cosh K_i,$$

where $K_i \equiv J_i/kT$ is the coupling parameter.

Obtain the spin–spin correlation function in terms of the partition function and show that for uniform interaction strength it takes the form:

$$G_n(r) \equiv \langle S_n S_{n+r} \rangle = \tanh^r K.$$

where K is the coupling constant for the model.

Comment on the possibility of long-range order appearing in the cases: (a) $T > 0$; and (b) $T = 0$.

2. Show that the transfer matrix method may be extended to the case of an Ising ring located in a constant field B and hence obtain the free energy per lattice site in the form:

$$f = -J - \frac{1}{\beta} \ln[\cosh \beta B \pm (e^{2\beta J} \sinh^2 \beta B + e^{2\beta J})^{\frac{1}{2}}],$$

where all the symbols have their usual meaning.

[Hint: When generalizing the transfer matrix to the case of nonzero external field remember that it must remain symmetric in its indices.]

Obtain an expression for the specific magnetization $m = \langle S \rangle$, and discuss its dependence on the external field B comment on your results.

3. The RG recursion relations for a one-dimensional Ising model in an external field B may be written as;

$$x' = \frac{x(1+y)^2}{(x+y)(1+xy)}; \qquad y' = \frac{y(x+y)}{(1+xy)},$$

where $x = e^{-4J/kT}$ and $y = e^{-B/kT}$. Verify the existence of fixed points as follows: $(x^*, y^*) = (0, 0)$; $(x^*, y^*) = (0, 1)$ and a line of fixed points $x^* = 1$ for $0 \leq y^* \leq 1$. Discuss the physical significance of these points and sketch the system point flows in the two-dimensional parameter space bounded by $0 \leq x \leq 1$ and $0 \leq y \leq 1$.

4. Verify that summing over alternate spins on a square lattice results in a new square lattice rotated through $45°$ relative to the original lattice and with a scale factor of $b = \sqrt{2}$.

Also verify that the effect of such a decimation on the Ising model is to change the original partition function involving only pairs of nearest neighbors to a form involving nearest-neighbors, next-to-nearest neighbors and the product of four spins taken "round a square."

3

MEAN-FIELD THEORIES FOR SIMPLE MODELS

In this chapter we deal with two microscopic mean-field theories, the Weiss theory of magnetism and the Debye–Hückel theory of electrolytes, respectively. Both of these theories are also self-consistent theories, and in our exposition we are careful to emphasize the difference between the mean-field and self-consistent assumptions. We conclude the chapter with a discussion of the Landau theory of critical phenomena. This is a macroscopic mean-field theory but its self-consistent nature is less obvious. We conclude by noting that the Van der Waals equation of state for a nonideal gas (which is derived as an application of renormalized perturbation theory Section 4.5) can also be interpreted as a macroscopic mean-field theory for the gas–liquid phase transition. However, we shall not pursue that here. Further discussion can be found in the books [29] and [32].

3.1 The Weiss theory of ferromagnetism

The Weiss theory dates from 1907, before the formulation of quantum mechanics, but we shall present a slightly modernized version which acknowledges the existence of quantum physics.

3.1.1 The ferro-paramagnetic transition: theoretical aims

When a piece of ferromagnetic material is placed in a magnetic field $\mathbf{B}$, a mean magnetization $\mathbf{M}$ is induced in the material which is proportional[16] to $\mathbf{B}$. Then, taking one coordinate axis along $\mathbf{B}$, we can work with the scalars B and M which we assume to be in the same direction.

The relationship between the applied magnetic field and the resulting magnetization is given by the *isothermal susceptibility*, as defined by the relation:

$$\chi_T \equiv \left(\frac{\partial M}{\partial B} \right)_T. \tag{3.1}$$

Note this is an example of an *response function*. In a fluid, for instance, the analogous response function would be the *isothermal compressibility*.

Accordingly, our general theoretical aim will be to obtain an expression relating magnetization to the applied field. However, in order to have a specific objective, we will seek a value for the critical temperature T_c, above which spontaneous magnetization cannot exist.

---

[16] We are assuming here that the magnetic material is isotropic.

3.1.2   The mean magnetization

The basic model of the magnetic material is much the same as used in Chapter 2 and consists of N spins on a lattice. This time we assume that each has magnetic moment μ_0 (up or down), corresponding to the spin S_i at the lattice site labeled by i, taking its permitted values of $S_i = \pm 1$. The instantaneous state at any one lattice site is therefore given by

$$\mu = \pm\mu_0.$$

We may define the magnetization M,

$$M = N\bar{\mu},$$

where $\bar{\mu}$ is the average value of the magnetic moment at a lattice site.

It is helpful to consider two extreme cases:

- If all spins are oriented at random then $\bar{\mu} = 0$ and so $M = 0$, and hence there is no net magnetization.

- If all spins are lined up then $\bar{\mu} = \mu_0$ and so the net magnetization is $M_\infty = N\mu_0$, which is the largest possible value and is often referred to as the *saturation value*.

In between these extremes there is an *average magnetization* appropriate to the temperature of the system:

$$\bar{\mu} = \sum_{\text{states}} P(\mu)\mu, \tag{3.2}$$

where $P(\mu)$ is the probability of any particular value μ of magnetic moment. This dependence on temperature is illustrated qualitatively in Fig. 3.1.

3.1.3   The molecular field B'

We assume that any spin experiences an effective magnetic field B_E, which is made up of an externally applied field B and a molecular field B' due to spin–spin interactions.

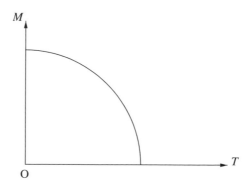

FIG. 3.1 Magnetization as a function of temperature.

This is the mean-field approximation

$$B_E = B + B'. \tag{3.3}$$

At thermal equilibrium the probability of any value of the magnetic moment is given in terms of the associated energy by eqns (2.1) and (2.2), and identifying the "magnetic energy" as $H = \mu B_E$, we have

$$P(\mu) = e^{-\mu B_E/kT} \Big/ \sum_{\text{states}} e^{-\mu B_E/kT}, \tag{3.4}$$

and hence the mean value of the individual magnetic moments is

$$\bar{\mu} = \sum_{\text{states}} \mu e^{-\mu B_E/kT} \Big/ \sum_{\text{states}} e^{-\mu B_E/kT}. \tag{3.5}$$

The possible states of the individual magnetic moments are given by $\mu = \pm\mu_0$, thus the expression for the mean becomes

$$\bar{\mu} = \frac{\mu_0 e^{-\mu_0 B_E/kT} - \mu_0 e^{+\mu_0 B_E/kT}}{e^{-\mu_0 B_E/kT} + e^{+\mu_0 B_E/kT}}, \tag{3.6}$$

and so

$$\bar{\mu} = \mu_0 \tanh\left[\frac{\mu_0}{kT}(B + B')\right]. \tag{3.7}$$

Or, in terms of the total magnetization of the specimen, we may write this as

$$M = N\bar{\mu} = N\mu_0 \tanh\left[\frac{\mu_0}{kT}(B + B')\right]. \tag{3.8}$$

But $N\mu_0 = M_\infty$ is the saturation value and hence we have

$$M = M_\infty \tanh\left[\frac{\mu_0}{kT}(B + B')\right], \tag{3.9}$$

which gives the magnetization at any temperature T as a fraction of the saturation magnetization, in terms of the applied field B and the unknown molecular field B'. This means, of course, that we only have one equation for two unknowns, M and B'.

3.1.4   The self-consistent assumption: $B' \propto M$

We are interested in the case where there is permanent magnetization which can be detected even when the external field B has been set to zero. Under these circumstances, the molecular field and the magnetization must be related to each other in some way. The self-consistent step which can close eqn (3.9) is to assume that B' is a function of M, and the simplest such assumption is $B' \propto M$. This is such an important step that we highlight it as:

This is the self-consistent assumption: $B' \propto M$.

We can identify the constant of proportionality in such a relationship as follows. For any one spin at a lattice site,

- Let z be the number of neighboring spins,
- Let z_+ be the number of neighboring spins up,
- Let z_- be the number of neighboring spins down.

Hence we may write

$$\frac{z_+ - z_-}{z} = \frac{M}{M_\infty},$$

and so

$$z_+ - z_- = z \frac{M}{M_\infty}. \tag{3.10}$$

On this picture the average energy of interaction of one spin with its neighbors is

$$\Delta E = \pm \mu_0 B'$$

and, from a microscopic point of view, we can express this in terms of the quantum-mechanical exchange interaction as

$$\Delta E = J(z_+ - z_-).$$

Equating these two expressions gives us

$$\mu_0 B' = J(z_+ - z_-). \tag{3.11}$$

From this result, and using eqn (3.10) for $(z_+ - z_-)$, we obtain an expression for the molecular field as

$$B' = \frac{J}{\mu_0}(z_+ - z_-) = \frac{J}{\mu_0} z \frac{M}{M_\infty}. \tag{3.12}$$

Lastly, we substitute for B' into equation for M:

$$\frac{M}{M_\infty} = \tanh\left[\frac{\mu_0}{kT}\left(B + \frac{Jz}{\mu_0} \cdot \frac{M}{M_\infty}\right)\right], \tag{3.13}$$

and obtain a closed equation for the magnetization of the system. For spontaneous magnetization, we have $B = 0$, and so

$$\frac{M}{M_\infty} = \tanh\left[\frac{Jz}{kT} \cdot \frac{M}{M_\infty}\right]. \tag{3.14}$$

We may solve this for the critical temperature above which spontaneous magnetization cannot occur. We note that the result depends on the coupling strength J and the number of nearest neighbors (in other words, the lattice type) but not on μ_0. The simplest method is graphical. The spontaneous magnetization is given by plotting the graph of

$$X = \tanh\left(\frac{zJ}{kT}X\right) \tag{3.15}$$

and looking for the intersection with the straight line

$$X = M_s/M_\infty,$$

where M_s is the spontaneous magnetization.

3.1.5   Graphical solution for the critical temperature T_c

Let us anticipate the fact that eqn (3.15) can be solved for the critical temperature T_c and rewrite it as:

$$X = \tanh\left(\frac{T_c}{T}X\right) \tag{3.16}$$

where

$$T_c = \frac{zJ}{k}. \tag{3.17}$$

We note that in general T_c depends on the lattice type (simple cubic, body-centered cubic etc.), through the parameter z, the strength of the interaction J and the Boltzmann constant k.

Now, for the case $T = T_c$, eqn (3.16) reduces to

$$X = \tanh X,$$

and for small values of X, this becomes

$$X = X$$

and the only possible solution is $X = 0$, corresponding to there being no mean magnetization. In general this is true for $T_c/T \leq 1$ and so the only possibility of an intersection at nonzero X is for $T_c/T > 1$, as shown in Fig. 3.2. We can summarize the situation as follows:

- $T > T_c$ $X = 0$, $M = 0$: disordered phase;
- $T < T_c$ $X \neq 0$, $M = 0$: ordered phase.

As the transition from a disordered to an ordered phase is from a more symmetric to a less symmetric state, such transitions are often referred to as *symmetry-breaking*.

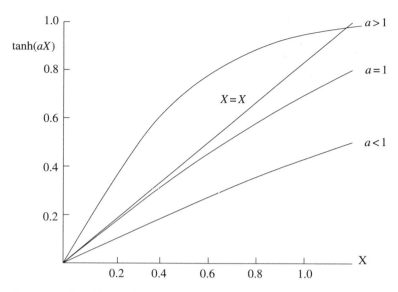

FIG. 3.2 Graphical solutions of eqn (3.16). Note that here $a = T_c/T$.

3.2   The Debye–Hückel theory of the electron gas

We can also apply the concept of the self-consistent field theory to the problem of a plasma or electrolyte, where the interaction between pairs of particles is the Coulomb potential.

The theory dates back to 1922 and, like the Weiss theory of ferromagnetism, is an ancient theory which is still close to the frontiers of many-body physics even today. It is not perhaps quite as important as the Weiss theory, but is of general relevance to the perturbation treatment of the electron gas at high temperatures or to electrolytes in the classical regime. It is also of relevance in rheology where it can be used to describe the mutual interactions of particles suspended in a fluid.

Although we have already taken a brief look at this topic in Section 1.2.1, it is worth stating our theoretical objective here, as follows:

We wish to calculate the electrostatic potential at a point r due to an electron at the origin while taking into account the effect of all the other electrons in the system.

3.2.1   The mean-field assumption

We shall discuss an idealized version of the problem in which N electrons are free to move in an environment with spatially uniform positive charge, chosen such that overall the system is electrically neutral. Both negative and positive charge densities are numerically equal to en_∞, where e is electronic charge and n_∞ is the number density when the electrons are spread out uniformly.

Consider the case where one electron is at $r = 0$. We wish to know the probability $p(r)$ of finding a second electron a distance r away. At thermal equilibrium eqns (2.1) and (2.2) apply, so this is given by

$$p(r) = \frac{e^{-W(r)/kT}}{\mathcal{Z}}, \tag{3.18}$$

where as usual $\mathcal{Z}$ is the partition function and here $W(r)$ is the renormalized interaction energy, which takes into account the collective effect of all N electrons. We expect that (like the *bare* Coulomb form) the *dressed* interaction will satisfy

$$W(r) \to 0 \quad \text{as} \quad r \to \infty. \tag{3.19}$$

It should be noted that $W(r)$ is analogous to the molecular field introduced in the Weiss theory and that (3.18) and (3.19) together amount to a *mean-field approximation*. Then, the probability P_r (say) of finding a second electron in the spherical shell between r and $r + dr$ around the first electron is just

$$P_r = P(r) \times 4\pi r^2 dr = \frac{1}{\mathcal{Z}} e^{-W(r)/kT} \times 4\pi r^2 dr, \tag{3.20}$$

and so the number of electrons in the shell is

$$N_r = N P_r = \frac{N}{\mathcal{Z}} e^{-W(r)/kT} \times 4\pi r^2 \, dr. \tag{3.21}$$

Hence, the *number density* $n(r)$ of electrons in the shell is given by

$$n(r) = \frac{N P_r}{4\pi r^2 \, dr} = \frac{N}{\mathcal{Z}} e^{-W(r)/kT}. \tag{3.22}$$

Now consider the ratio of this density to that at some other $r = R$, thus:

$$\frac{n(r)}{n(R)} = \frac{e^{-W(r)/kT}}{e^{-W(R)/kT}}. \tag{3.23}$$

Further, let us take $R \to \infty$, and using (3.19), we have

$$e^{-W(R)/kT} \to 1,$$

and so eqn (3.23) may be written as

$$\frac{n(r)}{n(\infty)} = e^{-W(r)/kT}. \tag{3.24}$$

Or, in terms of the uniform number density introduced at the beginning of this section: $n_\infty = n(\infty)$, we may rearrange this result into the form:

$$n(r) = n_\infty e^{-W(r)/kT}. \tag{3.25}$$

3.2.2 The self-consistent approximation

Debye and Hückel (1923) proposed that ϕ should be determined *self-consistently* by making a "*continuum approximation*" and solving Poisson's equation (from electrostatics), thus:

$$\nabla^2\phi = -4\pi\rho(r), \tag{3.26}$$

where $\rho(r)$ is the electron charge density. In this case, the electron charge density may be taken as

$$\rho(r) = en(r) - en_\infty,$$

and the Poisson equation becomes

$$\nabla^2\phi(r) = -4\pi en_\infty\{e^{-e\phi(r)/kT} - 1\}, \tag{3.27}$$

where we substituted

$$W = e\phi(r), \tag{3.28}$$

into the right-hand side of eqn (3.25) for $n(r)$, and $\phi(r)$ is the self-consistent field potential.

3.2.3 The screened potential

If we restrict ourselves to $W \ll kT$ (i.e. the high-temperature case), and expand out the exponential to first order, Poisson's equation further becomes

$$\nabla^2\phi = \frac{4\pi e^2 n_\infty \phi}{kT}, \tag{3.29}$$

with solution readily found to be

$$\phi = \left(\frac{e}{r}\right) exp\{-r/l_D\}, \tag{3.30}$$

where l_D is the Debye length and is given by

$$l_D = \left[\frac{4\pi e^2 n_\infty}{kT}\right]^{-1/2}. \tag{3.31}$$

Equation (3.30) represents a *screened potential* (Fig. 3.3). Physically, the Debye length is interpreted as the radius of the *screening cloud* of electrons about any one electron. This can also be interpreted as "charge renormalization," in the following sense

$$e \rightarrow e \times exp\{-r/l_D\}.$$

Note that it is necessary to consider the circumstances under which the cloud of discrete electrons can be regarded as a continuous charge density.

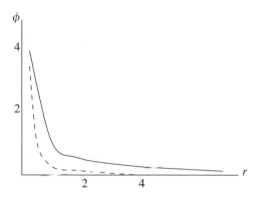

FIG. 3.3 Comparison of the Coulomb potential (full line) with a screened potential (dashed line).

3.2.4 Validity of the continuum approximation

The continuum approximation should be valid for the case where the distance between particles is much smaller than the Debye length. That is

$$l_D \gg N^{-1/3}; \quad \text{or} \quad l_D^3 \gg N^{-1},$$

and from eqn (3.31)

$$8\pi^{3/2} e^3 N^{1/2} \beta^{3/2} \ll 1,$$

where $\beta \equiv 1/kT$.

3.3 Macroscopic mean-field theory: the Landau model for phase transitions

As a preliminary to the Landau model, we state our theoretical aims: we wish to calculate the *critical exponents* of the system.

3.3.1 The theoretical objective: critical exponents

We have already met a critical exponent in Section 2.2.2, where we discussed the subject of percolation: this was associated with the correlation length and defined by eqn (2.45). Now, bearing in mind the analogy between probability in percolation and temperature in thermal systems, it will not be too surprising that we can define some critical exponents in terms of temperature. Here we shall introduce four critical exponents: those associated respectively with the heat capacity C_B, the magnetization M, the susceptibility X and the equation of state, which is the relationship between the applied field B and the magnetization. The defining relationships, which are no more than an arbitrary way of correlating experimental data, may be listed as follows:

$$C_B \sim \left| \frac{T - T_c}{T_c} \right|^{-\alpha} ; \qquad (3.32)$$

$$M \sim - \left(\frac{T - T_c}{T_c} \right)^{\beta} ; \qquad (3.33)$$

$$\chi_T \sim \left| \frac{T - T_c}{T_c} \right|^{-\gamma} ; \qquad (3.34)$$

and

$$B \sim |M|^{\delta} \, \mathrm{sgn} M, \qquad (3.35)$$

where sgn is the signum, or sign, function.[17] These relationships define the exponents α, β, γ, and δ. There are others, but we shall defer a more general discussion to Chapter 7 in the third part of this book.

3.3.2 Approximation for the free energy F

This theory is restricted to symmetry-breaking transitions, where the free energy F and its first derivatives vary continuously through the phase transition. We shall consider a ferromagnet in zero external field as an example. Let us assume that F is analytic in $\mathbf{M}$ near the transition point, so that we may expand the free energy in powers of the magnetization, as follows:

$$F(T, \mathbf{M}) = F_0(T) + A_2(T)M^2 + A_4(T)M^4 + \cdots . \qquad (3.36)$$

We note that only even terms occur in the expansion, as F is a scalar and can only depend on scalar products of $\mathbf{M}$. Referring to Fig. 3.4, we see that in broad qualitative terms, there are only four possible "shapes" for the variation of F with M, depending on the signs of the coefficients $A_2(T)$ and $A_4(T)$.

We may reject two of these cases immediately on purely physical grounds. That is, both cases with $A_4 < 0$ show decreasing F with increasing M. This is unstable behavior. Thus, for *global stability*, we have the requirement:

$$A_4 > 0.$$

Refer now to the two left hand graphs, where in both cases we have $A_4 > 0$. We shall consider the two cases separately:

Case 1: $A_2 > 0$: Here F has a minimum F_0 at $M = 0$. There is no permanent magnetization, so this can be interpreted as the paramagnetic phase, hence we may assume that this case corresponds to $T > T_c$.

---

[17] In fact $\mathrm{sgn} M = \mathbf{M}/|\mathbf{M}|$ in this case.

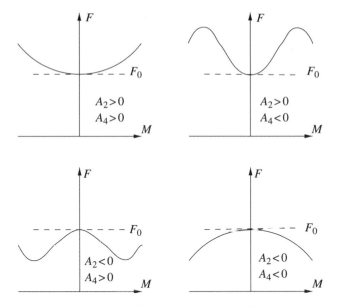

FIG. 3.4 Possible variations of the free energy F with the magnetization M.

Case 2: $A_2 < 0$: Here F has a maximum at $M = 0$, but two minima at $\pm M$, corresponding to permanent magnetization in one or other direction. Therefore we may interpret this as being the ferromagnetic phase and assume that this case corresponds to $T < T_c$.

Thus we conclude that $T = T_c$ corresponds to $A_2 = 0$.

Now let us reconsider these two cases from a mathematical point of view. The conditions for F to be minimized are:

$$\left(\frac{\partial F}{\partial M}\right)_T = 0; \qquad \left(\frac{\partial^2 F}{\partial M^2}\right)_T > 0.$$

Accordingly, we differentiate the expression for F, as given by (3.36) to obtain:

$$\left(\frac{\partial F}{\partial M}\right)_T = 2A_2(T)M + 4A_4(T)M^3, \tag{3.37}$$

and re-examine our two cases:

Case 1: Here we have $A_2 > 0$, $T > T_c$ and clearly,

$$\left(\frac{\partial F}{\partial M}\right)_T = 0 \quad \text{if and only if} \quad M = 0,$$

as M and M^3 have the same sign, and A_2 and A_4 are both positive, therefore no cancellations are possible.

Case 2: Here we have $A_2 < 0$, $T < T_c$. As **M** can point in one of two directions, we temporarily restore vector notation. It follows that

$$\left(\frac{\partial F}{\partial M}\right)_T = 2A_2\mathbf{M} + 4A_4M^2\mathbf{M} = 0.$$

Now A_2 can change sign so we write this as:

$$-2|A_2|\mathbf{M} + 4A_4M^2\mathbf{M} = 0,$$

or

$$\mathbf{M} = \pm(|A_2|2A_4)^{1/2}\widehat{\mathbf{M}},$$

where $\widehat{\mathbf{M}}$ is a unit vector in the direction of **M**. The fact that A_2 changes sign at $T = T_c$, implies that if we expand the coefficient $A_2(T)$ in powers of the temperature then it should take the form:

$$A_2(T) = A_{21}(T - T_c) + A_{23}(T - T_c)^3 + \text{higher order terms}, \qquad (3.38)$$

where we have set $A_{20} = A_{22} = 0$.

We may summarize all this as follows:

For $T \geq T_c$:
$$F(T, M) = F_0(T),$$

and thus corresponds to $M = 0$ which is the paramagnetic phase.
For $T \leq T_c$:
$$F(T, M) = F_0(T) + A_2(T)M^2 + \cdots,$$

and corresponds to
$$M^2 = -A_2/2A_4,$$

with
$$A_2 = A_{21}(T - T_c).$$

Evidently this is the ferromagnetic phase.

Equation (3.36) for the free energy may now be written as

$$F(T, M) = F_0(T) - (A_{21}^2/2A_4)(T - T_c)^2 + \cdots. \qquad (3.39)$$

Note that the equation of state may be obtained from this result by using the relationship $B = -\partial F/\partial M)_T$.

3.3.3 Values of critical exponents

Equilibrium magnetization corresponds to minimum free energy. From eqns (3.37) and (3.38):

$$\frac{dF}{dM} = 2A_2M + 4A_4M^3 = 0$$

$$= 2A_{21}(T - T_c)M + 4A_4M^3,$$

and so

$$M = 0 \quad \text{or} \quad M^2 \sim (T - T_c).$$

Hence

$$M \sim (T - T_c)^{1/2} \sim \theta_c^{1/2},$$

and from (3.33) we identify the exponent β as:

$$\beta = \tfrac{1}{2}.$$

To obtain γ and δ, we add a magnetic term due to an external field B; thus:

$$F = F_0 + A_{21}(T - T_c)M^2 + A_4M^4 - BM,$$

and hence

$$\frac{dF}{dM} = -B + 2A_{21}(T - T_c)M + 4A_4M^3 = 0.$$

This can be re-arranged as:

$$B = 2A_{21}(T - T_c)M + 4A_4M^3. \tag{3.40}$$

For the critical isotherm, $T = T_c$ and so $B \sim M^3$; or:

$$\delta = 3.$$

Lastly, as

$$\chi = \left.\frac{\partial M}{\partial B}\right)_T,$$

we differentiate both sides of equation (3.40) for equilibrium magnetization with respect to B:

$$1 = 2A_{21}(T - T_c)\left.\frac{\partial M}{\partial B}\right)_T + 12A_4M^2\left.\frac{\partial M}{\partial B}\right)_T,$$

and so,

$$\chi = \left(2A_2T_c\theta_c + 12A_4M^2\right)^{-1} = 1,$$

and from (3.33):

$$\gamma = 1.$$

These values may be compared with the experimental values: $\beta = 0.3\text{–}0.4$, $\delta = 4\text{–}5$, and $\gamma = 1.2\text{–}1.4$.

3.4 Exercises

Solutions to these exercises, along with further exercises and their solutions, will be found at: www.oup.com

1. On the basis of the Weiss molecular field theory, show that the Ising model cannot exhibit a spontaneous magnetization for temperatures greater than $T = T_c$, where the Curie temperature is given by $T_c = J/k$.

2. An isolated parallel plate capacitor has a potential difference V between its electrodes, which are situated at $x = \pm a$. The space between the electrodes is occupied by an ionic solution which has a dielectric constant of unity. Obtain an expression for $n(x)$, the space charge distribution, which exists after the system has reached thermal equilibrium. For the sake of simplicity you may assume that the potential difference between the plates is so small that $eV \ll kT$.
   [Hint: take $x = 0$ as a plane of symmetry where:

   1. $n_+(0) = n_-(0) = n_0$ (say)

   2. the potential equals zero.]

3. Using the Landau model for phase transitions obtain values for the critical exponents β, γ, and δ.

PART II

RENORMALIZED PERTURBATION THEORIES

4

PERTURBATION THEORY USING
A CONTROL PARAMETER

We have given a formal (if simplified) treatment of perturbation theory in Section 1.3, and we shall presently make use of this when we apply it to nonlinear differential equations in Chapter 5. However, in the present chapter we can cheat a little by simply expanding out the exponential form in the partition function.

As we saw in Section 1.2.7, we can make the many-body partition function more tractable by expanding out the interaction term in powers of the density (low-density expansions) or in powers of $1/T$ (high-temperature expansions). In this context the temperature and the density are regarded as *control parameters* since they control the strength of the interaction or coupling.

We shall begin with the temperature and make a high-temperature expansion of the partition function for a one-dimensional Ising model of a magnet. Strictly this is not really an example of renormalization, as there is no actual renormalized quantity or quasi-particle in the theory. But, it is a nice example of the use of diagrams and an effective renormalization does emerge. The technique is also of interest for the way in which it can be extended beyond its apparent range of validity.

In order to demonstrate the use of the density as a control parameter we consider a model for a real gas with various assumptions about the shape of the intermolecular potential. We show that it is possible to obtain "low-density" corrections to the equation of state for an ideal gas.

The combination of these two topics provides us with an opportunity to reinterpret the phenomonological Debye–Hückel theory of Section 3.2 in terms of perturbation theory.

For the purposes of this chapter we shall need the Taylor series for an exponential function:

$$e^x = 1 + x + \frac{x^2}{2!} + \frac{x^3}{3!} \cdots \frac{x^s}{s!} + \cdots = \sum_{s=0}^{\infty} \frac{x^s}{s!},$$

along with that for a natural logarithm, thus:

$$\ln(1 + x) = x - \frac{x^2}{2} + \frac{x^3}{3} + \cdots, \tag{4.1}$$

and the binomial expansion

$$(a + x)^n = a^n + na^{n-1}x + \frac{n(n-1)}{2!}a^{n-2}x^2 + \cdots. \tag{4.2}$$

4.1   High-temperature expansions

As usual, our starting point is the partition function,

$$Z = \sum_{\text{states}} \exp\{-\beta H\},$$

where H is either the energy or the Hamiltonian. We may expand out the exponential as a power series in

$$\beta \equiv 1/kT,$$

thus:

$$\exp\{-\beta H\} = 1 - \beta H + \beta^2 H^2/2 + \cdots,$$

and in this way we can obtain all the thermodynamic quantities as power series in β. We may note the following general points.

- For large T, only a few terms will be needed, because the "expansion parameter" is small.

- For $T \to T_c$, we shall need lots of terms. This is the main limitation of the method but, as we shall see, there are ways of extracting more information despite having made a finite-order truncation of the series.

We shall work with zero external field. This means that we can evaluate thermodynamic quantities, such as the order parameter and the heat capacity. However, in principle we shall need $B \neq 0$, if we want to obtain the susceptibility. This requirement naturally complicates terms at each order. Hence usually susceptibilities are worked out with fewer terms than specific heats.[18]

4.2   Application to a one-dimensional magnet

In principle, we can apply the technique to any model but we shall choose the Ising model. We can also work the technique out for any number of dimensions but as usual we begin with the case $d = 1$ as an easy example.

4.2.1   High-temperature expansion for the "Ising ring"

As in Section 2.1, we consider a one-dimensional chain of spins (or dipoles), but this time we take the boundary conditions to have the periodic form

$$S_{N+1} \equiv S_1, \tag{4.3}$$

as illustrated in Fig. 4.1. We start with the identity:

$$e^{\pm A} = \cosh A \pm \sinh A, \tag{4.4}$$

---

[18] We can evade this problem by using linear response theory (see Section 7.4) to work out the zero-field susceptibility.

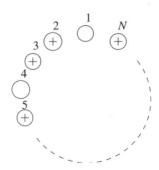

F IG. 4.1 A one-dimensional chain of dipoles with periodic boundary conditions.

and observe that if $\eta = \pm 1$, then we can equally well write it as

$$e^{\eta A} = \cosh A + \eta \sinh A. \tag{4.5}$$

In this particular case

$$\eta \equiv S_i S_j = \pm 1, \tag{4.6}$$

from which

$$\exp\{K S_i S_j\} = \cosh K + S_i S_j \sinh K, \tag{4.7}$$

where $K = \beta J$ is the coupling constant, as before. For the partition function Z, we again have

$$Z = \sum_{\{S\}} \exp\{-\beta H\} = \sum_{\{S\}} \prod_{\langle i,j \rangle} \exp\{K S_i S_j\}, \tag{4.8}$$

although it should be noted that we have reverted to the angle brackets around the indices i and j, as used in eqn (2.7), to indicate that the sum in the Hamiltonian is restricted to nearest neighbors. In turn this means that the product is over all nearest-neighbor pairs of lattice sites.
   There are P such nearest-neighbor pairs, where (for periodic boundary conditions)

$$P = Nz/2, \tag{4.9}$$

with N being the number of lattice sites and z the number of nearest neighbors to any lattice site (also known as the *coordination number*). It should also be noted that the formula for P is valid for the Ising model in any number of dimensions but (through its dependence on z) depends on lattice type. For example, in the present case for $d = 1$, we have $P = 2N/2 = N$.
   Now we use the identity given as eqn (4.5), in order to rewrite the expression for the partition function as

$$Z = [\cosh K]^P \sum_{S} \prod_{\langle i,j \rangle} (1 + S_i S_j \tanh K), \tag{4.10}$$

where we have factored out $[\cosh K]^P$ from the repeated product.

We shall carry out the analysis for the case $N = 3$ then generalize to any number of spins. For convenience, we set:

$$v = \tanh K. \tag{4.11}$$

Then, we multiply out the product operator up to $P = 3 \times 2/2 = 3$, thus:

$$\frac{Z_{N=3}}{\cosh^3 K} = \sum_{S_1=-1}^{1} \sum_{S_2=-1}^{1} \sum_{S_3=-1}^{1} \times \{1 + v(S_1 S_2 + S_2 S_3 + S_3 S_1)$$

$$+ v^2(S_1 S_2 S_2 S_3 + S_1 S_2 S_3 S_1 + S_2 S_3 S_3 S_1) + v^3(S_1 S_2 S_2 S_3 S_3 S_1)\}, \tag{4.12}$$

where we have taken the factor $\cosh^3 K$ across to the left hand side for sake of convenience. Note that the expansion is in powers of

$$v \equiv \tanh K \equiv \tanh(J/kT), \tag{4.13}$$

and is therefore an example of a high-temperature expansion.

Now we do the sums over the spins states $S_1, S_2, S_3 = \{S\}$ and write the results out as follows:

$$v^0: 2 \times 2 \times 2 = 2^P = 2^3 = 8. \tag{4.14}$$

$$v^1: \sum_{S_1=-1}^{1} S_1 S_2 = S_2 - S_2 = 0, \quad \sum_{S_2=-1}^{1} S_1 S_2 = 0, \quad \text{and so on.} \tag{4.15}$$

$$v^2: \sum_{S_1} (S_1 S_2)(S_2 S_3) = S_2^2 S_3 - S_2^2 S_3 = 0 \quad \text{and so on.} \tag{4.16}$$

$$v^3: \sum_{S_1} \sum_{S_2} \sum_{S_3} S_1 (S_2 S_2)(S_3 S_3) S_1 = 2^3 = 8. \tag{4.17}$$

It should be noted that for v^3, all sums over $\sum_{S=-1}^{+1}$ are non-vanishing because each variable is squared. This property may be expressed as:

$$\sum_{S_i=-1}^{1} S_i^r = 2 \quad \text{if } r \text{ is even,}$$

$$= 0 \quad \text{if } r \text{ is odd.} \tag{4.18}$$

Thus in the present case, only v^0 and v^3 contribute, hence:

$$Z_3 = \cosh^3 J (8 + 8v^3) = 2^3 (\cosh^3 K + \sinh^3 K). \tag{4.19}$$

4.2.2 Formulation in terms of diagrams

We assign a one-to-one correspondence between terms in the expansion for Z and graphs on the lattice. For the particular example of $N = 3$ being considered here, this may be seen from a comparison of eqns (4.12)–(4.15) with Fig. 4.2, where in both cases terms are ordered in powers of v.

With this in mind, we may make the following points:

- Working out coefficients in the expansion is replaced by enumerating graphs on the lattice.

- The topology of the graphs allows us to work out the coefficients in the expansion.

- Only graphs with even vertices contribute a non-vanishing coefficient. By an even vertex we mean one with an even number of bonds emerging.

Hence only the graph of order v^0 and the graph of order v^N (with N bonds) will contribute. Thus for an Ising ring with N-spins, we have

$$Z_N = \cosh^N K (2^N + 2^N v^N) \equiv 2^N (\cosh^N K + \sinh^N K). \qquad (4.20)$$

In general for any dimension of lattice, the Ising partition function may be written as

$$Z_N = (\cosh K)^P 2^N \sum_{r=0}^{P} g(r) v^r, \qquad (4.21)$$

where $g(r)$ is the number of closed graphs that we can draw on the lattice, such that $g(0) = 1$.

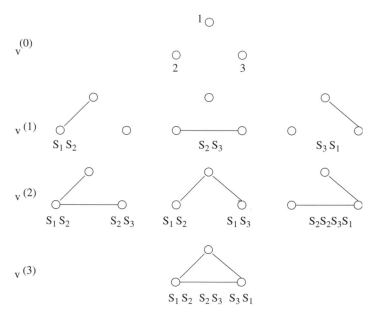

FIG. 4.2 Diagrams for high-temperature expansion of the Ising ring.

In practice, the calculation of the partition function for the Ising model is, by this method, equivalent to the problem of counting closed graphs on a lattice.

4.2.3   Behavior near the critical point

The expansion method is fine for $T \gg T_c$. But we really want $T \to T_c$ and there we have to reckon with nonanalytic behavior. We can improve the expansion method as follows. We have previously encountered the idea of critical exponents in Section 3.3.1. Now we give a more general definition of critical exponents, in the form:

$$F(\theta_c) = A\theta_c^{-s}(1 + \theta_c^y + \cdots), \tag{4.22}$$

for $T \to T_c$, where the reduced temperature is given by $\theta_c = (T - T_c)/T_c$ and F stands for any relevant macroscopic variable. We should note that the exponents are chosen such that $s > 0$ to give a singularity for temperatures approaching the critical temperature and $y > 0$ in order to ensure analytic behavior at temperatures much greater than T_c. As, in this section, we are interested in behavior near the critical point, we shall ignore the corrections required to give analytic behavior away from it. To do this, we simply set $y = 0$ and absorb the correction term into the value of the prefactor A. Accordingly, we write F as

$$F(\theta_c) = A \left(\frac{T}{T_c}\right)^{-s} \left(1 - \frac{T_c}{T}\right)^{-s} = A \left(\frac{\beta}{\beta_c}\right)^{s} \left(1 - \frac{\beta}{\beta_c}\right)^{-s}, \tag{4.23}$$

where $\beta \equiv 1/kT$. Then we expand $(1 - \beta/\beta_c)^{-s}$, using the binomial expansion, to obtain

$$F(\theta_c) = A \left(\frac{\beta}{\beta_c}\right)^{s} \left[1 + s\left(\frac{\beta}{\beta_c}\right) + \frac{s(s+1)}{2}\left(\frac{\beta}{\beta_c}\right)^{2} + \cdots \right.$$

$$\left. + \frac{s(s+1)\cdots(s+(n-1))}{n!}\left(\frac{\beta}{\beta_c}\right)^{n} + \cdots \right] \equiv \sum_{n} a_n \beta^{s+n}. \tag{4.24}$$

Now the convergence of this series will be limited by the presence of singularities. For instance, a simple pole on the real axis could suggest the presence of a critical point. We make use of the ratio test to examine the convergence. The radius of convergence r_c is given by:

$$r_c^{-1} = \lim_{n \to \infty} \frac{a_n}{a_{n-1}}.$$

Then we take the ratio of coefficients of β^n and β^{n-1} in the expansion:

$$\frac{a_n}{a_{n-1}} = \frac{1}{\beta_c} + \frac{(s-1)}{n\beta_c}, \tag{4.25}$$

and carry out the following procedure:

- Plot a_n/a_{n-1} against n.
- The intercept $\approx 1/\beta_c$ gives the critical temperature.
- The slope $\simeq (s-1)/\beta_c$ gives the critical exponent.

Further discussion of this approach can be found in the book [29]. See Fig. 9.5 of that reference where results are given for $F = X$, the susceptibility.

4.3   Low-density expansions

We shall consider a gas in which the molecules interact but we will "weaken" the interaction (or coupling) by restricting out attention to low densities. Accordingly we shall formulate the general theory on the assumption that interactions between particles will lead to perturbations of the "perfect gas" solution. Moreover, to stay within the confines of the present book, we shall only be able to treat the case of the "slightly imperfect gas" as a specific example of the method. With this in mind, it may be helpful to begin by considering the problem from the macroscopic point of view and try to anticipate the results of the microscopic theory, even if only qualitatively.

We know that the perfect gas law is consistent with the neglect of interactions between the molecules, and indeed also fails to allow for the fraction of the available volume which the molecules occupy. Thus, in general terms, we expect the perfect gas law to be a good approximation for a gas which is not too dense and not too cold. For a system of N molecules, occupying a fixed volume V, the perfect gas law, usually written as

$$PV = NkT, \tag{4.26}$$

tells us the pressure P. However, if we rewrite this in terms of the number density $n = N/V$, then we can assume that this must be the limiting form (at low densities) of some more complicated law which would be valid at larger densities. Thus,

$$P = nkT + O(n^2), \tag{4.27}$$

where for increasing values of number density we would expect to have to take into account higher-order terms in n. Formally, it is usual to anticipate that the exact form of the law may be written as the expansion

$$PV = NkT[B_1(T) + B_2(T)n + B_3(T)n^2 + \cdots]. \tag{4.28}$$

This is known as the virial expansion and the coefficients are referred to as the virial coefficients, thus:

$B_1(T)$: the first virial coefficient, which is equal to unity;
$B_2(T)$: the second virial coefficient;
$B_3(T)$: the third virial coefficient;

and so on, to any order. It should be noted that the coefficients depend on temperature because, for a given density, the effective strength of the particle interactions will depend on the temperature. It should also be noted that the status of eqn (4.28), on the basis of the reasoning given, is little more than that of a plausible guess. In the next section, we shall begin the process of seeing to what extent such a guess is supported by microscopic considerations.

4.4   Application to a "slightly imperfect" gas

Now we turn our attention to the microscopic picture, and consider N interacting particles in phase space, with the energy (or the Hamiltonian) as given by eqn (1.20). Although we shall base our approach on the classical picture, we shall divide phase space up into cells of volume $V_0 = h^3$. This allows us to take over the partition function for a quantum assembly to a classical description of the microstates. Equation (1.6) generalizes to

$$Z = \frac{1}{N!} \sum_{\text{cells}} e^{-E(\mathbf{X})/kT}, \tag{4.29}$$

where $\mathbf{X} \equiv (\mathbf{q}, \mathbf{p})$ is the usual "system point" in phase space, the sum over discrete microstates has been replaced by a sum over cells, and the factor $1/N!$ is required for the classical limit to take the correct form.

The cell size is small, being of the magnitude of the cube of Planck's constant h, so we can go over to the continuum limit and replace sums by integrals, thus:

$$\sum_{\text{cells}} \rightarrow \frac{1}{h^3} \int d\mathbf{p} \int d\mathbf{q}.$$

Hence eqn (4.29) can be written as:

$$Z = \frac{1}{N! h^{3N}} \int d\mathbf{p}_1 \cdots \int d\mathbf{p}_N \int d\mathbf{q}_1 \cdots \int d\mathbf{q}_N \times e^{-E(\mathbf{q},\mathbf{p})/kT}. \tag{4.30}$$

Note that the prefactor $1/N! h^{3N}$ guarantees that the free energy is consistent with the quantum formulation. Also note that we take the number of degrees of freedom to be determined purely by the translational velocities and exclude internal degrees of freedom such as rotations and vibrations of molecules. From now on in this chapter, we use Φ or ϕ for potential energy in order to avoid confusion with V for volume.

We can factor out the integration with respect to the momentum $\mathbf{p}$, by writing the exponential as

$$e^{-E(\mathbf{p},\mathbf{q})/kT} = e^{-\sum_{i=1}^{N} p_i^2/2mkT} \times e^{-\Phi(\mathbf{q})/kT},$$

and so write the total partition function for the system as the product of Z_0, the partition function for the perfect gas, with another function Q, thus

$$Z = Z_0 Q, \tag{4.31}$$

where

$$Z_0 = \frac{V^N}{N!} \left(\frac{2\pi mkT}{h^2}\right)^{3N/2},$$ (4.32)

and the *the configurational partition function* or, more usually, *configurational integral* Q is given by

$$Q = \frac{1}{V^N} \int d\mathbf{q}_1 \cdots \int d\mathbf{q}_N e^{-\Phi(\mathbf{q})/kT}.$$ (4.33)

We shall restrict our attention to the important general case of two-body potentials where

$$\Phi(\mathbf{q}) = \sum_{i<j=1}^{N} \phi(|\mathbf{q}_i - \mathbf{q}_j|) \equiv \sum_{i<j=1}^{N} \phi_{ij},$$ (4.34)

and hence the function $\Phi(\mathbf{q})$ will be written as the double sum over ϕ_{ij} from now on.

Evaluation of (4.33) for Q is difficult in general, and depends very much on the form of the two-body potential ϕ_{ij}. For instance, for molecules with radius $\sim b$, the hard-sphere potential is

$$\phi_{HS}(r) = \infty \quad \text{for } r < 2b,$$

$$= 0 \quad \text{for } r > 2b,$$ (4.35)

where we have taken the interparticle separation to be r. This potential is illustrated in Fig. 4.3.

Or, the more realistic Lennard–Jones (or "six -twelve") potential is given by

$$\phi_{LJ}(r) = 4\varepsilon[(b/r)^{12} - (b/r)^6]$$ (4.36)

where ε is related to binding energy, and this is illustrated schematically in Fig. 4.4. Evidently, if the temperature of the gas (and hence the kinetic energy of the molecules) is sufficiently low,

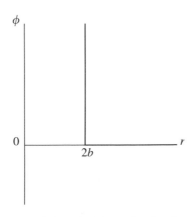

FIG. 4.3 The potential equivalent to hard-sphere interactions.

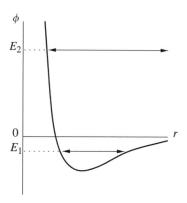

FIG. 4.4 A schematic impression of the Lennard–Jones potential.

a bound state may occur, as shown in the figure for an interparticle energy of E_1. However, if the temperature is high (interparticle energy labeled by E_2 in the figure), then the use of a hard-sphere potential might be a satisfactory approximation.

Other forms, such as the Coulomb potential, can be considered as required, but usually the configuration integral can only be evaluated approximately. In the next section we consider the use of the perturbation expansion in terms of a "book-keeping" parameter. This is introduced as an arbitrary factor, just as if it were the usual "small quantity" in perturbation theory; and, just as if it *were* the perturbation parameter, it is used to keep track of the various orders of terms during an iterative calculation. However, unlike the conventional perturbation parameter, it is not small and in fact is put equal to unity at the end of the calculation.

4.4.1 Perturbation expansion of the configuration integral

If the potential is, in some sense, weak (and we shall enlarge on what we mean by this at the end of the section), then we can expand out the exponential in (4.33) as a power series and truncate the resulting expansion at low order. In general, for any exponential we have the result given at the beginning of the chapter, and expanding the exponential in eqn (4.33) in this way gives us

$$Q = V^{-N} \int d\mathbf{q}_1 \cdots \int d\mathbf{q}_N \sum_{s=o}^{\infty} - \left(\frac{\lambda}{kT} \right)^s \frac{1}{s!} \left(\sum_{i<j=1}^{N} \phi_{ij} \right)^s , \qquad (4.37)$$

where λ is a "book-keeping" parameter ($\lambda = 1$). Any possibility of low-order truncation depends on integrals being well-behaved and this in turn depends very much on the nature of ϕ. Also, combinatorial effects increase with order λ^s, as follows:

$s = 0$:

$$Q_0 = V^{-N} \int d\mathbf{q}_1 \cdots \int d\mathbf{q}_N = 1, \qquad (4.38)$$

where,

$$\int d\mathbf{q}_1 = V \cdots \int d\mathbf{q}_N = V.$$

$s = 1$:

$$(-kT)Q_1 = V^{-N} \int d\mathbf{q}_1 \cdots \int d\mathbf{q}_N \left(\sum_{i<j=1}^{N} \phi_{ij} \right)$$

$$= V^{-N} \int d\mathbf{q}_1 \cdots \int d\mathbf{q}_N (\phi_{12} + \phi_{13} + \phi_{23} + \phi_{14} + \cdots)$$

$$= V^{-2} \left(\int \int d\mathbf{q}_1 d\mathbf{q}_2 \phi_{12} + \int \int d\mathbf{q}_1 d\mathbf{q}_3 \phi_{13} + \cdots \right). \tag{4.39}$$

And so on. Noting that Q_1 is made up of many identical terms, each of which is a double integral over the same pairwise potential, it follows that we need to evaluate only one of these integrals, and may then multiply the result by the number of pairs which can be chosen from N particles. Hence

$$Q_1 = -\tfrac{1}{2}N(N-1)V^{-1} \int \frac{\phi_{12}}{kT} d\mathbf{r}_{12}, \tag{4.40}$$

where we have made the change of variables $\mathbf{r}_{12} = |\mathbf{q}_1 - \mathbf{q}_2|)$, and the integration with respect to the centroid coordinate $\mathbf{R} = (\mathbf{q}_1 + \mathbf{q}_2)/2$ cancels one of the factors $1/V$.

Higher orders get more complicated and, as we shall see, diagram methods can be helpful. But the real problem is unsatisfactory behavior when we attempt to take the thermodynamic limit:

$$Lt \; N/V \to n, \quad \text{as } N, V \to \infty.$$

The expansion fails this test as, at any order s, there are various dependences on n, so that it does not take the form expected in eqn (4.28). (In mathematical terms, the expansion is *inhomogeneous*.)

This problem is not as serious as it might first appear, although our way of dealing with it may look rather like a trick. First we recall that the object of calculating the partition function is to calculate the free energy F (and hence all thermodynamical properties). To do this we use the bridge equation $F = -kT \ln \mathcal{Z}$ (see eqn (1.125)), which tells us that $F \sim \mathcal{Z}$. Hence our trick is to work with $\ln Q$ rather than Q (in other words, with the free energy due to interactions) to get a new series. In practice this amounts to a rearrangement of the perturbation expansion such that one finds an infinite series of terms associated with n, n^2, n^3, and so on. Each of these infinite series must be summed to give a coefficient in our new expansion in powers of n. We shall discuss some aspects of doing this in the next section, along with the helpful introduction of *Mayer functions*.

However, we close here by reconsidering what we mean by saying that the potential is "weak." We obtain one immediate clue from the above problem with the perturbation expansion, which is effectively (as is usual in many-body problems) in terms of the interaction strength. Intuitively, we can see that if the density is low, on average the particles will be far

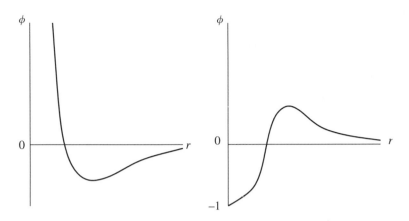

FIG. 4.5 The Mayer function f corresponding to a realistic choice of interparticle potential ϕ.

apart and hence the contribution of the interaction potential to the overall potential energy will be small. Also, we note that the potential energy (just like the kinetic energy) always appears divided by the factor kT. But, unlike the kinetic energy, the potential energy is purely configurational and does not increase with temperature and so for large temperatures the argument of the exponentials will be small. Thus, for either low densities or high temperatures the exponentials can be expanded and truncated at low order. In the first part of this chapter, where we consider critical phenomena, in the nature of things (e.g. a gas becoming a liquid) the density cannot realistically be treated as small. In these circumstances however, we saw that it can be useful to use the temperature as a *control parameter* and the interaction divided by kT is referred to as the *coupling*.

4.4.2   The density expansion and the Virial coefficients

In real gases, it is postulated that higher-density corrections to the perfect gas equation take the form given by eqn (4.28). Here we shall use statistical mechanics to explore the general method of calculating the virial coefficients, and although we shall not give a complete treatment, we shall highlight some of the difficulties involved. In the following section we shall then calculate the second virial coefficient B_2 explicitly.

From eqns (4.33) and (4.34), we may write the configurational integral as

$$Q = \frac{1}{V^N} \int d\mathbf{q}_1 \cdots \int d\mathbf{q}_N \, e^{-\sum_{i<j} \phi_{ij}/kT} = \frac{1}{V^N} \int d\mathbf{q}_1 \cdots \int d\mathbf{q}_N \prod_{i<j} e^{-\phi_{ij}/kT}. \quad (4.41)$$

Now we introduce the *Mayer functions* f_{ij}, which are defined (Fig. 4.5) such that

$$f_{ij} = e^{-\phi_{ij}/kT} - 1. \quad (4.42)$$

These possess the useful property that:

$$\text{as} \quad r \to 0, \quad f_{ij} \to -1 \quad \text{for} \quad \phi_{ij} \to \infty,$$

and changes the repeated product of (4.41) into an expansion in which the terms correspond to order of interaction: that is, two-body interactions, three-body interactions, and so on.

Upon substitution of (4.42), eqn (4.41) for the configurational integral becomes:

$$Q = \frac{1}{V^N} \int d\mathbf{q}_1 \cdots \int d\mathbf{q}_N \prod_{i<j}(1 + f_{ij})$$

$$= \frac{1}{V^N} \int d\mathbf{q}_1 \cdots \int d\mathbf{q}_N \left[1 + \sum_{i<j} f_{ij} + \sum_{i<j}\sum_{k<l} f_{ij} f_{kl} + \cdots \right]. \tag{4.43}$$

Note three points about this:

1. f_{ij} is negligibly small in value unless the molecules making up the pair labeled by i and j are close together. Hence, for nonnegligible values, f_{12} requires molecules 1 and 2 to collide; $f_{12} f_{34}$ requires molecules 1 and 2 to collide simultaneously with the collision between molecules 3 and 4; $f_{12} f_{23}$ requires a triple collision of molecules 1, 2, and 3; and so on.

2. The terms in eqn (4.43) involve *molecular clusters*. For this reason the multiple integrals in (4.43) are known as *cluster integrals*.

3. The expansion given in eqn (4.43) is known as the *virial cluster expansion*.

The graphical representation of the types of integral which occur in the expansion in (4.43) is shown in Fig. 4.6 for two-particle and three-particle clusters. Each circle corresponds to a particle, along with the corresponding integration over its position coordinate; while the line joining any pair of circles labeled by i and j stands for the interaction f_{ij}. Before considering the different orders, we shall find it helpful to characterize the terms of (4.43) by means of coefficients c_l, such that

$$c_l = \int d\mathbf{q}_1 \cdots \int d\mathbf{q}_{l-1} \overset{*}{\sum} \left[\prod_{i<j} f_{ij} \right], \tag{4.44}$$

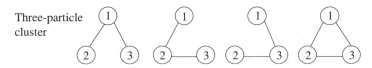

FIG. 4.6 Graphical representation of the integrals for two- and three-particle clusters.

where l is the number of particles in a cluster and

$$\sum{}^{*} \equiv \text{the sum of over all products consistent with a single cluster.} \qquad (4.45)$$

The significance of c_l will emerge as we consider succesively larger particle clusters.

4.4.3 The two-particle cluster

This is the lowest-order, and also the simplest, cluster, as it can only be formed in one way. The basic form is illustrated in Fig. 4.6. For this case, the index in eqn (4.43) is $l = 2$ and the corresponding coefficient c_2 is given by

$$c_2 = \int d\mathbf{q}_1 \, f_{12}(\mathbf{q}_1) \equiv I_2. \qquad (4.46)$$

It should be noted that only one integral appears, as the other one can be done directly by a change of variables to cancel a factor $1/V$. We have discussed this point previously when deriving eqn (4.40).

The overall contribution to the configuration integral is easily worked out in this case, as we just multiply by the number of pairs of particles, which is $N(N-1)/2$. This is what we shall do in the next section in order to calculate the second virial coefficient.

4.4.4 The three-particle cluster

This is the case $l = 3$ and as we may see from Fig. 4.6, there are four possible arrangements. The first three give the same contribution, so we shall just consider the first of these as typical and multiply our answer by a factor of three.

Considering the first of these terms, on the extreme left hand side of the figure, we may integrate over either $\mathbf{q}_1$ or $\mathbf{q}_2$ and obtain I_2, just as in the two-particle case. Integrating again over the volume of one of the remaining two particles also gives I_2 and hence the contribution of this type of arrangement is I_2^2, and with three such arrangements in all, we have

$$c_3 = 3I_2^2 + \int d\mathbf{q}_1 \int d\mathbf{q}_2 \, f_{12} f_{23} f_{13}, \qquad (4.47)$$

where the last term corresponds to the closed triangle in Fig. 4.5. We note that the linkages between the interactions are such that it is not possible to factor this into simpler forms and hence we give it its own name and write it as follows:

$$I_3 = \int d\mathbf{q}_1 \int d\mathbf{q}_2 \, f_{12} f_{23} f_{13}, \qquad (4.48)$$

and hence

$$c_3 = 3I_2^2 + I_3. \qquad (4.49)$$

We refer to I_3 and I_2 as *irreducible integrals* and in general we may expect that the numerical value of any c_l will be determined by sums and products of such integrals.

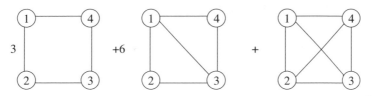

FIG. 4.7 Graphical representation of the irreducible integrals for a four-particle cluster.

4.4.5 The four-particle cluster

We shall not provide a full analysis of the four-particle case, but will merely concentrate on the irreducible integrals. In this instance, we have $l = 4$ and the three possibilities for irreducible integrals at this order are shown in Fig. 4.7.

The corresponding mathematical forms may be written down as follows:

$$I_{40} = \int d\mathbf{q}_1 \int d\mathbf{q}_2 \int d\mathbf{q}_3 \, f_{12} f_{23} f_{34} f_{14};$$ (4.50)

$$I_{41} = \int d\mathbf{q}_1 \int d\mathbf{q}_2 \int d\mathbf{q}_3 \, f_{12} f_{23} f_{34} f_{14} f_{13};$$ (4.51)

$$I_{42} = \int d\mathbf{q}_1 \int d\mathbf{q}_2 \int d\mathbf{q}_3 \, f_{12} f_{23} f_{34} f_{14} f_{13} f_{24};$$ (4.52)

and it can be shown by repeating the analysis of the preceding sections that

$$c_4 = 16I_2^2 + 12I_2 I_3 + I_4,$$ (4.53)

where

$$I_4 = 3I_{40} + 6I_{41} + I_{42}.$$ (4.54)

The full analysis goes on to obtain the general term and hence the general form of the configuration integral to any order in particle clusters. The procedure is both complicated and highly technical, and ultimately depends on the model taken for the interaction potential. For those who wish to pursue it further, we shall give some guidance in the section on further reading at the end of this chapter. However, much of the intrinsic physical interest can be seen by calculating the lowest nontrivial correction to perfect gas behavior and this we shall do in the next section.

4.4.6 Calculation of the second virial coefficient B_2

We shall work only to first order in f_{ij}. That is,

$$Q = \frac{1}{V^N} \int d\mathbf{q}_1 \cdots \int d\mathbf{q}_N \left[1 + \sum_{i<j} f_{ij} \right].$$ (4.55)

Now, evidently $f_{12} = f_{13} = \cdots = f_{23}$, so we shall take f_{12} as representative. Also, there are $N(N-1)/2$ pairs. Hence, to first order in the interaction potential, we have

$$Q = \frac{1}{V^N} \int d\mathbf{q}_1 \cdots \int d\mathbf{q}_N \left[1 + \frac{N(N-1)}{2} f_{12} \right],$$

and so

$$Q = \frac{1}{V^N} \left[V^N + V^{N-2} \int d\mathbf{q}_1 \int d\mathbf{q}_2 \frac{N(N-1)}{2} f_{12} \right]$$

$$= 1 + V^{-2} \int d\mathbf{q}_1 \int d\mathbf{q}_2 \frac{N(N-1)}{2} f(|\mathbf{q}_1 - \mathbf{q}_2|). \tag{4.56}$$

Next we change variables to work in the relative and centroid coordinates, $\mathbf{r} = \mathbf{q}_1 - \mathbf{q}_2$ and $\mathbf{R} = \frac{1}{2}(\mathbf{q}_1 + \mathbf{q}_2)$, respectively: this is illustrated in Fig. 4.8. Then, assuming spherical symmetry of the interaction potential, we obtain

$$Q = 1 + V^{-2} \int d\mathbf{R} \frac{N(N-1)}{2} \int d\mathbf{r} f(r)$$

$$= 1 + \frac{N(N-1)}{2V} \int f(r) d\mathbf{r}$$

$$= 1 + \frac{N^2}{2V} I_2, \tag{4.57}$$

where I_2 is the cluster integral

$$I_2 = \int d\mathbf{r} f(r) = \int d\mathbf{r} [e^{-\phi(r)/kT} - 1], \tag{4.58}$$

and, in the last step, we have substituted for the Mayer function to obtain our result for the second virial coefficient in terms of the interaction potential.

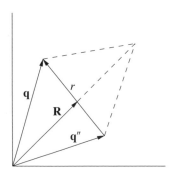

FIG. 4.8 Change to centroid and difference coordinates.

Now we resort to two tricks. First, we rewrite eqn (4.57) for Q as the leading terms in an expansion

$$Q = 1 + N\left(\frac{NI_2}{2V}\right) + \cdots. \tag{4.59}$$

Second, we note that the free energy $F \sim \ln Q$ must be extensive, so we must have $\ln Q \sim N$. It follows that the most likely form of the sum of the series on the right hand side of (4.59) is

$$Q = \left(1 + \frac{NI_2}{2V}\right)^N. \tag{4.60}$$

So, from this result, the bridge equation for F and (4.31) for $\mathcal{Z}$ we obtain the following expression for the free energy:

$$F = -kT \ln \mathcal{Z} = -kT \ln \mathcal{Z}_0 - kT \ln Q$$

$$= F_0 - kT \ln Q = F_0 - NkT \ln\left(1 + \frac{NI_2}{V2}\right)$$

$$= F_0 - \frac{NkT}{2}\left(\frac{N}{V}\right)I_2, \tag{4.61}$$

where we used the Taylor series for $\ln(1+x)$, as given at the beginning of this chapter and truncated it at small x: in this case, number density. We may obtain the equation of state from the usual thermodynamic relationship $P = (-\partial F/\partial V)_{T,N}$. Thus

$$P = \frac{NkT}{V} - \frac{NkT}{V}\left(\frac{N}{V}\right)\frac{1}{2}I_2 = \frac{NkT}{V}\left[1 - \frac{I_2}{2}\left(\frac{N}{V}\right)\right]. \tag{4.62}$$

Then, comparison with the expansion of (4.28) yields

$$B_2 = -\tfrac{1}{2}I_2, \tag{4.63}$$

as the second virial coefficient.

Lastly, we should note that the procedure just followed, although on the face of it *ad hoc*, nevertheless constitutes an effective renormalization, equivalent to a partial summation of the perturbation series.

4.5 The Van der Waals equation

The Van der Waals equation takes the form

$$\left(P + \frac{a}{V^2}\right)(V - b) = NkT,$$

where a/V^2 represents the effect of mutual attraction between molecules, and b is the "excluded volume" due to the finite size of the molecules. The equation is based on a model where $\phi(r)$ is taken as corresponding to a "hard sphere" potential for $r \leq d$, but is weakly attractive for $r > d$.

From eqns (4.63) and (4.58) we have

$$B_2 = -\frac{1}{2}I_2 = -\frac{1}{2} \int dr \left[e^{-\phi(r)/kT} - 1 \right],$$

and, on the basis of our assumptions about $\phi(r)$, we may make the simplification

$$e^{-\phi(r)/kT} \simeq 0 \qquad\qquad \text{for } r < d;$$

$$\simeq 1 - \phi/kT, \quad \text{for } r > d.$$

Then, dividing the range of integration into two parts, we obtain

$$B_2 = \frac{1}{2} \int_0^d 4\pi r^2 dr + \frac{1}{2} \int_d^\infty 4\pi r^2 \frac{\phi(r)}{kT} dr = B - A/kT,$$

where

$$B = \frac{2\pi d^3}{3} = 4v_0,$$

where $v_0 \equiv$ volume of a molecule, and

$$A = -2\pi \int_d^\infty r^2 \phi(r) dr.$$

Now, from (4.62) and these results, the pressure is given by

$$P = \frac{NkT}{V} \left[1 + \frac{N}{V} \left(B - \frac{A}{kT} \right) \right].$$

Let us define the Van der Waals constants as:

$$b = 4Nv_0,$$

which is the total excluded volume of N molecules, and

$$a = N^2 A,$$

which is the total effect of interactions between all possible pairs. Then the equation for the pressure becomes

$$P = \frac{NkT}{V} + \frac{NkT}{V} \cdot \frac{b}{V} - \frac{a}{V^2};$$

or:

$$P + \frac{a}{V^2} \simeq \frac{NkT}{V} \cdot \frac{1}{1 - b/V},$$

where the interpretation in terms of a first-order truncation of the binomial expansion is justified for small values of b/V. Then multiplying across, and cancelling as appropriate, yields

$$\left(P + \frac{a}{V^2} \right)(V - b) = NkT,$$

as required.

4.6   The Debye–Hückel theory revisited

In Section 3.2 we met the Debye–Hückel theory, which was a phenomenological treatment of the long-range forces, and saw that it was essentially a self-consistent or mean-field theory. However, one can also apply the perturbation methods of the present chapter to the problem of a plasma or electrolyte, where the Coulomb potential is involved, and in 1950 Mayer showed that the Debye–Hückel (1923) result of a screened potential could be obtained by summing over a restricted class of diagrams (or terms) in perturbation theory. We shall only give a brief outline of that theory here.

Previously, we considered a system of many electrons, each free to move in a medium which carried a uniform positive charge, thus ensuing electrical neutrality. For a more funda-mental treatment, it is helpful to think of the summations in eqn (4.43) for the configuration integral to be over different species (and charge) of ion. Then we impose the constraint that the sum over all ions is such as to ensure electrical neutrality. However, in applying perturb-ation theory, our first problem is to decide which form of potential to substitute for ϕ_{ij} in the individual terms of the expansion on the left hand side of eqn (4.43).

The difficulty lies in the fact that we are dealing with particles which are free to move and which possess electrical charges of various magnitudes and which can be of either negative or positive sign. Accordingly, we start with the free-field or infinite-dilution case and assume that, for large enough distances r_{ij} between ions labeled by i and j, the Coulomb potential applies, and so we write the potential energy for a pair of ions in the form

$$\phi_{ij} = \frac{z_i z_j e^2}{r_{ij}} + \phi_{ij}^*, \tag{4.64}$$

where z_i, z_j are the charges on the ions labeled i, j, respectively, the dielectric coefficient has been taken as unity; and ϕ_{ij}^* is a correction which ensures correct short-range behavior. In fact, the usual expedient is to assume that the correction term has the form of a hard-sphere potential, but we shall not take that any further and instead will concentrate on long-range behavior.

We do this because the real difficulty is that the long-range $1/r$ tail in the potential can cause the integrals to diverge as $r \to \infty$. We may investigate the nature of this problem as follows. Suppose we pick a particular pair of ions, one of type a and one of type b and expand the function f_{ba}, as defined by eqn (4.42), for large distances (i.e. small values of the exponent), thus:

$$f_{ba} \simeq \frac{z_a z_b e^2}{kT} \times \frac{1}{r_{ba}}, \tag{4.65}$$

at first order. If we now try to work out the second virial coefficient, $B_{2,ba}$,

$$B_{2,ba} = \frac{1}{2} \int_0^\infty 4\pi r_{ba}^2 f_{ba} dr_{ba}, \tag{4.66}$$

then it is clear that the integral diverges at the upper limit if we substitute the first-order term as given by eqn (4.65) for the potential.

In fact, this is not a serious problem at first order. This is because the requirement of overall electrical neutrality ensures that the sum over the various values of the indices b and a gives zero before the integration is carried out. However, the next term in the expansion of the exponential involves the squares of the charges and therefore does not in general vanish and hence the divergence in the integrals then becomes serious.

This problem can be circumvented by *renormalising* the expansions, with an upper limit $r = R$, to cut off each integral, and then letting $R \to \infty$ at the end of the calculation. However, Mayer was guided by the Debye–Hückel theory, and modified the trial potential from (4.65) to the following

$$\phi_{ij} = \frac{z_i z_j e^2}{r_{ij}} e^{-\alpha r_{ij}} + \phi_{ij}^*, \tag{4.67}$$

where α is an arbitrary constant, analogous to the Debye length. This form of the potential allows the integrals to converge, which in turn allows one to interchange the order of summations and integrations in the theory. If this is done carefully, the integrals of sums of products are found to converge, even in the limit $\alpha \to 0$.

This allows a well-behaved perturbation theory to be set up, and it is found that terms corresponding to diagrams which are fully connected but have no internal lines can be summed to recover the Debye–Hückel result. We show some of these diagrams (up to sixth order) in Fig. 4.9. Mayer referred to these as "cycles," but they are often referred to nowadays as "ring diagrams." It should be appreciated that when one does this sort of partial summation, a great many other terms of unknown properties are being neglected simply because their diagrammatic representation belongs to a different class on topological grounds.

This is as far as we shall take this topic in this part of the book. However, it is worth remarking in summary that we have touched on many of the characteristic features of many-body or field theory; in that there are usually divergences at short or long distances (and sometimes both!) and that renormalization often involves a partial summation of the perturbation series. We shall return to these matters in the third part of the book when we consider renormalization group theory.

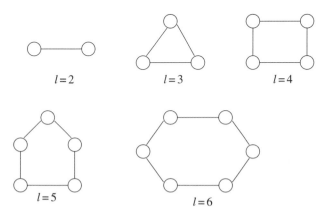

$l=2$ $\qquad$ $l=3$ $\qquad$ $l=4$

$l=5$ $\qquad$ $l=6$

FIG. 4.9 Irreducible cluster diagrams contributing to Debye–Hückel theory up to sixth order.

Further reading

A more extensive discussion of the high-temperature expansion can be found in [29]. In order to give a proper treatment of the virial-type expansion for the pressure in a fluid of interacting particles it is necessary to use the Grand Canonical Ensemble, rather than just the Canonical Ensemble, as used in the present book. Fuller treatments of this type will be found in [12] and [26]. Some of the original papers are worth consulting and [18] is of particular interest in that it has the first appearance of diagrams (even although little use is made of them!) and thus possesses additional historic interest. In [19] and [27], the theory of Deybe and Hückel is revisited and one also sees a marked development in the use of diagrams.

4.7 Exercises

Solutions to these exercises, along with further exercises and their solutions, will be found at: www.oup.com

1. Use the high-temperature expansion to show that the partition function of the (open) Ising linear chain takes the form

$$Z_N = 2^N \cosh^{N-1} K,$$

and explain the differences between this and the result for the Ising ring.

2. Use the method of high-temperature expansion to obtain the spin–spin correlation $\langle S_n S_m \rangle$ of the Ising model as

$$\langle S_m S_n \rangle = Z_N^{-1} \cosh^P K \, 2^N \sum_{r=1}^{P} f_{mn}(r) v^r,$$

where P is the number of nearest neighbor pairs, $v \equiv \tanh K$ and $f_{mn}(r)$ is the number of graphs of r lines, with even vertices except at sites m and n.
   Show that for the case of the Ising linear chain, this reduces to

$$\langle S_m S_n \rangle = v^{|n-m|},$$

and draw the corresponding graph.

3. If one can assume that the interparticle potential $\phi(r)$ is large (and positive) on the scale of kT, for $r < d$; and is small for $r > d$, where d is the molecular diameter, then it may be shown that the second virial coefficient can be written as

$$B_2 = B - A/kT.$$

Obtain explicit expressions for the constants a and b, and show that the resulting equation of state may be reduced to the van der Waal's equation.

4. If a gas of interacting particles is modeled as hard spheres of radius a, show that the second virial coefficient takes the form:

$$B_2 = \frac{2\pi a^3}{3}.$$

Given that the third virial coefficient may be written as:

$$B_3 = \tfrac{1}{3} \int d^3r \int d^3r' f(|\mathbf{r}|) f(|\mathbf{r}'|) f(|\mathbf{r} - \mathbf{r}'|),$$

where $f(r) = e^{-\beta \Phi(r)} - 1$, show that this is related to the second virial coefficient B_2 by

$$B_3 = \tfrac{5}{8} B_2^2,$$

for a system of hard spheres.

5. A gas consisting of N classical point particles of mass m occupies a volume V at temperature T. If the particles interact through a two-body potential of the form:

$$\phi(r_{ij}) = \frac{A}{r_{ij}^n},$$

where A is a constant, $r_{ij} = |\mathbf{q}_i - \mathbf{q}_j|$ and n is positive, show that the canonical partition function is a homogeneous function, in the sense

$$Z(\lambda T, \lambda^{-3/n} V) = \lambda^{3N(1/2 - 1/n)} Z(T, V),$$

where λ is an arbitrary scaling factor.

6. Prove the identity

$$\frac{\partial}{\partial x} e^{-\beta H} = - \int_0^\beta e^{-(\beta - y)H} \frac{\partial H}{\partial x} e^{-yH} dy,$$

to second order in β by equating coefficients in the high-temperature (small β) expansion of each side of the relation, where H is an operator.

7. Show that at high temperatures, the heat capacity of a quantum assembly can be written as

$$C_V = \frac{1}{kT^2} \left\{ \frac{Tr(H^2)}{Tr(1)} - \frac{[Tr(H)]^2}{[Tr(1)]^2} + 0(\beta) \right\},$$

where $\beta = 1/kT$.

8. Show that the use of the Van der Waals equation,

$$\left(P + \frac{a}{V^2} \right) (V - b) = NkT,$$

to describe phase transitions in a fluid system leads to the following values for the critical parameters:

$$P_c = a/27b^2, \quad V_c = 3b, \quad NkT_c = 8a/27b.$$

Hence show that the Van der Waals equation may be written in the universal form

$$\left(\tilde{p} + \frac{3}{\tilde{v}^2}\right)(3\tilde{v} - 1) = 8\tilde{t},$$

where

$$\tilde{p} = P/P_c, \quad \tilde{v} = V/V_c, \quad \tilde{t} = T/T_c.$$

9. By re-expressing the Van der Waals equation in terms of the reduced variables; $P = (P - P_c)/P_c$, $v = (V - V_c)/V_c$ and $\theta_c = (T - T_c)/T_c$, obtain values for the critical exponents γ and δ. Comment on the values which you obtain.

5

CLASSICAL NONLINEAR SYSTEMS
DRIVEN BY RANDOM NOISE

In this chapter we at last make use of the formal perturbation theory as presented in Section 1.3.5. We apply it to the study of classical nonlinear systems such as fluid turbulence. It has been recognized since the 1950s that turbulence theory has a similar structure to quantum field theory and this idea has proved beneficial to research in turbulence. At the same time, there are essential differences betweem the two subjects and in recent years there has been some tendency to lose sight of that fact. This aspect becomes particularly important as growing activity in soft condensed matter draws increasingly on turbulence theory for certain basic procedures. For this reason we have added some cautionary words on the analogies between classical and quantum field theories at the end of the chapter.

Before closing this introductory section, we shall make a few remarks on the general structure of this chapter.

After discussing and formulating the problem of classical nonlinear fields, we present what we call a "toy version" of the equations and base our perturbation theory—and its subsequent renormalization—on this simplified system. However it should be emphasized that the simplification is purely notational. We wish to avoid becoming bogged down in algebraic detail and tensor indices. The procedures presented are the correct ones, and full algebraic detail can be easily restored. In Chapter 6 we shall examine some representative equations which arise at second-order renormalized level and discuss their ability to reproduce the physics of the system.

It is also worth pointing out that although in general the approach will be applicable to all the problems for which the generic equation (see eqn (5.2)) is relevant, we shall focus on the problem of fluid turbulence as a specific example of the perturbative approach.

5.1 The generic equation of motion

The evolution of various classical stochastic fields can be described by nonlinear partial differential equations. As such equations are inherently deterministic, it is necessary to add some kind of random noise in order to produce the requisite stochastic behavior of the field. Such an approach has its origins in the study of microscopic phenomena such as Brownian motion, where the molecular impacts on a colloidal particle can be modeled both as a random force (which "drives" the colloidal particle) and a dynamical friction which determines its response to the driving force. The result is the well-known (linear) Langevin equation.

However, in recent years there has been great interest in macroscopic, continuous fields ranging from fluid turbulence (including diffusion of heat and matter) which is described by the Navier–Stokes equations (NSE), through the development of shock waves (Burgers equation), to a variety of growth problems (the Kardar–Parisi–Zhang or KPZ equation). All these equations can be represented as special cases of a generic nonlinear Langevin equation.

In order to see this, we represent any of the fields by a general vector field $u_\alpha(\mathbf{x}, t)$, where α is a cartesian tensor index which can take values $\alpha = 0, 1, 2,$ and 3. This can include the usual three-dimensional fluid velocity field (where $\alpha = 1, 2,$ and 3), or the velocity field supplemented by a scalar contaminant (e.g. heat or matter) denoted by $u_0(\mathbf{x}, t)$, or a one-dimensional height function $u_1(\mathbf{x}, t) \equiv u(\mathbf{x}, t)$ on a d-dimensional substrate $\mathbf{x}$. With this in mind, we can write a general equation for all of these processes and, as most of the time, we work in Fourier *wave number* space so we introduce the Fourier transform $u_\alpha(\mathbf{k}, t)$ of the velocity field in d dimensions by means of the relationship

$$u_\alpha(\mathbf{x}, t) = \int d^d k \, u_\alpha(\mathbf{k}, t) e^{i\mathbf{k}\cdot\mathbf{x}}. \tag{5.1}$$

Then the generic equation of motion may be written as:

$$\left(\frac{\partial}{\partial t} + \nu k^2\right) u_\alpha(\mathbf{k}, t) = f_\alpha(\mathbf{k}, t) + \lambda M_{\alpha\beta\gamma}(\mathbf{k}) \int d^3 j \, u_\beta(\mathbf{j}, t) u_\gamma(\mathbf{k} - \mathbf{j}, t). \tag{5.2}$$

The noise f is arbitrary and is usually chosen to have a multivariate normal distribution, with autocorrelation of the form:

$$\langle f_\alpha(\mathbf{k}, t) f_\beta(\mathbf{k}', t') \rangle = 2W(k)(2\pi)^d D_{\alpha\beta}(\mathbf{k}) \delta(\mathbf{k} + \mathbf{k}') \delta(t + t'). \tag{5.3}$$

It is usual to group the two linear terms together on the left hand side of the equation of motion in order to make a single linear operator. However, it can be helpful in interpreting the equation to re-write it as

$$\frac{\partial}{\partial t} u_\alpha(\mathbf{k}, t) = f_\alpha(\mathbf{k}, t) - \nu k^2 u_\alpha(\mathbf{k}, t) + \lambda M_{\alpha\beta\gamma}(\mathbf{k}) \int d^3 j \, u_\beta(\mathbf{j}, t) u_\gamma(\mathbf{k} - \mathbf{j}, t).$$

Then we can write the equation in words as:

Rate of change of $\mathbf{u}(\mathbf{k}, t)$ with time = Gain from forcing

+ Loss due to "damping"

+ Gain or loss due to coupling with other modes.

The last term poses the theoretical challenge. The effect of Fourier transformation is to turn the nonlinear term into a coupling term. In other words, the different spatial frequencies or wave number modes are coupled together. This also leads to problems when we try to obtain an equation for the mean value of $\mathbf{u}(\mathbf{k}, t)$, and we shall discuss that shortly after some remarks about the various physical systems described by eqn (5.2).

5.1.1 The Navier–Stokes equation: NSE

If we take $\alpha = 1, 2$, or 3, then eqn (5.2) is the NSE for incompressible viscous fluid motion. Here ν is the coefficient of kinematic viscosity of the fluid, we take the fluid density to be unity (that is, we work in "kinematic" units) and the operator $M_{\alpha\beta\gamma}(\mathbf{k})$ takes account of pressure gradients and maintains the incompressibility property as a constraint. We shall defer consideration of the specific form of $M_{\alpha\beta\gamma}(\mathbf{k})$ for turbulence until the next chapter. One can also, as mentioned above, incorporate a scalar contaminant into the problem by taking $\alpha = 0$ and using u_0 to represent temperature or some diffusing substance but we shall not pursue that here.

We note that $\lambda = 0$ (linear system) or $\lambda = 1$ (nonlinear system). That is, λ is also a control parameter. If we scaled variables in a suitable way, we could replace λ by a Reynolds number. Hence the perturbation expansion in λ is *effectively* in powers of a Reynolds number.

Lastly, $W(k)$ is a measure of the rate at which the stirring force does work on the fluid. The rate at which the force does work on the fluid must equal the energy dissipation rate:

$$\int W(k)\, \mathrm{d}k = \varepsilon. \tag{5.4}$$

5.1.2 The Burgers equation

If we take

$$u_\alpha(\mathbf{k}, t) \equiv u(k, t),$$

and set

$$\lambda M_{\alpha\beta\gamma}(\mathbf{k}) = k,$$

then eqn (5.2) reduces to the Burgers equation. This is a one-dimensional analogue of the NSE and was originally posed by Burgers as a simplified equation for the study of turbulence. It was not very helpful in that role but has since become of considerable importance in other fields such as nonlinear waves, the dynamics of growing interfaces and large-scale structures of the universe. In recent years it has also provided some support for the derivation of the KPZ equation, as discussed in the next sub-section.

5.1.3 The KPZ equation

If we take

$$u_\alpha(\mathbf{k}, t) \equiv h(\mathbf{k}, t)$$

and set

$$M_{\alpha\beta\gamma}(\mathbf{k}) = k,$$

where ν is a material constant associated with surface tension effects and λ is some other material constant of the system being studied, eqn (5.2) now reduces to the KPZ equation. It should be noted that $h(\mathbf{k}, t)$ is a height rather than a velocity.

There does not appear to be any general agreement on which specific systems are described by the KPZ equation but there is a widespread consensus that the equation is worthy of study in the context of nonlinear growth processes.

5.2   The moment closure problem

The fundamental problem of turbulence is encountered when we attempt a statistical formulation. To begin, let us express it symbolically. We write the equation of motion, (5.2) in a very simplified form as:

$$L_0 u = Muu,$$

where L_0 is the linear operator on the left hand side of eqn (5.2) and M is an obvious abbreviation of the operator on the right hand side.

Then we average term by term to obtain an evolution equation for the mean velocity $\bar{u} = \langle u \rangle$:

$$L_0 \langle u \rangle = M \langle uu \rangle. \tag{5.5}$$

Thus our equation for the mean velocity contains the second moment of the velocities on the right hand side.

Now we only know the relation between moments if we know the probability distribution $P(u)$, and as this is not the case, we must resort to finding an equation for the unknown second moment. To obtain an equation for $\langle uu \rangle$ we multiply the equation of motion through by u and again average term by term, thus:

$$L_0 \langle uu \rangle = M \langle uuu \rangle, \tag{5.6}$$

which introduces a further new unknown $\langle uuu \rangle$.

In order to form an equation for this new unknown, we must multiply through the equation of motion by uu and again average to obtain

$$L_0 \langle uuu \rangle = M \langle uuuu \rangle, \tag{5.7}$$

and so on, to any order.

Our problem now is clear:

- to solve (5.5), we need the solution of (5.6),

- to solve (5.6), we need the solution of (5.7),

- to solve (5.7), we need solution of …;

and so on … *ad infinitum*. This is the moment closure problem.

5.3   The pair-correlation tensor

By definition the general correlation tensor is:

$$Q_{\alpha\beta}(\mathbf{x}, x'; t, t') = \langle u_\alpha(\mathbf{x}, t) u_\beta(\mathbf{x}', t') \rangle, \tag{5.8}$$

where the correlation is between two velocities taken at different positions and at different times, and $\langle \ldots \rangle$ denotes the operation of taking averages. We restrict our attention to the case of spatial homogeneity and this implies that it takes the form

$$Q_{\alpha\beta}(\mathbf{x}, x'; t, t') = Q_{\alpha\beta}(\mathbf{x} - \mathbf{x}'; t, t'),$$
$$= Q_{\alpha\beta}(\mathbf{r}; t, t'). \tag{5.9}$$

We wish to work in wave number space, so taking the Fourier transform yields:

$$\langle u_{\alpha}(\mathbf{k}, t)u_{\beta}(\mathbf{k}', t')\rangle = \delta(\mathbf{k} + \mathbf{k}')Q_{\alpha\beta}(\mathbf{k}; t, t'), \tag{5.10}$$

and, if the field is stationary, we have the further simplification

$$Q_{\alpha\beta}(\mathbf{k}; t, t') = Q_{\alpha\beta}(\mathbf{k}; t - t'),$$
$$= Q_{\alpha\beta}(\mathbf{k}; \tau). \tag{5.11}$$

For the specific case of isotropic turbulence this can be written as:

$$Q_{\alpha\beta}(\mathbf{k}; t, t') = \delta_{\alpha\beta}Q(k; t, t') - \frac{k_{\alpha}k_{\beta}}{k^2}Q(k; t, t'),$$
$$= D_{\alpha\beta}(\mathbf{k})Q(k; t, t'). \tag{5.12}$$

where the projection operator is given by:

$$D_{\alpha\beta}(\mathbf{k}) = \delta_{\alpha\beta} - \frac{k_{\alpha}k_{\beta}}{k^2}. \tag{5.13}$$

Note that this form satisfies the continuity condition for incompressible fluid motion in both tensor indices.

5.4   The zero-order "model" system

We base our perturbation approach on a model. Take $\lambda = 0$, and equation (5.2) becomes:

$$\left(\frac{\partial}{\partial t} + \nu k^2\right)u_{\alpha}(\mathbf{k}, t) = f_{\alpha}(\mathbf{k}, t). \tag{5.14}$$

Choose the noise term $\mathbf{f}(\mathbf{k}, t)$ to have Gaussian statistics. The zero-order ($\lambda = 0$) velocity field is the solution of above equation, thus:

$$\mathbf{u}^{(0)}(\mathbf{k}, t) = \int dt' e^{-\nu k^2(t-t')}\mathbf{f}(\mathbf{k}, t') \equiv \int dt' G^{(0)}(\mathbf{k}; t - t')f(\mathbf{k}, t'). \tag{5.15}$$

This is *stirred* fluid motion, valid only in the limit of zero Reynolds number. Because $\mathbf{f}$ is Gaussian, so also is $\mathbf{u}^{(0)}$.

When we set $\lambda = 1$ the full equation is restored. The mixing effect of the nonlinear term couples together modes with different wave numbers. The physical effect is to induce an *exact*, non-Gaussian velocity field $\mathbf{u}(\mathbf{k}, t)$. This may be written as a series in powers of λ:

$$\mathbf{u}(\mathbf{k}, t) = \mathbf{u}^{(0)}(\mathbf{k}, t) + \lambda\mathbf{u}^{(1)}(\mathbf{k}, t) + \lambda^2\mathbf{u}^{(2)}(\mathbf{k}, t) + \cdots \tag{5.16}$$

The coefficients $\mathbf{u}^{(1)}, \mathbf{u}^{(2)}, \ldots$ are calculated iteratively in terms of $\mathbf{u}^{(0)}$.

5.5   A toy version of the equation of motion

Let us now introduce a simplified notation which is rather more realistic than that used in Section 5.2: the equation of motion (5.2) can be written in the reduced form

$$L_{0k}u_k = f_k + \lambda M_{kjl}u_j u_l, \tag{5.17}$$

where

$$L_{0k} \equiv \left(\frac{\partial}{\partial t} + \nu k^2\right), \tag{5.18}$$

and subscripts stand for all the variables (wave vector, tensor index, and time). This result is what we have referred to as the "toy version of the equations" in our opening remarks to this chapter. In this simplified notation the zeroorder solution (5.15) becomes

$$u_k^{(0)} = G_k^{(0)} f_k, \tag{5.19}$$

and the perturbation expansion (5.16) becomes

$$u_k = u_k^{(0)} + \lambda u_k^{(1)} + \lambda^2 u_k^{(2)} + \cdots . \tag{5.20}$$

5.6   Perturbation expansion of the toy equation of motion

The Gaussian model is soluble because we can factor moments of the $\mathbf{u}^{(0)}$ to all orders. We can summarize this as:

1. All odd-order moments $\langle u^{(0)}u^{(0)}u^{(0)}\rangle$ etc., are zero.
2. All even-order moments $\langle u^0 u^{(0)}u^{(0)}u^{(0)}\rangle$ etc., can be expressed as products of the second-order moments $\langle u^{(0)}u^{(0)}\rangle$.

As we shall see the resulting perturbation expansion for the exact second-order moment is $\langle uu\rangle$ wildly divergent. Nevertheless, we go ahead anyway!

5.6.1   The iterative calculation of coefficients

For convenience, we invert the operator on the left hand side of eqn (5.17) to write:

$$u_k = u_k^{(0)} + \lambda G_k^{(0)} M_{kjl}u_j u_l, \tag{5.21}$$

where we have used (5.19). Now substitute the perturbation expansion (5.20) for **u** into the above equation and multiply out the right hand side, thus:

$$
\begin{aligned}
u_k^{(0)} + \lambda u_k^{(1)} + \lambda^2 u_k^{(2)} + \cdots = \; & u_k^{(0)} + \lambda G_k^{(0)} M_{kjl} \\
& \times [u_j^{(0)} + \lambda u_j^{(1)} + \lambda^2 u_j^{(2)} + \cdots] \\
& \times [u_l^{(0)} + \lambda u_l^{(1)} + \lambda^2 u_l^{(2)} + \cdots].
\end{aligned}
\tag{5.22}
$$

Equate terms at each order in λ, thus:
Order λ^0

$$
u_k^{(0)} = G_k^{(0)} f_k.
\tag{5.23}
$$

Order λ^1

$$
u_k^{(1)} = G_k^{(0)} M_{kjl} u_j^{(0)} u_l^{(0)}.
\tag{5.24}
$$

Order λ^2

$$
u_k^{(2)} = 2 G_k^{(0)} M_{kjl} u_j^{(0)} u_l^{(1)}.
\tag{5.25}
$$

Order λ^3

$$
u_k^{(3)} = 2 G_k^{(0)} M_{kjl} u_j^{(0)} u_l^{(2)} + G_k^{(0)} M_{kjl} u_j^{(1)} u_l^{(1)}.
\tag{5.26}
$$

And so on, to any order. Then we can substitute successively for $u^{(1)}, u^{(2)}, \ldots$, in terms of $u^{(0)}$, so that all coefficients are expressed in terms of G_k^0, u_k^0 and M_{kjl} only.

5.6.2 Explicit form of the coefficients

First order is already in the correct form:

$$
u_k^{(1)} = G_k^{(0)} M_{kjl} u_j^{(0)} u_l^{(0)}.
\tag{5.27}
$$

At second order, substitute (5.27) for $u^{(1)}$ into (5.25) for $u^{(2)}$:

$$
u_k^{(2)} = 2 G_k^{(0)} M_{kjl} u_j^{(0)} G_l^{(0)} M_{lpq} u_p^{(0)} u_q^{(0)}.
\tag{5.28}
$$

At third order, substitute (5.27) for $u^{(1)}$ and (5.28) for $u^{(2)}$, respectively, into (5.26) for $u^{(3)}$:

$$
\begin{aligned}
u_k^{(3)} = \; & 4 G_k^{(0)} M_{kjl} u_j^{(0)} G_l^{(0)} M_{lpq} u_p^{(0)} G_q^{(0)} M_{qrs} u_r^{(0)} u_s^{(0)} \\
& + G_k^{(0)} M_{kjl} G_j^{(0)} M_{jpq} u_p^{(0)} u_q^{(0)} G_l^{(0)} M_{lrs} u_r^{(0)} u_s^{(0)};
\end{aligned}
\tag{5.29}
$$

and so on, for higher orders.

5.6.3 Equation for the exact correlation

The exact second moment or pair correlation is then given up to $\mathcal{O}(\lambda^4)$ by

$$
\begin{aligned}
Q_k = \langle u_k u_{-k} \rangle = \langle u_k^{(0)} u_{-k}^{(0)} \rangle + \langle u_k^{(0)} u_{-k}^{(2)} \rangle \\
+ \langle u_k^{(1)} u_{-k}^{(1)} \rangle + \langle u_k^{(2)} u_{-k}^{(0)} \rangle + \mathcal{O}(\lambda^4),
\end{aligned}
\tag{5.30}
$$

the first step being by definition and the second following from the substitution of (5.20) for the velocity field.[19]

Substituting in (5.27), (5.28), ... for the coefficients $\mathbf{u}^{(1)}$, $\mathbf{u}^{(2)}$, ..., yields.

$$
\begin{aligned}
Q_k = Q_k^{(0)} + 2 G_k^{(0)} M_{-kjl} M_{lpq} G_l^{(0)} \langle u_k^{(0)} u_j^{(0)} u_p^{(0)} u_q^{(0)} \rangle \\
+ G_k^{(0)} M_{kjl} M_{-kpq} G_k^{(0)} \langle u_j^{(0)} u_l^{(0)} u_p^{(0)} u_q^{(0)} \rangle \\
+ 2 G_k^{(0)} M_{kjl} M_{lpq} G_l^{(0)} \langle u_{-k}^{(0)} u_j^{(0)} u_p^{(0)} u_q^{(0)} \rangle \\
+ \mathcal{O}(\lambda^4).
\end{aligned}
\tag{5.31}
$$

5.6.4 Factorizing the zero-order moments

We won't go higher than fourth-order moments: we take the first such term as an example. We use a property of Gaussian statistics:

$$
\begin{aligned}
\langle u_k^{(0)} u_j^{(0)} u_p^{(0)} u_q^{(0)} \rangle = \langle u_k^{(0)} u_j^{(0)} \rangle \langle u_p^{(0)} u_q^{(0)} \rangle \\
+ \langle u_k^{(0)} u_p^{(0)} \rangle \langle u_j^{(0)} u_q^{(0)} \rangle \\
+ \langle u_k^{(0)} u_q^{(0)} \rangle \langle u_j^{(0)} u_p^{(0)} \rangle,
\end{aligned}
\tag{5.32}
$$

where the permutation of the indices on the right hand side should be noted. For isotropic, homogeneous fields we have[20] (for example) that

$$
\langle u_k^{(0)} u_{k'}^{(0)} \rangle = \delta_{kk'} D_k Q_k^{(0)}.
\tag{5.33}
$$

Hence (5.32) can be written as:

$$
\begin{aligned}
\langle u_k^{(0)} u_j^{(0)} u_p^{(0)} u_q^{(0)} \rangle = \delta_{kj} \delta_{pq} D_k D_p Q_k^{(0)} Q_p^{(0)} \\
+ \delta_{kp} \delta_{jq} D_k D_j Q_k^{(0)} Q_j^{(0)} \\
+ \delta_{kq} \delta_{jp} D_k D_j Q_k^{(0)} Q_j^{(0)}.
\end{aligned}
\tag{5.34}
$$

---

[19] Remember: odd-order moments vanish in the Gaussian case.

[20] Remember: this is just eqn (5.12) in the abbreviated notation.

Recall $l = |\mathbf{k} - \mathbf{j}|$, thus $\delta_{kj} M_{lqp} = 0$ as $k = j$ implies $l = 0$, and so the first term gives zero. Second and third terms are identical (with dummy variables swapped), hence we have

$$\langle u_k^{(0)} u_j^{(0)} u_p^{(0)} u_q^{(0)} \rangle = 2\delta_{kq}\delta_{jp} D_k D_j Q_j^{(0)} Q_k^{(0)}. \tag{5.35}$$

With these results, eqn (5.31) for $Q(k)$ becomes

$$Q_k = Q_k^{(0)} + 4G_k^{(0)} M_{-kjl} M_{lkj} G_l^{(0)} D_k D_j Q_j^{(0)} Q_k^{(0)}$$

$$+ 2G_k^{(0)} M_{kjl} M_{-klj} G_k^{(0)} D_j D_l Q_j^{(0)} Q_l^{(0)}$$

$$+ 4G_k^{(0)} M_{kjl} M_{jlk} G_j^{(0)} D_l D_k Q_l^{(0)} Q_k^{(0)} + \mathcal{O}(\lambda^4). \tag{5.36}$$

We can combine all the M's and D's into a simple coefficient: $L(k, j, l)$. Hence:

$$Q_k = Q_k^{(0)} + G_k^{(0)} L(k, j, l) G_l^{(0)} Q_j^{(0)} Q_k^{(0)}$$

$$+ G_k^{(0)} L(k, j, l) G_k^{(0)} Q_j^{(0)} Q_l^{(0)} + G_k^{(0)} L(k, j, l) G_j^{(0)} Q_l^{(0)} Q_k^{(0)}$$

$$+ \mathcal{O}(\lambda^4), \tag{5.37}$$

where the coefficient takes the form:

$$L(k, j, l) = 4M_{kjl} M_{lkj} D_k D_j = 2M_{kjl} M_{-klj} D_j D_l. \tag{5.38}$$

A fuller version of this will be given in Chapter 6.

5.7 Renormalized transport equations for the correlation function

Let us remind ourselves what we are trying to do. In reduced notation, we have the equation of motion:

$$L_{0k} u_k = f_k + M_{kjl} u_j u_l. \tag{5.39}$$

In order to obtain an equation for the pair correlation, we multiply through by u_{-k} and average:

$$L_{0k} Q_k = \langle f_k u_{-k} \rangle + M_{kjl} \langle u_j u_l u_{-k} \rangle. \tag{5.40}$$

The first term on the right hand side is the *input term*, due to the noise. This can be worked out exactly by perturbation theory. The second term on the right hand side is the coupling due to the nonlinearity. This can only be treated approximately.

Nevertheless we can use the exact calculation of the input term in order to suggest a way of introducing an exact response function. This we shall do in the following subsections. Then we will examine the underlying ideas in terms of the topological properties of an expansion in diagrams.

It should be emphasized, however, that the resulting theory is only one of a number of candidates, but we shall defer this particular issue until Chapter 6.

5.7.1 Introduction of an exact response function

We begin by considering the relationship between the zero-order second moment of the velocity field and the noise, with a view to adopting an analogous (or postulated) relationship between the exact second moment of the velocity field and the noise. That is:

$$Q_k^0 = \langle u_k^{(0)} u_{-k}^{(0)} \rangle = \langle G_k^0 f_k G_k^0 f_{-k} \rangle = G_k^0 G_k^0 W_k \tag{5.41}$$

where W_k is the autocorrelation of the stirring force (or noise term), as defined by (5.3), in reduced notation. Note that the zero-order response function is statistically sharp and can be taken outside the average.

Now, substituting (5.41) into (5.37), and operating with L_{0k} from the left (recalling that $L_{0k} G_k^0 = \delta(t - t') \equiv 1$ in our abbreviated notation) gives us (with some re-arrangement)

$$L_{0k} Q_k = G_k^0 W_k + L(k, j, l) G_l^0 Q_j^0 Q_k^0$$
$$+ L(k, j, l) G_j^0 Q_l^0 Q_k^0$$
$$+ L(k, j, l) G_k^0 Q_j^0 Q_l^0 + \mathcal{O}(\lambda^4). \tag{5.42}$$

It is tempting to try to repeat the trick used in (5.41) to relate the velocity correlation to the noise; but, if we look at eqn (5.42), we see that only two of the three second-order terms have a factor Q_k^0. The other term has the convolution $Q_j^0 Q_l^0 \equiv Q_j^0 Q_{k-j}^0$ and this distinction between two kinds of terms persists to all orders in λ.

Accordingly, grouping the two types of terms separately, and substituting (5.41) for Q_k^0, along with the contraction property $G_k^0 G_k^0 = G_k^0$, we have

$$L_{0k} Q_k = \{ G_k^0 + L(k, j, l) G_l^0 Q_j^0 G_k^0$$
$$+ L(k, j, l) G_j^0 Q_l^0 G_k^0 + \mathcal{O}(\lambda^4) \} W_k$$
$$+ L(k, j, l) G_k^0 Q_j^0 Q_l^0 + \mathcal{O}(\lambda^4). \tag{5.43}$$

Then, if we identify the expansion inside the curly brackets as the *exact* response function G_k, we can rewrite equation (5.43) as

$$L_{0k} Q_k = G_k W_k + L(k, j, l) G_k^0 Q_j^0 Q_l^0 + \mathcal{O}(\lambda^4), \tag{5.44}$$

where

$$G_k = G_k^0 + L(k, j, l) G_l^0 Q_j^0 G_k^0$$
$$+ L(k, j, l) G_j^0 Q_l^0 G_k^0 + \mathcal{O}(\lambda^4). \tag{5.45}$$

Lastly, we can operate on each term of (5.45) from the left with L_{0k} to obtain:

$$L_{0k} G_k = \delta(t - t') + L(k, j, l) G_l^0 Q_j^0$$
$$+ L(k, j, l) G_j^0 Q_l^0 + \mathcal{O}(\lambda^4). \tag{5.46}$$

Thus, in eqns (5.44) and (5.46), we now have equations for exact (or renormalized) correlation and response functions, Q_k and G_k. Of course, the right hand sides are expansions to all orders in λ. Bearing in mind that $\lambda = 1$ and the nonlinear term is generally large, these expansions are wildly divergent. To have any chance of succeeding, we must find a way of summing certain classes of terms to all orders.

5.7.2   RPT equations for the exact correlation and response functions

One method of partial summation is to make the replacements:

$$Q^0 \rightarrow Q; \quad G^0 \rightarrow G,$$

on the right hand side of eqns (5.44) and (5.46) and truncate at second-order to obtain:

$$L_{0k} Q_k = G_k W_k + L(k, j, l) G_k Q_j Q_l; \tag{5.47}$$

$$L_{0k} G_k = \delta(t - t') + L(k, j, l) G_l Q_j + L(k, j, l) G_j Q_l. \tag{5.48}$$

These replacements can be justified in two different ways:

1. Reversion of power series, as discussed in the next section.

2. Use of diagrams to show that the original (primitive) power series can be recovered from the renormalized perturbation expansion. We shall discuss this in Section 5.9.

However, the important thing is that we make *both* replacements. That is what makes it a renormalization process.

5.8   Reversion of power series

The method of reversion (or, sometimes, inversion) of power series has been used in other comparable problems in theoretical physics. For example, the renormalization of the expansion for the free energy, which we discussed in Section 4.4, can be treated in this way. To be precise, the expansion of the configuration integral in terms of cluster functions can be reverted into a power series in the Mayer functions.

The general method can readily be explained, as follows. Consider a pair of real variables x and y, which are connected by the power series

$$y = ax + bx^2 + cx^3 + dx^4 + \cdots \tag{5.49}$$

We now wish to invert this relationship and express x in terms of y. We begin by supposing that x is small enough for us to neglect its square. Then we have the immediate result

$$y = ax \quad \text{or} \quad x = y/a.$$

Evidently this is the lowest-order approximation to the general result which we are seeking. It can be made the basis of an iteration. We anticipate the required result by writing

$$x = Ay + By^2 + Cy^3 + Dy^4 + \cdots, \tag{5.50}$$

where it follows immediately that the first unknown coefficient is given by

$$A = 1/a. \tag{5.51}$$

The second coefficient is found by going to second order. That is, we assume that x is not quite so small and that we need to include its square: thus

$$\begin{aligned} y = ax + bx^2 &= a(Ay + By^2) + b(Ay + By^2)^2 \\ &= aAy + (aB + A^2b)y^2 + O(y^3) \\ &= y + (aB + A^2b)y^2 + O(y^3), \end{aligned} \tag{5.52}$$

where we have substituted from (5.50) for x and, in the first term on the right hand side, from (5.51) for A. Consistency then requires that the term of order y^2 on the right hand side of (5.52) vanishes and hence

$$B = -b/a^3. \tag{5.53}$$

Clearly this iteration can be carried on to any order but we shall not pursue that here. The method can also be extended to functional power series and specifically to the present renormalization problem as follows.

(a) We begin with the primitive power series for Q and G in terms of Q_0 and G_0.

(b) We revert these primitive expansions to obtain Q_0 and G_0 as power series in Q and G.

(c) Then substitute these new expansions for each Q_0 and G_0 factor in the primitive expansions for the triple moments.

(d) Lastly, multiply out and collect terms of each order.

The result of all this is line renormalization of the primitive perturbation series.

5.9 Formulation in Wyld diagrams

The perturbation theory has been presented in Section 5.7 in a simplified way, so that the underlying structure is not obscured too much by algebraic detail. It is possible to reveal even more of this structure by working with diagrams, rather like those used in Chapter 4 (the Mayer diagrams) or the Feynman diagrams to be used in Chapter 10. For Navier–Stokes turbulence these are known as "Wyld" diagrams.

We shall present a rather abbreviated version of the original treatment, as even in diagrams turbulence theory can quickly get out of hand.

Essentially we shall follow the steps in the analysis just given, but now we replace the mathematical terms by pictures known as graphs or diagrams. We favor the latter usage and will stick with it.

The first step is shown in Fig. 5.1. We represent the zeroorder velocity $u^{(0)}$ by a short, full line. Then a correlation is obtained by joining two full lines to represent $Q^0 = \langle u^{(0)} u^{(0)} \rangle$

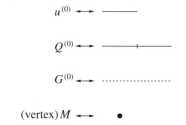

$$u^{(0)} \longleftrightarrow \text{———}$$

$$Q^{(0)} \longleftrightarrow \text{———|———}$$

$$G^{(0)} \longleftrightarrow \text{·················}$$

$$(\text{vertex}) M \longleftrightarrow \bullet$$

FIG. 5.1 Basic elements in diagrams.

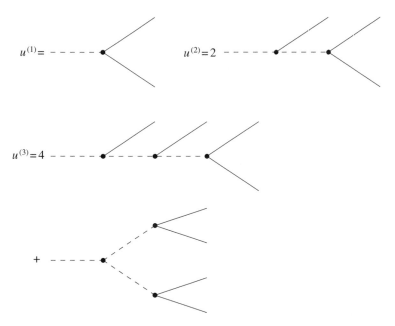

FIG. 5.2 Diagrammatic representation of the coefficients in eqns (5.27)–(5.29).

by longer full line, with a tick in the middle to remind us that two lines have been joined. G^0 is represented by the same length of dotted line and the operator M by a solid circle. We have not specified vector or other labels. If one wishes, one can add such labels to the diagrams and the price one pays is messy diagrams. If one is unsure about the labels, then a detailed comparison can be made of a diagram with its corresponding term in the equations of Section 5.6.

We can begin such a comparison with Fig. 5.2, where the coefficients $u^{(1)}$, $u^{(2)}$, and $u^{(3)}$ are shown in graphical form and may be compared with their algebraic equivalents in eqns (5.27)–(5.29). Taking $u^{(1)}$ as an example, a comparison with (5.27) shows that the emergent full lines are labeled by wave vectors $\mathbf{j}$ and $\mathbf{l}$, while the dotted line is labeled by $\mathbf{k}$.

The reason for the name "vertex," for the M_k element is now obvious! Moreover, the requirement that the wave vectors $\mathbf{k}, \mathbf{j}$, and $\mathbf{l}$ must form a triangle, such that

$$\mathbf{k} + \mathbf{j} + \mathbf{l} = 0,$$

ensures wave number conservation at a vertex. This applies to *all* vertices. So if the labels on two lines joining at a vertex are fixed then the label of the third line (dotted or full) is also fixed by the triangle requirement.

5.9.1   Diagrams in the expansion for the exact correlation

The basic procedure for obtaining the expansion for the exact propagator is given in Section 5.6.4. The first step in diagrams is the same: we insert the expansion with coefficients given by Fig. 5.2 into each of the u_k in the relation $Q_k = \langle u_k u_{-k} \rangle$, multiply out and then average products against the Gaussian distribution.

The second step is to evaluate higher-order moments as either zero (odd-order in $u^{(0)}$) or factored into products of pair correlations (even-order in $u^{(0)}$). This is a purely combinatorial process and that is where the basic strength of the diagrams lies.

In Fig. 5.3 we introduce elements for the exact or renormalized quantities, using thicker lines for the exact u, Q, and G; and an open circle for the renormalized vertex. To begin with, we shall only need the exact element for u, as we write down an expansion for Q in terms of entirely zeroorder quantities.

The general procedure can be summarized as follows:

1. Note that each term on the right hand side of eqn (5.30) for Q_k is (to all orders) a product of two coefficients.

2. Accordingly each diagram on the right hand side of the diagrammatic expansion for Q_k must consist of a product of two of the diagrams given in Fig. 5.2.

3. Think of the diagrams in Fig. 5.2 as being "trees" lying on their sides. Each "tree" has a "trunk" made up of dotted lines (response functions) and "branches" which are full lines $u^{(0)}$.

4. The two stages of first averaging and second evaluating higher-order moments in terms of pair-correlations can be accomplished in one step by placing pairs of tree diagrams

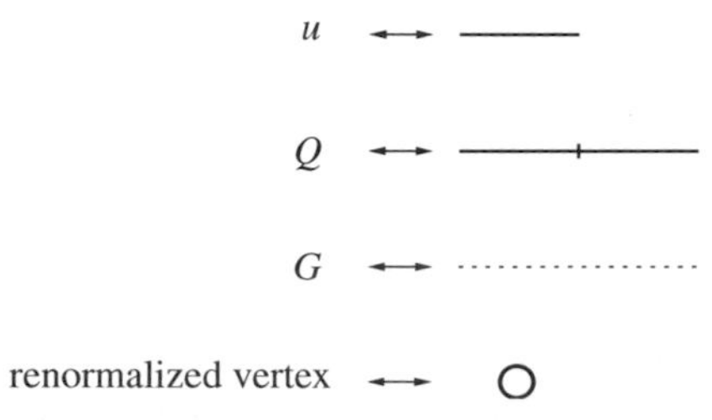

FIG. 5.3 Renormalized elements in diagrams.

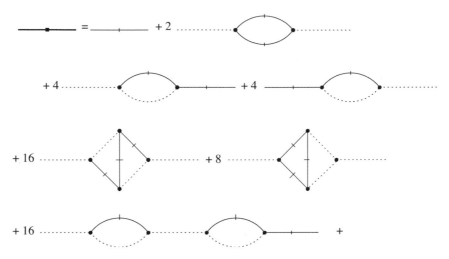

FIG. 5.4 Expansion in diagrams for the exact correlation function.

with their branches facing each other and then joining up branches in all possible ways. Remember: each pair of joined branches corresponds to a Q^0.

5. Warning! If you join up the two branches from one vertex, the result is zero! This is because one is simultaneously trying to satisfy the homogeneity requirement both for Q and also for the vertex and this can only be done for $k = 0$.

The result of all this is set out in Fig. 5.4, with only three of the fourth-order terms shown explicitly. This figure is the diagrammatic equivalent of eqn (5.31).

5.9.2 The renormalized response function

We can now use the diagrams to give a topological interpretation of the introduction in Section 5.7.1 of an exact response function. Corresponding to the grouping of terms in (5.43) into two categories, we can identify two classes of diagram:

A. *Reducible diagrams*: these are diagrams which can be split into two parts by cutting a single Q^0 line. In Fig. 5.4, the reducible diagrams are Q^0 itself (the zeroorder), the second and third diagrams at second order and the first and third diagrams at fourth order. (There are, of course, many other diagrams at fourth order, as well as those which we have shown explicitly.)

B. *Irreducible diagrams*: these are all the other diagrams, and the first diagram of the second order in Fig. 5.4 is a good example. Clearly if we cut one of the two Q^0 lines joining the two vertices, then the diagram is still connected by the remaining one.

The reducible diagrams correspond to the first category of terms in eqn (5.43) and clearly must correspond to the exact response function as they can all be written as a diagram which

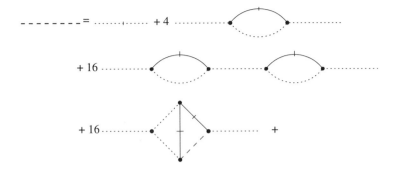

FIG. 5.5 Expansion in diagrams for the renormalized response function.

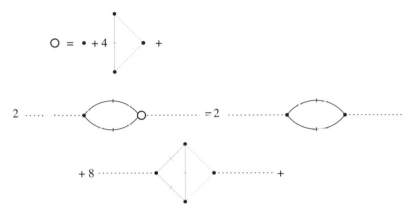

FIG. 5.6 Effect of renormalizing the vertex.

connects like a response function, and which is connected to a noise term. This leads us to the following identification:

The diagram corresponding to the exact response function is equal to the sum of the diagrams (to all orders) which connect like a response function.

The diagrammatic expansion for the exact response function is shown in Fig. 5.5.

5.9.3 Vertex renormalization

We have just seen that it is possible to identify diagram parts which connect like a response function and hence to identify a possible perturbation expansion for an exact or renormalized response function.

We can make a similar identification for vertices. For example, a connected diagram part with three vertices will connect just like a point vertex. So that we can infer the existence of a renormalized vertex expansion, as illustrated in Fig. 5.6. In the same figure, we show that if we replace one vertex of the irreducible second-order diagram of Fig. 5.5 by a renormalized vertex then some of the higher-order terms can be generated by its expansion.

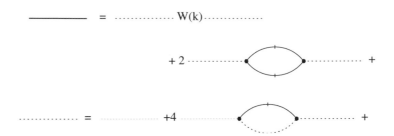

FIG. 5.7 Diagrams corresponding to integral equations for the exact correlation and response functions.

Vertex renormalization in turbulence theory has its controversial aspects and this will not be pursued here. However, it is worth making the pragmatic point that the effects of this type of renormalization are felt at third-order and above, whereas turbulence theories invariably are based on second-order truncations of the renormalized expansions and in practice the issue need not arise.[21]

5.9.4 Renormalized expansions for the exact correlation and response functions

The key to renormalization is now seen to be the irreducible diagrams. The procedure is as follows:

1. In the expansions for the exact quantities Q and G (see Figs 5.4 and 5.5, respectively) identify the irreducible diagrams.

2. Replace all elements in the irreducible diagrams by their exact or renormalized forms.

3. Write down all these modified diagrams in order, thus generating a *renormalized* perturbation expansion.

The result is illustrated in Fig. 5.7, where we have truncated the expansion at second order. We shall discuss the algebraic forms of equations at this order, along with an indication of their ability to predict turbulence, in the next chapter.

5.9.5 Comparison with quantum field theory

A truncation, at second order, as shown in Fig. 5.7, is often referred to as a "one loop" approximation. The term comes from quantum field theory and may perhaps be misleading when applied to turbulence theory. Accordingly, it may be helpful if we make a few remarks here on the subject.

Let us begin by defining a "loop." We will take it to be any closed circuit in a diagram, where we can get back to our starting point without retracing our steps. Thus, referring to Fig. 5.4, we see that each of the second-order diagrams possesses one loop, whereas each of

---

[21] We shall modify this statement a little when we consider renormalization group approaches in Part III of this book.

TABLE 5.1 Comparison of quantum field theory and turbulence theory

	Observable	NOT observable
Quantum field theory	G	G_0
Turbulence theory	G^0, Q	G, Q^0

the three fourth-order diagrams which we have chosen to show, posesses two loops. This is particularly obvious in the case of the third diagram at fourth order and evidently we could make the interpretation:

$$\text{second order} \equiv \text{one loop},$$

$$\text{fourth order} \equiv \text{two loops};$$

and so on.

However, as we saw in Sections 1.4.4 and 1.4.3, in quantum field theory there is a connection between integration of G_0 with respect to wave number and the problem of divergences. We shall enlarge on this aspect in Part III, when we will introduce Feynman diagrams and it will be seen that these divergences arise when one considers what is called a "loop integral." At this stage we shall merely make some cautionary remarks about the differences between the two kinds of expansion: scalar field theory on the one hand, turbulence theory on the other.

The first thing to note about scalar field theory is that G_0 as the bare propagator does not correspond to an observable particle. It is the renormalized propagator G which represents the observable particles. Thus, in shorthand, we say that G_0 is not an observable, whereas G is.

The situation in turbulence is rather more complicated in that we have two bare quantities G^0 and Q^0 which are jointly renormalized to give us the exact forms G and Q (Table 5.1). Here, in contrast to quantum field theory, the bare response function G^0 is an observable but the exact form is not. This is really rather obvious. We know the mathematical form of G^0. That is, from (5.15) we can infer that

$$G^0(k, t - t') = e^{\nu k^2 (t-t')}.$$

So, as ν can be measured for any Newtonian fluid by means of a simple deterministic experiment, it follows that G^0 is an observable. However, even if we assumed that the renormalized response G had the same exponential form as G^0, we would still have to measure a renormalized (turbulence) viscosity which (unlike ν) would vary from one realization to another. Accordingly we would have to settle for some mean value.[22]

Perhaps rather surprisingly the converse situation applies when we consider the correlation function. The exact correlation Q is the observable here, as it is obtained by measuring the

---

[22] However, in this connection see the remarks about turbulence theories as mean-field theories in the next chapter.

correlation of real fluid velocities. In contrast Q^0 is the correlation of zeroorder velocities *which cannot exist*. If we were to stir a real fluid with Gaussian stirring forces, then inter-mode coupling would rapidly establish a non-Gaussian velocity field. Thus Q^0 is not an observable.

These conclusions are summarized in the table. We close by pointing out that the loop integral which leads to divergences in quantum field theory occurs as a *product* in the perturbation expansion and involves a single G^0 being integrated over all wave numbers.

In contrast, the "loop integrals" of the perturbation theory of the NSE are *convolutions* and where a G^0 occurs it is always convolved with an unobservable Q^0. Accordingly, it is quite impossible to make any statement about divergences in the primitive perturbation series for turbulence. Q^0 does not exist. It is a purely Gaussian fiction and has no direct relationship to any turbulence quantity. For this reason we may conclude that there are no inherent divergences in the expansion of the NSE of the kind which crop up in quantum field theory.

6

APPLICATION OF RENORMALIZED PERTURBATION THEORIES TO TURBULENCE AND RELATED PROBLEMS

In this chapter we discuss two specific theories of turbulence. They are the direct-interaction approximation (or DIA, for short) and the local energy transfer (or LET) theory. They are both examples of second-order renormalized perturbation theories, as discussed in the previous chapter, and hence share the same equation for the pair correlation of velocities, but differ in the way in which renormalization is accomplished by the introduction of a second (response or propagator) function.

DIA is the oldest of the modern turbulence theories and relies on the introduction of a renormalized response function $G(k; t, t')$, which relates velocity fields to the stirring forces or noise. The derivation of this theory, which closely parallels the treatment of the previous chapter, dates back to the 1950s and it has long been known that DIA successfully reproduces all the symmetries of the problem and gives good qualitative and quantitative predictions of isotropic turbulence freely decaying from some arbitrary initial state. However, it has also long been known that the DIA is incapable of describing turbulence at very large values of the Reynolds number.[23] In particular, it does not yield the Kolmogorov power-law for the energy spectrum in the inertial range of wave numbers.

In contrast, the LET theory *is* compatible with the Kolmogorov spectrum. It is based on the introduction of a propagator $H(k; t, t')$ with connects Fourier modes at different times. The governing equation for the propagator is obtained on the basis of a hypothesis that it can be determined by a local (in wave number) energy balance.

Of course $G(k; t, t')$ in DIA is also a propagator: the difference between the two theories lies in the way the governing equation for the renormalized response function or propagator is obtained. In the case of DIA, this is done by considering the rate at which the stirring forces do work on the fluid. So although $G(k; t, t')$ (DIA) and $H(k; t, t')$ (LET) are the same kind of thing, it is helpful to give them different symbols in order to remind ourselves that they belong to different theories.

The organization of the rest of this chapter is as follows. In Section 6.1 we give some background information about isotropic turbulence, and in Section 6.2 we state the governing equations for the two theories, DIA and LET. In Sections 6.3 and 6.4 we examine the physics of turbulent energy transfer and dissipation, using the results obtained from numerical computation of the DIA and LET theories, to illustrate new concepts as they arise. This allows us to provide a general phenomenological treatment of the physics of turbulence, a subtle and

---

[23] DIA was later applied in a mixed Eulerian–Langrangian coordinate system, with some success, but this was at the price of great complication, and we shall not pursue that here.

interesting subject which deserves to be better known. Moreover, comparisons with results from "computer" and other experiments allow us to give some impression of the ability of renormalized perturbation theories (RPTs) to make theoretical predictions about turbulence. The experimental tests presented in Sections 6.3 and 6.4 are, respectively, the free decay of isotropic turbulence at moderate Reynolds number and stationary turbulence at high Reynolds number.

6.1   The real and idealized versions of the turbulence problem

"Real" turbulence occurs in pipes, jets, wakes, boundary-layers, and other more complicated configurations. It is driven by a force which is deterministic and often constant in space and time (e.g. a pressure gradient or a mean rate of shear). The resulting turbulence is due to *instability*. The chaotic phenomenon which we observe is accompanied by energy and momentum flows in both x- and k-space.

To make the turbulence problem look as much as possible like other problems in physics, we consider an idealized "turbulence in a box." The situation is artificial and if we wish to study a statistically stationary field we have to apply a random forcing to the Navier–Stokes equation (NSE) in order to achieve this. If we apply the forcing only to low-k modes (large eddies) we can hope that behavior at high-k (small eddies) is like that in turbulent shear flows (i.e. universal). The forcing we choose is multivariate normal (or Gaussian) and we have to specify its strength.

To do this, we consider an infinitely repeating box of incompressible fluid turbulence of side L with periodic boundary conditions. Our turbulence is *isotropic* (statistically invariant under rotation and reflection of axes) *homogeneous* (statistically invariant under translation of axes) and sometimes *stationary* (statistically invariant with respect to time).

6.1.1   Stationary isotropic turbulence

If the turbulence is stirred at low wave numbers the resulting kinetic energy is transferred through the wave number modes, to be dissipated at a rate ε. As we have previously noted in Section 5.1.1, the rate at which the force does work on the fluid must equal the dissipation rate:

$$\int_0^\infty W(k)\, \mathrm{d}k = \varepsilon. \tag{6.1}$$

We can see that this is so by deriving a general result for *all* fluid motion where the fluid is acted upon by a force $\mathbf{f}(\mathbf{x}, t)$. Consider a fluid of density ρ and kinematic viscosity ν occupying a volume V. Total energy of the fluid motion is given by:

$$E = \frac{1}{2} \sum_\alpha \int_V \rho u_\alpha^2 \, \mathrm{d}v. \tag{6.2}$$

We can obtain an equation for E directly from the NSE. This is usually done in $\mathbf{x}$-space. The balance equation is found to take the form:

$$\frac{\mathrm{d}E}{\mathrm{d}t} = \int_u \rho u_\alpha f_\alpha \, \mathrm{d}V - \int_V \rho \varepsilon \, \mathrm{d}V, \tag{6.3}$$

where ε is the energy dissipation per unit mass of fluid per unit time. A steady state exists when $u_\alpha f_\alpha = \varepsilon$. In other words, the rate at which work is done by the force equals the dissipation rate. When the external forces are zero, the kinetic energy of the fluid flow dies away at a rate given by ε.

The nonlinear and pressure terms do no net work on the system. Their role is limited to moving energy from one place to another (transport in x-space), or from one length scale to another (transport in k-space). The latter phenomenon is the well-known energy cascade.

6.1.2   Freely-decaying isotropic turbulence

By definition $\varepsilon = -\mathrm{d}E/\mathrm{d}t$ for freely decaying turbulence. The energy balance can be rewritten as:

$$\frac{\mathrm{d}E}{\mathrm{d}t} = -\varepsilon = -\int_0^\infty 2\nu k^2 E(k, t)\,\mathrm{d}k. \tag{6.4}$$

Hence the dissipation rate is also given by:

$$\varepsilon = \int_0^\infty 2\nu k^2 E(k, t)\,\mathrm{d}k. \tag{6.5}$$

The factor of k^2 ensures that dissipation is a high$-k$ (or small eddy) effect. The region in $k-$space where dissipation occurs is characterized by the Kolmogorov dissipation wave number:

$$k_\mathrm{d} = (\varepsilon/\nu^3)^{1/4}. \tag{6.6}$$

This result is obtained by dimensional analysis and this can be seen as follows. We work in kinematic units where quantities like energy are expressed "per unit mass of fluid". Thus kinetic energy per unit mass of fluid has dimensions of velocity squared. Energy dissipation is just rate of change of energy with respect to time and so we have:

$$\text{dimensions of } \varepsilon = L^2 T^{-3}.$$

The dimensions of kinematic viscosity are those of a diffusivity, thus:

$$\text{dimensions of } \nu = L^2 T^{-1}.$$

Obviously we can obtain a group which is dimensionless with respect to time if we divide ε by ν^3 and then taking the fourth root gives the desired result, in accord with eqn (6.6).

6.1.3   Length scales for isotropic turbulence

In turbulent shear flows, the relevant large length scale is the diameter of the pipe or the width of the jet. The relevant short length scale is usually taken to be l_d, which is the Kolmogorov dissipation length scale and is the inverse of the Kolmogorov dissipation wave number k_d as given by eqn (6.6).

For isotropic turbulence, it can be helpful to work with the integral length scale L, which is defined as follows.

$$L(t) = \frac{3\pi}{4} \int_0^\infty k^{-1} E(k, t) \mathrm{d}k / E(t).$$ (6.7)

This quantity is used to normalize some of the numerical results presented in Section 6.3.

A key quantity is the Taylor microscale λ. For decaying isotropic turbulence this is

$$\lambda(t) = \left[5E(t) / \int_0^\infty k^2 E(k, t)\, \mathrm{d}k \right]^{1/2}.$$ (6.8)

The formula can, of course, be used for stationary turbulence, where E and $E(k)$ do not depend on time. However, even in freely decaying turbulence the microscale λ can become independent of time when the energy $E(t)$ and the energy spectrum $E(k, t)$ take on the same time dependence. In fact, as we shall see, this does occur and is one of a number of possible indications that the turbulence has become well developed; that is, universal in character and independent of the way in which it was generated.

We can characterize the turbulence by the Taylor–Reynolds number R_λ, thus:

$$R_\lambda = \lambda u / v,$$ (6.9)

where u is the root-mean-square velocity and v is the viscosity of the fluid.

6.1.4   Numerical simulation of turbulence

Although stationary isotropic turbulence is a well defined concept, it can only exist in nature or the laboratory as an approximation. As a result, for many years most investigations of how well theories performed were limited to freely decaying turbulence. This could at least be produced in a laboratory by passing an irrotational flow through a grid or mesh. The resulting "grid-generated" turbulence would decay (in space) downstream from the grid; and, by changing to a set of coordinates moving with the free stream velocity, this behavior could be transformed to a decay in time. It is perhaps worth mentioning that for regions away from the boundaries of the apparatus, measurements have established that such turbulent velocity fields are both homogeneous and isotropic.

More recently the development of numerical simulations of the NSE have provided an alternative, "computer experiment" with which theories can be compared. The pioneering work in this field was limited to freely decaying turbulence at rather low values of the Reynolds number. However, nowadays results are available for forced (stationary) turbulence at large Reynolds numbers.

As we shall be presenting some results obtained by numerical simulations it may be helpful to provide here a few remarks about the way in which such simulations are carried out.

We assume that the turbulent fluid occupies a cubical box. First we divide our box up into a three-dimensional lattice or grid. It is usual to work with both the real-space (x-space) lattice and the reciprocal (k-space) lattice. An initial velocity field is set up by using a random number generator to assign values to each mesh point, subject to the constraints that the fluid is incompressible and that the initial energy spectrum $E(k, 0)$ takes some arbitrarily prescribed functional form.

The discretized NSE is then used to calculate new values of the velocity field at some slightly later time, and the calculation is stepped forward in time in this way until the field is deemed to be fully developed.

It should be emphasized that the initial distribution of velocities is (as determined by the choice of random number generator) Gaussian or normal, and as the turbulence simulation evolves, the effect of the inter-mode coupling is to take the distribution away from the Gaussian form. Hence, one measure of the success of a simulation is the behavior of the *skewness factor*, which should rise from zero (the Gaussian value) at $t = 0$ and reach some plateau, (indicating non-Gaussian behavior) when the turbulence is fully developed.

We shall discuss this aspect further when considering the actual results. For the moment it is worth raising the subject of *experimental error*. This can arise because a different choice of initial (random) velocity field may lead to the chosen initial spectrum being more or less well represented. Accordingly this leads to some uncertainty about the initial conditions. This can be assessed statistically by running the simulation a number of times with different initial velocity fields.

A different but related problem can arise with stationary simulations, which require some form of forcing to counteract the effects of viscous dissipation. This can be managed in various ways but invariably the resulting "stationary" simulation will fluctuate about some mean value. Again this leads to uncertainty but the degree of uncertainty can readily be determined by the usual statistical methods.

6.2   Two turbulence theories: the DIA and LET equations

As pointed out in Section 5.1.1, the M coefficient which turns (5.2) into the NSE equation, takes account of pressure gradients and maintains incompressibility as a constraint. It is given by

$$M_{\alpha\beta\gamma}(\mathbf{k}) = (2i)^{-1}\left\{k_\beta D_{\alpha\gamma}(\mathbf{k}) + k_\gamma D_{\alpha\beta}(\mathbf{k})\right\}, \tag{6.10}$$

where $D_{\alpha\beta}(\mathbf{k})$ is defined by eqn (5.13). The second-order renormalized equation for the two-time correlation function $Q(k, t, t')$ takes the form:

$$\left[\frac{\partial}{\partial t} + \nu k^2\right] Q(k, t, t')$$

$$= \int d^3 j\, L(\mathbf{k}, \mathbf{j}) \left[\int_0^{t'} dt''\, G(k, t', t'') Q(j, t, t'') Q(|\mathbf{k} - \mathbf{j}|, t, t'')\right.$$

$$\left. - \int_0^t dt''\, G(j, t, t'') Q(k, t'', t') Q(|\mathbf{k} - \mathbf{j}|, t, t'')\right], \tag{6.11}$$

where the coefficient $L(\mathbf{k}, \mathbf{j})$ is given by:

$$L(\mathbf{k}, \mathbf{j}) = \frac{\left[\mu(k^2 + j^2) - kj(1 + 2\mu^2)\right](1 - \mu^2)kj}{k^2 + j^2 - 2kj\mu}. \tag{6.12}$$

The statistical closure is completed by an equation for the response function $G(k, t, t')$. The basic ansatz of DIA is that there is a response function such that

$$\delta u_\alpha(\mathbf{k}, t) = \int_{-\infty}^{t} \hat{G}_{\alpha\beta}(\mathbf{k}, t, t')\delta f_\beta(\mathbf{k}, t')\, dt', \tag{6.13}$$

and that this *infinitesimal response* function can be renormalized. The resulting response equation is

$$\left[\frac{\partial}{\partial t} + \nu k^2\right]G(k, t, t')$$
$$+ \int d^3\mathbf{j}L(\mathbf{k}, \mathbf{j})\int_{t'}^{t} dt'' G(k, t'', t')G(j, t, t'')Q(|\mathbf{k} - \mathbf{j}|, t, t'') = 0, \tag{6.14}$$

where the ensemble-averaged response function $G(k, t, t')$ is given by

$$\langle \hat{G}_{\alpha\beta}(\mathbf{k}, t, t')\rangle = D_{\alpha\beta}(\mathbf{k})G(k, t, t'). \tag{6.15}$$

Essentially, eqns (6.11) and (6.14) are just eqns (5.47) and (5.48), with the full forms of $M_{\alpha\beta\alpha}$ and $D_{\alpha\beta}$ put in, and all the algebra worked out!

Instead of G, the LET theory introduces a *renormalized propagator* H such that:

$$Q_{\alpha\beta}(\mathbf{k}, t, t') = H_{\alpha\gamma}(\mathbf{k}, t, t')Q_{\gamma\beta}(\mathbf{k}, t, t). \tag{6.16}$$

Then the LET equations consist of eqn (6.11), with G re-named H, thus:

$$\left[\frac{\partial}{\partial t} + \nu k^2\right]Q(k, t, t')$$
$$= \int d^3j L(\mathbf{k}, \mathbf{j})\left[\int_{0}^{t'} dt'' H(k, t', t'')Q(j, t, t'')Q(|\mathbf{k} - \mathbf{j}|, t, t'')\right.$$
$$\left. - \int_{0}^{t} dt'' H(j, t, t'')Q(k, t'', t')Q(|\mathbf{k} - \mathbf{j}|, t, t'')\right], \tag{6.17}$$

and eqn (6.16). Alternatively we can derive a general propagator equation for H and this takes the form

$$\left[\frac{\partial}{\partial t} + \nu k^2\right] H(k, t, t')$$

$$+ \int d^3j L(\mathbf{k}, \mathbf{j}) \int_{t'}^{t} dt'' H(k, t'', t') H(j, t, t'') Q(|\mathbf{k} - \mathbf{j}|, t, t'')$$

$$= \frac{1}{Q(k, t', t')} \int d^3j L(\mathbf{k}, \mathbf{j}) \left[\int_{0}^{t'} dt'' Q(|\mathbf{k} - \mathbf{j}|, t, t'')\right.$$

$$\left.\{H(k, t', t'') Q(j, t, t'') - Q(k, t', t'') H(j, t, t'')\}\right]. \tag{6.18}$$

The governing equations for DIA can be recovered by putting the right hand side of the LET equation for the exact propagator equal to zero, and setting $H = G$. In fact the LET equations are computed using (6.16) rather than (6.18), but this step allows us to see the existence of an extra term which cancels the "infrared divergence" in the DIA at large Reynolds numbers. Further discussion of this can be found in the books [14] and [20].

6.2.1 DIA and LET as mean-field theories

It is of interest to show that both the DIA and the LET theory belong to the class of mean-field theories as discussed in Chapter 3. We may see this as follows. In order to derive the DIA equations one must take the step

$$\langle \hat{G}(t - t') u(t) u(t') \rangle = \langle \hat{G}(t - t') \rangle \langle u(t) u(t') \rangle \tag{6.19}$$

$$= G(t - t') Q(t - t'). \tag{6.20}$$

This is a mean-field approximation.

LET begins with the postulate

$$u_\alpha(\mathbf{k}, t) = \hat{H}_{\alpha\beta}(\mathbf{k}, t, t') u_\beta(\mathbf{k}, t'). \tag{6.21}$$

Multiplying through by $u_\alpha(-\mathbf{k}, t')$ and averaging

$$\langle u_\alpha(\mathbf{k}, t) u_\gamma(-\mathbf{k}, t') \rangle = \langle \hat{H}_{\alpha\beta}(\mathbf{k}, t, t') u_\beta(\mathbf{k}, t') u_\gamma(-\mathbf{k}, t') \rangle. \tag{6.22}$$

If, as in DIA, we take the propagator and velocity field to be uncorrelated, we can write

$$Q_{\alpha\gamma}(\mathbf{k}, t, t') = H_{\alpha\beta}(\mathbf{k}, t, t') Q_{\beta\gamma}(\mathbf{k}, t', t'), \tag{6.23}$$

where

$$H_{\alpha\beta}(k, t, t') = \langle \hat{H}_{\alpha\beta}(\mathbf{k}, t, t') \rangle. \tag{6.24}$$

and, again, this is a mean-field approximation.

6.3 Theoretical results: free decay of turbulence

This is an initial value problem and we integrate eqns (6.16) and (6.17) for the LET theory forward in time from the initial conditions that $E(k, 0)$ is prescribed and $H(k, 0, 0) = 1$. Similar procedures were carried out for DIA; and, so far as possible both theories had the same initial spectrum as the corresponding numerical simulation.

6.3.1 The energy spectrum $E(k, t)$

From eqn (5.12), we have

$$Q_{\alpha\beta}(\mathbf{k}, t, t') = D_{\alpha\beta}(\mathbf{k}) Q(k, t, t), \tag{6.25}$$

where $Q(k, t, t')$ is called the *spectral density*. For total turbulent energy (per unit mass of fluid), we set $\alpha = \beta$ and sum over $\alpha = 1, 2$, and 3. The *energy spectrum* $E(k, t)$ is related to the spectral density by

$$E(k, t) = 4\pi k^2 Q(k, t) \equiv 4\pi k^2 Q(k, t, t). \tag{6.26}$$

It is also related to the total turbulent energy E per unit mass of fluid by:

$$E = \int_0^\infty E(k) \, dk. \tag{6.27}$$

Typical results for the energy spectrum are shown in Fig. 6.1, where DNS stands for *direct numerical simulation*. Note that the DNS results are plotted with error bars. It should

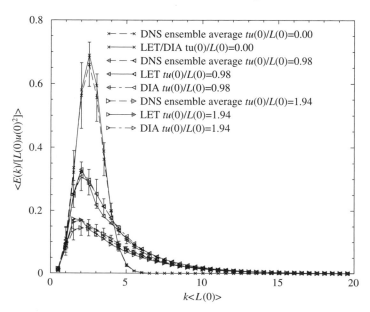

FIG. 6.1 Energy spectra for $R_\lambda(t = 0) \simeq 95$ freely decaying turbulence.

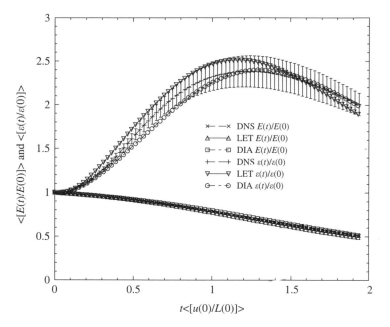

FIG. 6.2 Total energy per unit mass and dissipation rate per unit mass (upper group of curves) for $R_\lambda(t=0) \simeq 95$ freely decaying turbulence.

be noted that, as well as decaying with time, the spectra also spread out, indicating the transfer of energy to higher wave numbers due to inter-mode coupling. Evidently both theories agree well with the results of the computer experiment, in a quantitative sense as well as showing the expected qualitative behavior.

In Fig. 6.2 we see that the total energy shows the expected monotonic decline, with very close agreement between the three sets of results, whereas the dissipation rate increases greatly and passes through a peak, before declining with increasing time. This behavior is due to the build up of the turbulence, with its efficient energy transfer and dissipation, from the initial Gaussian random fluid motion. Although in this case the agreement is less close, both theories agree with the computer experiment to within the indicated experimental error.

The Taylor microscale and the skewness factor are both plotted in Fig. 6.3 and both graphs show how the turbulence develops with time to a self-preserving state. Evidently the agreement for the values of the microscale is very close, but in the case of the skewness, the qualitative agreement is a great deal better than the quantitative. The skewness factor is generally regarded as the most sensitive test of agreement between theories or between theory and experiment.

6.3.2 The energy transfer spectrum $T(k, t)$

We form the spectral energy-balance equation from the NSE. Multiplying each term of eqn (5.2) by $u_\alpha(-\mathbf{k}, \mathbf{t})$, averaging through according to (5.9)–(5.13), and invoking (6.10),

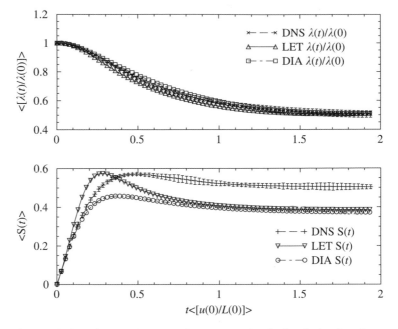

FIG. 6.3 Microscale length (upper group of curves) and velocity derivative skewness (lower group of curves) for $R_\lambda(t = 0) \simeq 95$ freely decaying turbulence.

we have

$$\left(\frac{\partial}{\partial t} + 2\nu_0 k^2\right) E(k, t) = W(k) + T(k, t), \qquad (6.28)$$

where the nonlinear inertial transfer term (or transfer spectrum, for short) $T(k, t)$ is to be defined in terms of the third-order moment $Q_{\beta\gamma\alpha}$, which is

$$Q_{\beta\gamma\alpha}(\mathbf{j}, \mathbf{k} - \mathbf{j}, -\mathbf{k}, t) = \langle u_\beta(\mathbf{j}, t)u_\gamma(\mathbf{k} - \mathbf{j}, t)u_\alpha(-\mathbf{k}, t)\rangle. \qquad (6.29)$$

We choose $\mathbf{k}$ as the polar axis and introduce

$$\mu \equiv \cos\theta_{kj} \qquad (6.30)$$

where θ_{kj} is the angle between the vectors $\mathbf{k}$ and $\mathbf{j}$. The nonlinear term takes the form

$$T(k, t) = \int_0^\infty dj \int_{-1}^1 d\mu \, T(k, j, \mu) \qquad (6.31)$$

with

$$T(k, j, \mu) = -8i\pi^2 k^2 j^2 \left\{ j_\gamma \, Q_{\beta\gamma\beta}(\mathbf{j}, \mathbf{k} - \mathbf{j}, -\mathbf{k}) \right.$$
$$\left. - k_\gamma \, Q_{\beta\gamma\beta}(-\mathbf{j}, \mathbf{j} - \mathbf{k}, \mathbf{k}) \right\}. \qquad (6.32)$$

Then the transfer spectrum can be defined by

$$T(k, t) = \int_0^\infty \mathrm{d}j\, T(k, j), \qquad (6.33)$$

with

$$T(k, j) = \int_{-1}^{1} \mathrm{d}\mu\, T(k, j, \mu). \qquad (6.34)$$

We have the antisymmetry

$$T(k, j, \mu) = -T(j, k, \mu). \qquad (6.35)$$

Or, alternatively,

$$T(k, j) = -T(j, k) \qquad (6.36)$$

In this formulation conservation of energy follows as:

$$\int_0^\infty \mathrm{d}k \int_0^\infty \mathrm{d}j\, T(k, j) = 0 = \int_0^\infty \mathrm{d}k\, T(k) = 0. \qquad (6.37)$$

This antisymmetry of $T(k, j)$, which is required for conservation of energy, may be clearly inferred from Fig. 6.4, where we plot results for $T(k, t)$. The initial value $T(k, 0) = 0$, for all k, when the velocity field is Gaussian. However, as time goes on, $T(k, t)$ develops, and its qualitative interpretation is quite straightforward.

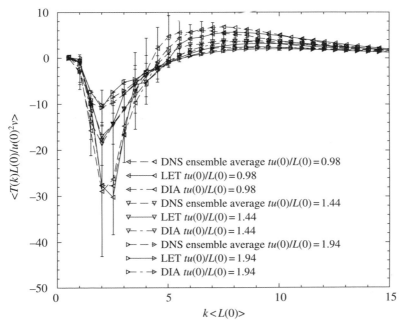

FIG. 6.4 Transfer spectra for $R_\lambda(t = 0) \simeq 95$ freely decaying turbulence.

At low wave numbers $T(k, t)$ is negative and absorbs energy, either from $\partial E(k, t)/\partial t$ (decaying turbulence) or from the stirring forces (stationary turbulence). This energy is transferred to high wave numbers where $T(k, t)$ is positive. Hence $T(k, t)$ behaves like a source of energy at high wave numbers. Here it emits the energy which it absorbed at low-wave numbers and the emitted energy is dissipated locally (in wave number) by the effects of the fluid's viscosity.

We shall enlarge on these points slightly in the next section where we discuss results for stationary turbulence.

6.4   Theoretical results: stationary turbulence

The application of renormalized perturbation theories such as DIA or LET to stationary turbulence has generally been limited to a few asymptotic results. In order to evaluate their behavior in sufficient detail such that comparisons can be made with computer experiments, it is necessary to add some forcing at low wave numbers in order to ensure stationarity. However, as we are dealing with purely ensemble-averaged quantities in a theory this means that the forcing must also be deterministic and, at best, related to the forcing used in the numerical simulation only at mean-field level. We should point out therefore, that the results presented here are possibly the first results obtained in this way from forced computations of LET and DIA. In view of their pioneering status, they should be viewed with some caution.

6.5   Detailed energy balance in wave number

In order to ensure stationarity, energy is added to the system at low wave numbers by a source term $W(k)$ which satisfies

$$\int_0^\kappa \mathrm{d}k\, W(k) = \varepsilon, \tag{6.38}$$

for some $\kappa \ll k_\mathrm{d}$, where k_d, as given by eqn (6.6), is the Kolmogorov dissipation wave number, and ε is the rate of energy input. Note that this is more restrictive than in eqn (6.1) in that the forcing acts only at low wave numbers.

Then $\mathrm{d}E(k, t)/\mathrm{d}t = 0$, and the energy balance of eqn (6.28) becomes:

$$T(k) = W(k) - 2\nu k^2 E(k) = 0; \tag{6.39}$$

or,

$$\int_0^\infty \mathrm{d}j\, T(k, j) = W(k) - 2\nu k^2 E(k) = 0. \tag{6.40}$$

At sufficiently high Reynolds numbers, we assume there is a wave number κ such that the input effects are below it and dissipation effects above it. That is, for a well-posed problem:

$$\int_0^\kappa W(k)\mathrm{d}k \simeq \varepsilon \simeq -\int_\kappa^\infty 2\nu k^2 E(k)\,\mathrm{d}k. \tag{6.41}$$

We can obtain two detailed energy balance equations by first integrating each term with respect to k from zero up to κ and then from infinity down to κ.

First,

$$\int_0^\kappa \mathrm{d}k \int_\kappa^\infty \mathrm{d}j\, T(k, j) + \int_0^\kappa W(k)\mathrm{d}k = 0, \qquad (6.42)$$

that is, energy supplied directly by input term to modes with $k \leq \kappa$ is transferred by the nonlinearity to modes with $j \geq \kappa$. Thus $T(k)$ behaves like a dissipation and absorbs energy.

Second,

$$\int_\kappa^\infty \mathrm{d}k \int_0^\kappa \mathrm{d}j\, T(k, j) - \int_\kappa^\infty 2\nu k^2 E(k)\, \mathrm{d}k = 0, \qquad (6.43)$$

that is, nonlinearity transfers energy from modes with $j \leq \kappa$ to modes with $k \geq \kappa$, where it is dissipated into heat. Thus in this range of wave numbers $T(k)$ behaves like a source and emits energy which is then dissipated by viscosity.

6.5.1   Scale-invariance and the Kolmogorov spectrum

We have discussed the turbulence energy balance in terms of the separation between input (stirring) effects and output (viscous effects). If we keep increasing the Reynolds number (i.e. we either reduce the viscosity ν or increase the rate of doing work on the fluid ε), then clearly the dissipation wave number k_d, as given by (6.6), will take ever larger values. This raises the possibility that we can separate the input and output effects by such a large distance in k-space, that there is an intermediate range of wave numbers where only the inertial transfer term, $T(k)$, acts.

Kolmogorov's hypotheses about this possibility were essentially the following:

1. The turbulence energy transfer is local in wave number: that is, the *cascade picture* is assumed to be valid.

2. For sufficiently large wave numbers there exists an intermediate range of wave numbers where the energy spectrum does not depend on the details of the input—only on its rate ε, nor does it depend on the viscosity ν.

With these assumptions, dimensional analysis immediately yields[24]:

$$E(k) = \alpha \varepsilon^{2/3} k^{-5/3}. \qquad (6.44)$$

The prefactor α is generally known as the Kolmogorov constant. This result for the energy spectrum has received ample experimental confirmation with the value of the prefactor given by $\alpha = 1.620 \pm 0.168$.

Another way of looking at this is in terms of scale invariance. Let us introduce the *transport power* $\Pi(\kappa, t)$. For some given wave number $k = \kappa$, the transport power is the rate

---

[24] This can be verified by using the information given following eqn (6.6).

at which energy is transferred from modes with $k \leq \kappa$ to modes with $k \geq \kappa$. It is given by:

$$\Pi(\kappa, t) = \int_{\kappa}^{\infty} T(k, t) \, dk. \qquad (6.45)$$

By the antisymmetry of $T(k, j)$:

$$\Pi(\kappa, t) = \int_{\kappa}^{\infty} dk \int_{0}^{\kappa} dj \, T(k, j). \qquad (6.46)$$

Also by antisymmetry of $T(k, j)$:

$$\Pi(\kappa, t) = -\int_{0}^{\kappa} T(k, t) \, dk. \qquad (6.47)$$

A criterion for the existence of an inertial range can be obtained by integrating both sides with respect to k. We find that for any K_I in the inertial range, we have the condition

$$\int_{K_I}^{\infty} dk \, T(k) = -\int_{0}^{K_I} dk \, T(k) = \varepsilon. \qquad (6.48)$$

One point worth making is that if $E(k, t)$, with time dimension T^{-2}, is expressed in terms of the *rate* of energy transfer ε which has time dimension T^{-3}, then

$$E(k, t) \sim \varepsilon^{2/3}$$

is inevitable. If there is scale invariance, as defined by (6.48), then eqn (6.44) is also inevitable.

6.5.2 Theoretical results for the energy spectrum at large Reynolds numbers

In Fig. 6.5 we show the theoretical predictions of the energy spectrum as a function of wave number scaled on the dissipation wave number k_d (see eqn (6.6)). In addition to the DNS results, the theoretical spectra can be compared with the *ad hoc* spectrum due to Qian (1984), as this is generally regarded as being a good fit to experimental data.

The actual spectra are divided by $\varepsilon^{2/3} k^{-5/3}$ and from (6.45) we can see that a plateau region of the graph could correspond to the Kolmogorov pre-factor α. This form of spectral plot is known as a *compensated spectrum*. For a Taylor–Reynolds number of $R_\lambda \simeq 232$ we would expect to see a small amount of inertial range and this is just about perceptible here around $k/k_d \sim 10^{-1}$.

It should be borne in mind that k_d is based on dimensional analysis and therefore only gives a rough idea of the position of the boundary between inertial and dissipation ranges of wave number. From experimental results, one would in fact expect a cross-over at wave numbers of the order of $0.1k_d - 0.2k_d$.

In Fig. 6.6 we show spectra obtained from the LET theory at two values of the Taylor–Reynolds number compared with some representative experimental results. Here we show

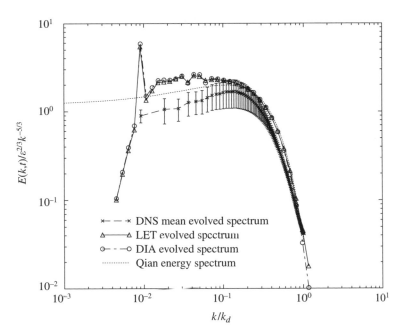

FIG. 6.5 Comparison of the compensated energy spectrum at $R_\lambda(t_{\text{Evolved}}) \simeq 232$ in forced turbulence with the *ad hoc* energy spectrum due to Qian (1984).

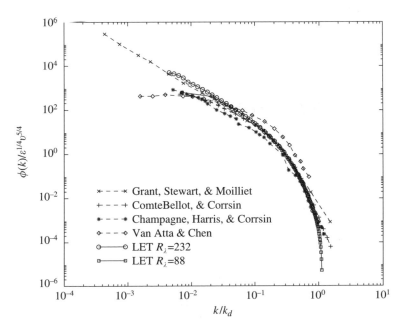

FIG. 6.6 Comparison of the LET one-dimensional energy spectrum for $R_\lambda(t_{\text{Evolved}}) \simeq 88$ and 232 in forced turbulence with some experimental results.

a one-dimensional projection of the spectrum $\phi(k, t)$ because this is what experimentalists actually measure. It can be related to the usual energy spectrum by the formula

$$\phi(k, t) = \frac{1}{2} \int_k^\infty \left\{ 1 - \frac{k^2}{p^2} \right\} p^{-1} E(p, t) \, dp, \qquad (6.49)$$

which holds for isotropic turbulence.

Obviously one may conclude that the LET theory agrees as well with experiment as the experiments agree with each other. Unfortunately, not only is the turbulence problem very difficult, but the experimental side is relatively underdeveloped when compared with physics as a whole. This means that decisive comparisons are rarely, if ever, possible.

6.6   Application to other systems

The turbulence theories which we have discussed in this chapter could provide a method of tackling other problems in which the equation of motion is a nonlinear Langevin equation with some form of noise input. However, this approach is relatively undeveloped, and the only investigations in this area known to us are the application of DIA to the KPZ equation by Bouchaud and Cates and the application of LET (along with assorted RG methods) to the long-time, large-scale properties of the noisy, driven diffusion equation by Prakash, Bouchaud and Edwards. Recently, the statistical closure of Edwards (1964) has been extended to time-dependent problems in both fluid turbulence and the wider class of problems associated with noisy, driven diffusion equations. References may be found in [24], along with details of the numerical calculations of the LET and DIA theories.

We return to this topic in the context of RG at the end of Section 11.3, where some more general references are given for background reading.

PART III

RENORMALIZATION GROUP (RG)

7

SETTING THE SCENE: CRITICAL PHENOMENA

In the third and last part of this book we concentrate on renormalization group (or RG) in its various forms. As the great success of RG has been in the context of critical phenomena, it seems sensible to consider the general theoretical situation in that subject as it existed just prior to the coming of RG. Moreover, as the application of RG to critical phenomena was to some extent inspired by the concepts of scaling and "blocking the lattice," it will be helpful to examine these ideas also.

Accordingly we present a slightly more extended treatment of critical exponents. As the calculation of their numerical values is the main test of theories, this is clearly worthwhile. Then we give a more general discussion of theoretical models, although it is invariably the Ising model which interests us most. Then we show how scaling theory (based on the use of homogeneous functions) leads to relationships between critical exponents, so that in order to work out all six exponents for a system it is only necessary to work out any two of them.

We then close with a consideration of mean-field theory. This was the main theory of critical phenomena until it lost its crown to RG. A particularly interesting feature of this version of mean-field theory is that the combination of a variational principle with linear response theory allows us to work out the theoretical form of the pair-correlation, despite the fact that the Ising model is equivalent to an assumption that there are no correlations!

As well as calculating the mean-field values for the critical exponents, we also present two quite demanding pieces of mathematical analysis in Sections 7.7 and 7.8 in which we work out the form of the connected correlation function and also show that mean-field theory becomes exact in an infinite number of dimensions. These analyses could be omitted on a first reading, but the subject of the *upper critical dimension* in Section 7.9 is of such importance later on that it merits attention at this stage.

7.1 Some background material on critical phenomena

We have previously met four of the six critical exponents for the "para–ferro" magnetic transition in Section 3.3. Here we introduce the concept rather more formally and conclude by defining and summarizing all the exponents for magnetic and gas–liquid transitions.

7.1.1 Critical exponents

Critical points occur in a great variety of systems. The value of T_c depends on the details of the atomic or molecular interactions of the system and hence will vary widely from one system to another. However there is a considerable degree of similarity in the way systems approach a critical point: macroscopic variables like specific heat, magnetic susceptibility or correlation length either diverge or go to zero as $T \to T_c$. We can characterize this behavior by the introduction of critical exponents.

We may represent any macroscopic variable by $F(T)$ and introduce the reduced temperature θ_c by

$$\theta_c = \frac{T - T_c}{T_c}. \tag{7.1}$$

Then a critical exponent n can be defined for $\theta_c \approx 0$ (i.e. $T \approx T_c$) by

$$F(\theta_c) = A\theta_c^{-n}, \tag{7.2}$$

where A is a constant. We note that there are two broad cases as follows, depending only on the sign of the critical exponent:

1. Critical exponent n is positive, $F(\theta_c)$ diverges as $T \to T_c$.
2. Critical exponent n is negative, $F(\theta_c) \to 0$ as $T \to T_c$.

Actually F may be expected to behave analytically away from the fixed point. With this in mind, we can write an expression with a greater range of validity as

$$F(\theta_c) = A\theta_c^{-n}(1 + B\theta_c^y + \cdots), \tag{7.3}$$

where $y > 0$ for analytic behavior at large θ_c and B is a constant.[25]
More formally, the critical exponent n of $F(\theta_c)$ is defined to be:

$$n = -\lim_{\theta_c \to 0} \frac{\ln F(\theta_c)}{\ln \theta_c}. \tag{7.4}$$

Lastly, we should mention at this stage the idea of *universality*. The critical exponents are to a large extent universal, depending only on the symmetry of the Hamiltonian and its dimension, provided the interatomic forces are short range.

7.1.1.1 The para-ferromagnetic transition. When a piece of ferromagnetic material is placed in a magnetic field **B**, a mean magnetization **M** is induced in the material which is proportional[26] to **B**. Then, taking one coordinate axis along **B**, we can work with the scalars B and M (assumed to be in the same direction!).

It should be noted that M is zero in the disordered phase (spins oriented at random) and nonzero in the ordered phase (spins aligned), and so is an example of an *order parameter*.

---

[25] We have previously given this expression as eqn (4.22).
[26] We are assuming here that the magnetic material is isotropic.

The relationship between the applied magnetic field and the resulting magnetization is given by the isothermal susceptibility, as defined by the relation:

$$\chi_T \equiv \left(\frac{\partial M}{\partial B} \right)_T. \tag{7.5}$$

Note this is an example of a *response function*. For a fluid, the analogous response function would be the isothermal compressibility, which is a measure of the volume change (response) when the system is subject to a change in pressure (applied field).

7.1.2 Correlation functions

Correlations are an essential tool for the statistical analysis of all fluctuating systems and they play a particularly important part in the study of critical phenomena. We shall discuss them here for the case of a magnetic system consisting of "spins on a lattice." We later introduce *linear response theory*, with the emphasis on methods of calculating correlations. Here we shall concentrate on the physical aspects and in the process we shall introduce two further critical exponents.

7.1.2.1 Two-point correlations. The two-point correlation G is defined by:

$$G(\mathbf{x}_i, \mathbf{x}_j) \equiv \langle S_i S_j \rangle \tag{7.6}$$

where $\mathbf{x}_i$ is the position vector of the ith lattice site, with associated spin vector S_i. This is also known as the pair correlation and sometimes denoted by $G^{(2)}$. Usually we shall restrict our attention to materials which are spatially homogenous and isotropic. Then we may benefit from the following simplifications:

1. Homogeneity means that average values (including correlations) are translationally invariant. This implies that two-point correlations can only depend on the distance between the points and not on their absolute position in space, hence $G \equiv G(\mathbf{x}_i - \mathbf{x}_j)$.

2. Isotropy means that average values are rotationally invariant and hence two-point correlations cannot depend on the direction of the vector connecting the two points. This implies that $G \equiv G(|\mathbf{x}_i - \mathbf{x}_j|)$.

Then if we put $|\mathbf{x}_i - \mathbf{x}_j| = r$, we may write the correlation in the following convenient form:

$$G(r) = \langle S(0)S(r) \rangle, \tag{7.7}$$

Alternatively, we may also write it as:

$$G(i, j) \equiv G_{ij} = \langle S_i S_j \rangle, \tag{7.8}$$

and all these forms will be employed as appropriate to the circumstances.

7.1.2.2 Connected correlations. Below the critical temperature T_c, both correlations and mean values become large and this leads to large values of the correlation G_{ij} for all values of

i and j. We may deal with this problem by subtracting out the mean values and in this way we introduce the *connected pair correlation* $G_c(i, j)$, thus

$$G_c(i, j) = \langle S_i S_j \rangle - \langle S_i \rangle \langle S_j \rangle. \tag{7.9}$$

Alternatively, we may write this in a form analogous to (7.7), thus

$$G_c(r) = \langle S(0)S(r) \rangle - \langle S \rangle^2. \tag{7.10}$$

It should be noted that:

- Above T_c, $\langle S \rangle = 0$, hence $G_c(r) \equiv G(r)$.
- Below T_c, G_c is a measure of the fluctuations in the spins about a general alignment.

7.1.3 The correlation length

For temperatures near the critical temperature, it is found experimentally that the pair correlation decays exponentially with increasing separation of the measuring points, provided that this separation is not too small.

That is, for r large and the reduced temperature $\theta_c \ll 1$, it is found that

$$G_c(r) \sim e^{-r/\xi}, \tag{7.11}$$

where ξ has the dimension of a length and is interpreted as the correlation length. That is, it is the distance over which spin fluctuations are correlated (in the sense that correlations over greater distances are rare). As $T \to T_c$, then it is found that ξ grows without limit and this allows us to define another critical exponent ν, by means of the relation

$$\xi \sim \theta_c^{-\nu}, \quad \text{as } \theta_c \to 0. \tag{7.12}$$

At the critical point, it is also found empirically that the connected pair correlation takes the form

$$G_c(r) \sim \frac{1}{r^{d-2+\eta}}, \tag{7.13}$$

where d is the number of space dimensions and ν and η are constants.

In the Table 7.1 we list all six exponents for fluid systems as well as magnetic systems.

TABLE 7.1 Summary of critical exponents.

Exponent	Variable	Magnetic system	Fluid system
α	Heat capacity	$C_B \sim |\theta_c|^{-\alpha}$	$C_V \sim |\theta_c|^{-\alpha}$
β	Order parameter	$M \sim (-\theta_c)^\beta$	$(\rho_l - \rho_g) \sim (-\theta_c)^\beta$
γ	Response function	$\chi_T \sim |\theta_c|^{-\gamma}$	$\kappa_T \sim |\theta_c|^{-\gamma}$
δ	Critical isotherm	$B \sim |M|^\delta \text{ sgn } M$	$P - P_c \sim |\rho_l - \rho_g|^\delta$
ν	Correlation length	$\xi \sim |\theta_c|^{-\nu}$	$\xi \sim |\theta_c|^{-\nu}$
η	Pair correlation at T_C	$G^c(r) \sim 1/r^{d-2+\eta}$	$G^c(r) \sim 1/r^{d-2+\eta}$

7.1.4 Summary of critical exponents

7.1.4.1 Critical exponent inequalities. Using thermodynamic relationships it is possible to prove various inequalities among exponents, for example, Rushbrooke's inequality: $\alpha + 2\beta + \gamma \geq 2$. For some systems, these hold as equalities, but it is not worth doing a lot on this: better results come more easily from scaling theory (see Section 7.3) and even RG (see Section 8.3).

7.2 Theoretical models

We have previously introduced the idea of models for magnetic systems in Chapter 2 in an informal way. Now we introduce the idea more formally. This includes the formal use of the term "Hamiltonian," although we shall still mean by this the energy. Remember that for quantum systems, the Hamiltonian is an operator and the energy is its eigenvalue.

We begin by noting that the microscopic behavior of assemblies can often be regarded as "classical" rather than "quantum mechanical" for the following reasons:

- Thermal fluctuations are often much larger than quantum fluctuations.

- Classical uncertainty in the large-N limit overpowers the quantum uncertainty.

- The complexity of the microstructure of the assembly introduces its own uncertainty.

We can set up theoretical models which should be:

1. physically representative of the system to some reasonable degree of approximation;

2. soluble.

But, usually (2) is incompatible with (1) and attempting to reconcile the two normally involves some form of perturbation theory. In practice, one sacrifices some degree of physical "correctness" in order to be able to solve the model. Invariably, by "solve," we mean "obtain a good approximation to the partition function."

7.2.1 Generic Hamiltonian for D-dimensional spins on a d-dimensional lattice

If we restrict our attention to magnetic systems, then a generic model for the Hamiltonian H can be written in d-dimensions[27] in terms of a D-dimensional spin vector $S_i^{(D)}$, thus:

$$H = -\sum_{i,j} J_{ij} S_i^{(D)} S_j^{(D)} - \sum_i B_i S_i^{(D)} \tag{7.14}$$

where B_i is an externally applied field which is assumed for generality at this stage, to be spatially *nonuniform* and hence to vary with lattice position. The following points should be noted:

1. J_{ij} is a coupling parameter, which in general depends on the distance between sites i and j. For anisotropic assemblies it also depends on the orientation of the vector $\mathbf{r}_i - \mathbf{r}_j$.

---

[27] That is, we consider a lattice in an arbitrary number of dimensions d which may be greater than $d = 3$.

2. The dimension of the spin vector D is not necessarily equal to the dimension d of the lattice.

3. The theoretical treatment which we employ can be either classical or quantum mechanical.

Most of the theoretical models of critical phenomena can be regarded as either a special case of the above or can be generated from it by suitable procedures.

7.2.2 Examples of models of magnetism

The Heisenberg model (classical) $D = 3$. In this case the spin is a three-dimensional vector and the model can be regarded as the classical limit of the quantum-mechanical Heisenberg model. Instead of the quantum mechanical $(2S + 1)$ discrete spin orientations, there is a continuum of orientations: S is a classical vector. It is the most realistic model for temperatures near T_c. With $d = 3$ and $D = 3$, it models three-dimensional spin vectors on a three-dimensional lattice. It has not been solved analytically and its study relies on numerical computation.

   The XY model $D = 2$. This is just a version of the Heisenberg model with $D = 2$. The spins can point in any direction in a plane.

   The Ising model $D = 1$. This model is Boolean in character–the spins are either up or down. It may also be regarded as an example of a "two-state" model. In addition, the spin interactions are limited to pairs of nearest neighbors on the lattice.

   The Potts model (q-state). This is a generalization of the Ising model to have more states. For the particular case $q = 2$, it is identical to the Ising model.

   The Gaussian and spherical models. These were invented to try to make the Ising model more tractable but they are both very nonphysical. The Gaussian does not undergo a phase transition in any number of dimensions. But it is widely studied because it gives a starting point for perturbative calculations, as we shall see in Chapter 9. The spherical model is one of the very few exactly soluble models for a $d = 3$ lattice. It has been solved for $d = 1, 2$, and 3. Also its partition function is the same as that for the Heisenberg model, in the limit $D \to \infty$.

7.2.3 The Ising model

This is the most widely studied model in statistical field theory, including both the theory of critical phenomena and particle theory. So naturally it is the one upon which we shall concentrate here. We recall that it is generated from the generic model Hamiltonian of eqn (7.14) by specializing the dimension of the spin vector to the case $D = 1$. If we write

$$S_i^{(1)} \equiv S_i \quad \text{such that} \quad S_i = \pm 1,$$

then the Hamiltonian can be written for the generic Ising model as

$$H = -\sum_{\langle i,j \rangle} J_{ij} S_i S_j - \sum_i B_i S_i, \tag{7.15}$$

where (as previously encountered in Section 2.1.1) the restriction to nearest-neighbor pairs of spins in the double sum is indicated by the use of angle brackets to enclose the indices i and j. As we have seen in Section 2.1.2 this restriction can be indicated more explicitly, so that there is no need to remember that each pair of spins should only be counted once.

The Ising model is really a family of models, the individual members being determined by our choice of the dimensionality d. We begin by summarizing a few features of the model for each value of d.

1. $d = 1$. In this case we envisage a line of spins, as illustrated in Fig. 2.1. It presents a very simple problem and can be solved exactly. In Appendix C we give one of the methods of calculating the partition function for a one-dimensional Ising model. But, although its solution is of considerable pedagogical importance, the model does not exhibit a phase change.[28]

2. $d = 2$. In two dimensions, the Ising model can be thought of as an array of spins on a square lattice, with each lattice site having an associated spin vector at right angles to the plane of the array. The situation is illustrated schematically in Fig. 2.8, where a plus sign may be taken as denoting a spin pointing upwards and the absence of sign denoting a spin pointing downwards. This model is more realistic in that a phase transition appears in the thermodynamic limit. It was solved exactly by Onsager (1944) and this work is still regarded as a theoretical *tour de force*.

3. $d \geq 3$. These cases are more difficult to draw but at least the three-dimensional problem is easily visualized. One simply imagines a cubic lattice with the three main coordinate directions corresponding to the cartesian coordinate axes x, y, and z. Then we assume that spin vectors at each lattice site can point in the directions of $\pm z$. The Ising models for $d \geq 3$ cannot be solved exactly but they can be treated numerically, and numerical simulation of Ising models is a very active area of statistical physics. As we shall see in the next section, mean-field theory gives a reasonable approximation to the partition function for $d = 3$ and a very good approximation for $d = 4$.

7.3   Scaling behavior

Ideas of scaling and self-similarity had been a crucial element in the subject of fluid mechanics from the late nineteenth century onwards, but it was not until the mid-1960s that such ideas were formulated in the theory of critical phenomena. We have already touched on these topics in Section 1.5, and in particular on the concept of a homogeneous function in Section 1.5.5. Now we extend this idea to functions of two variables and this provides us with an elegant and economic way of obtaining relationships between critical exponents.

7.3.1   Generalized homogeneous functions

A generalized homogeneous function $f(x, y)$ is defined by the property:

$$f(\lambda^a x, \lambda^b y) = \lambda f(x, y), \qquad (7.16)$$

---

[28] It is sometimes said that there is a phase change at $T = 0$, but in any model the spins will be aligned at zero temperature, so arguably this is a rather trivial example of a phase transition.

for arbitrary numbers a and b. It can be shown that generalized homogeneous functions of two variables must be of the form:

$$f(x, y) = y^n F(x/y). \tag{7.17}$$

where F is a function of a single variable and is defined by

$$F(z) = f(z, 1). \tag{7.18}$$

Note that we can recover the ordinary homogeneous function by setting $a = b$ and $n = 1/a$.

7.3.2   The static scaling hypothesis

Let us consider the Gibbs free energy $G(T, B)$ for a magnetic system at a temperature T and subject to an imposed magnetic field B. We may write it in terms of the reduced temperature θ_c, as $G(\theta_c, B)$, near the critical temperature $T = T_c$.

Then the scaling hypothesis put forward by Widom in 1965 is just the assumption that $G(\theta_c, B)$ is a generalized homogeneous function. That is, we assume that there exist two parameters r and p such that:

$$G(\lambda^r \theta_c, \lambda^p B) = \lambda G(\theta_c, B) \tag{7.19}$$

The hypothesis leaves r and p undetermined. But, as we shall see, all the critical exponents can be expressed in terms of r and p and this leads to new relationships between the critical exponents.

7.3.3   Relations among the critical exponents

We begin by showing that the above equation implies a scaling relationship for the magnetization (or order parameter) M. Then, by considering two particular cases, we obtain expressions for the critical exponents β and δ.

In order to do this, we differentiate both sides of eqn (7.19), with respect to B to obtain

$$\lambda^p \frac{\partial G(\lambda^r \theta_c, \lambda^p B)}{\partial (\lambda^p B)} = \lambda \frac{\partial G(\theta_c, B)}{\partial B}. \tag{7.20}$$

Next, recalling that $M = (-\partial G/\partial B)_T$, we have

$$\lambda^p M(\lambda^r \theta_c, \lambda^p B) = \lambda M(\theta_c, B), \tag{7.21}$$

as the scaling relation for the magnetization.

Now we consider critical behavior, for two particular cases.

Case (1): $B = 0$ and $\theta_c \to 0$

Re-arrange (7.21) to obtain

$$M(\theta_c, 0) = \lambda^{p-1} M(\lambda^r \theta_c, 0). \qquad (7.22)$$

This equation is valid for all λ: hence it must hold for any arbitrarily chosen value of λ. Accordingly, we shall choose the value:

$$\lambda = (-1/\theta_c)^{1/r},$$

and so eqn (7.22) becomes:

$$M(\theta_c, 0) = (-\theta_c)^{(1-p)/r} M(-1, 0).$$

For $\theta_c \to 0^-$, we may define the scaling exponent β by the relationship

$$M(\theta_c, 0) \sim (-\theta_c)^{\beta}.$$

Then, comparison of the two immediately preceeding expressions for M gives:

$$\beta = \frac{1-p}{r}, \qquad (7.23)$$

which is the critical exponent β in terms of the scaling parameters r and p.

Case (2): $\theta_c = 0$ and $B \to 0$

In this case, eqn (7.22) becomes

$$M(0, B) = \lambda^{p-1} M(0, \lambda^p B).$$

This time we choose our arbitrary value of the scaling parameter to be

$$\lambda = B^{-1/p},$$

and so the scaling relation for the magnetization takes the form

$$M(0, B) = B^{(1-p)/p} M(0, 1).$$

For $B \to 0$, the critical exponent δ is defined by:

$$M(0, B) \sim B^{1/\delta},$$

and once again, comparison yields

$$\delta = \frac{p}{1-p}. \qquad (7.24)$$

Then, if we solve eqns (7.23) and (7.24) simultaneously, we obtain

$$r = \frac{1}{\beta} \cdot \frac{1}{\delta + 1};$$
$$p = \delta \cdot \frac{1}{\delta + 1}.$$
(7.25)

Then, if other critical exponents are expressed in terms of the parameters r and p, we may derive equations connecting the various sets of critical exponents by invoking the above equations. In general this may be done by forming other partial derivatives of the Gibbs function.

7.3.4 Relationship between β, γ, and δ

For instance, if we differentiate twice with respect to B, we may obtain a relationship for the magnetic susceptibility and hence for the exponent γ, as follows:

$$\lambda^{2p} \chi_T(\lambda^r \theta_c, \lambda^p B) = \lambda \chi_T(\theta_c, B),$$
(7.26)

where we used

$$\chi_T = \left(\frac{\partial^2 G}{\partial B^2} \right)_T.$$

Consider the particular case $B = 0$, and choose the arbitrary value of the scaling parameter to be

$$\lambda = (-\theta_c)^{-1/r},$$

then the scaling relation takes the form

$$\chi_T(\theta_c, 0) = (-\theta_c)^{-(2p-1)/r} \chi_T(-1, 0).$$

Now, the critical exponent γ is defined by the relation

$$\chi_T(\theta_c, 0) \sim (-\theta_c)^{-\gamma},$$

as $\theta_c \to 0$. Comparison of the two preceeding expressions yields:

$$\gamma = (2p - 1)/r.$$

Then, substituting for p and r from eqn (7.25) above gives:

$$\gamma = \beta(\delta - 1).$$
(7.27)

This is a typical example of the kind of equation one can obtain in this way which gives relationships among the critical exponents.

7.3.5 Magnetic equation of state

The scaling hypothesis leads to specific predictions about the magnetic equation of state. This is a relation between the magnetization M, the applied field B, and the temperature T. If we go back to eqn (7.21), and rearrange:

$$M(\theta_c, B) = \lambda^{p-1} M(\lambda^r \theta_c, \lambda^p B). \qquad (7.28)$$

Once again we make our arbitrary choice of scaling parameter: $\lambda = |\theta_c|^{-1/r}$, and the scaling relation for the magnetization becomes:

$$M(\theta_c, B) = |\theta_c|^{(1-p)/r} M\left(\frac{\theta_c}{|\theta_c|}, \frac{B}{|\theta_c|^{p/r}}\right). \qquad (7.29)$$

Then, from (7.23) which is $\beta = (1 - p)/r$ and an additional expression,

$$\beta\delta = p/r,$$

which is easily obtained by multiplying eqns (7.23) and (7.24) together, we find

$$M(\theta_c, B) = |\theta_c|^\beta M\left(\frac{\theta_c}{|\theta_c|}, \frac{B}{|\theta_c|^{\beta\delta}}\right); \qquad (7.30)$$

or, with some rearrangement,

$$\frac{M(\theta_c, B)}{|\theta_c|^\beta} = M\left(\frac{\theta_c}{|\theta_c|}, \frac{B}{|\theta_c|^{\beta\delta}}\right). \qquad (7.31)$$

If we plot M against B, then data falls on different isotherms: the relationship between these two quantities depends on temperature. If we divide M by $|\theta_c|^\beta$ and B by $|\theta_c|^{\beta\delta}$, then the data for various temperatures all collapses onto one curve.

A good illustration of this may be seen in the book ([29], p. 118, Fig. 11.4). It should be noted that Stanley gives many inequalities between critical exponents ([29], p. 61, Table 4.1), based on thermodynamics and convexity properties. Many of these turn up as equalities in ([29], p. 185, Table 11.1) based on scaling relations.

7.3.6 Kadanoff's theory of block spins

If we judge the Widom scaling hypothesis by results, it is clearly successful in giving us some insight into the universal nature of the critical exponents. Now we would like to obtain some insight into the nature of this hypothesis. The crucial idea in understanding it was put forward by Kadanoff in the mid-1960s and is based on the assumption that the underlying regularity of a lattice (in the case of a magnet) could be reflected in the larger scale structures of domains or correlated islands of magnetism.

We rely on the Ising model, in which (let us remind ourselves) we consider N spins, each of which can be either "up" or "down" on a d-dimensional lattice. Each spin S_i is labeled by its position in the lattice where $i = 1, 2, \ldots, N$. The interactions between spins are limited to pairs of nearest neighbors, each of which has an interaction of magnitude J.

The Hamiltonian (7.15) may be written in the form

$$H = -\frac{K}{\beta} \sum_{\langle i,j \rangle} S_i S_j - \frac{h}{\beta} \sum_{i=1}^{N} S_j, \tag{7.32}$$

where the angle brackets grouping the indices i and j indicate that the summation is restricted to "nearest neighbor pairs," and of course each pair must be counted once only: see Section 2.1.1. We have introduced a change of notation which will be very useful from now on and have set

$$K \equiv \beta J = \frac{J}{kT}, \tag{7.33}$$

and

$$h \equiv \beta B = \frac{B}{kT}, \tag{7.34}$$

where B is the externally applied magnetic field. Variables divided by kT are known generically as *reduced variables*. It is usual to refer to h as the *reduced magnetic field* and to refer to K as the *coupling constant*. This is implicit recognition of the fact that the effective coupling is controlled by the temperature. This will turn out to be a very important idea in coming chapters. Recall that in Section 1.2.7 we have interpreted the temperature as being a *control parameter* for the coupling constant.

The idea proposed by Kadanoff amounts to a coarse-graining of our microscopic description of the lattice. Starting with a lattice of points a distance a apart (where a is the lattice constant), we partition the lattice points in groups into cells of side La, where $L \gg 1$ but is otherwise arbitrary. This procedure results in a number of cells n given by

$$n = \frac{N}{L^d}, $$

and each cell contains L^d spins. This is illustrated for the case $d = 2$ in Fig. 7.1. Now consider the case where the system is near the critical point, with $T \to T_c$ such that the correlation length $\xi \gg La$, and so our arbitrarily chosen coarse-graining parameter will satisfy the condition $1 \ll L \ll \xi/a$. Then, label each cell with a value of the an index α, where ($\alpha = 1, 2, \ldots, n$), and associate with each cell a magnetic moment $\tilde{S}_\alpha$.

7.3.6.1 Assumption 1. Next, we assume that each *cell* moment $\tilde{S}_\alpha$ behaves like a *site* moment S_j. That is, it is either up or down. (This assumption will be valid for most cells, if $T \sim T_c$ and we are dealing with big clusters of correlated spins.)

This then implies that, if we write the system Hamiltonian H in terms of cell moments $\tilde{S}_\alpha$ we may expect it to have the same form as H based on site moments S_j, although it may have different values of K and h. Hence we associate analogous parameters $\tilde{K}$ and $\tilde{h}$ with our cell model. Actually as K depends on both J and T, we shall work with $\tilde{\theta}_c$ rather than $\tilde{K}$.

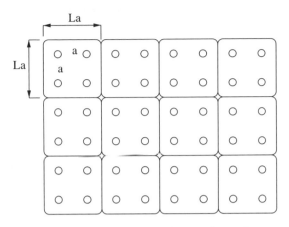

FIG. 7.1 "Blocking" spins in two dimensions.

7.3.6.2 Assumption 2. The next important step is to note that if the Hamiltonian H has the same *structure* in terms of both lattice sites and cells then we can assume that the form of the partition function will be the same, and hence (through the "bridge equation") that the thermodynamic potentials for the two models are also similar. Thus, taking the Gibbs potential as specific example, we may infer that the two descriptions are related by:

$$G(\tilde{\theta}_c, \tilde{h}) = L^d G(\theta_c, h). \tag{7.35}$$

7.3.6.3 Assumption 3. Now we need to relate $\tilde{\theta}_c$ to θ_c and $\tilde{h}$ to h. In order to do this, Kadanoff assumed that the "block" and "site" variables were related by:

$$\tilde{\theta}_c = L^y \theta_c; \tag{7.36}$$

$$\tilde{h} = L^x h, \tag{7.37}$$

and, substituting these forms into the equation for the Gibbs free energy, we obtain:

$$G(L^y \theta_c, L^x h) = L^d G(\theta_c, h). \tag{7.38}$$

7.3.6.4 Comparison with the Widom scaling hypothesis. This is of same form as the Widom scaling hypothesis, although, in that case λ was arbitrary and here L is only arbitrary within $1 \ll L \ll \xi/a$. Thus it can be written as:

$$G(L^{y/d} \theta_c, L^{x/d} h) = L G(\theta_c, h), \tag{7.39}$$

where previous scaling parameters can now be identified as

$$r = y/d; \tag{7.40}$$

$$p = x/d. \tag{7.41}$$

Thus in the absence of characteristic scales, we can assume that a coherent region on a given scale, has the scale of the coherent region below it in the hierarchy as its characteristic scale. This is the idea of *self-similarity* or *fractal structure*, as discussed in Section 1.5.

7.4 Linear response theory

This is a method for working out the response function (e.g. the magnetic susceptibility) of a system in thermal equilibrium. It relies on a clever mathematical trick but also has an underlying physical interpretation. For once we shall begin with the mathematical idea and come back to the physics later.

In statistical mechanics, we saw that we can work out the mean value of the energy, for instance, $\langle E \rangle$, either directly from the usual expectation value (as in eqn (1.129)) or indirectly by differentiating the partition function with respect to the temperature T, as in eqn (1.130). In fact, this idea is available as a general technique for any variable X (say) and the general algorithm may be stated as follows:

- Add a fictitious term $-XY$ to the energy (or Hamiltonian) of the system.

- Work out the partition function $\mathcal{Z}$ as a function of Y.

- Differentiate $\ln \mathcal{Z}$ with respect to Y and then put $Y = 0$.

That is,

$$\frac{1}{\beta}\frac{\partial \ln \mathcal{Z}}{\partial Y}\bigg|_{Y=0} = \frac{1}{\beta \mathcal{Z}}\frac{\partial}{\partial Y}\sum_i e^{-\beta(E_i - X_i Y)}\bigg|_{Y=0} = \frac{1}{\mathcal{Z}}\sum_i X_i e^{-\beta(E_i - X_i Y)}\bigg|_{Y=0} = \langle X \rangle. \tag{7.42}$$

So the two methods are equivalent. If we use

$$-\frac{1}{\beta}\ln \mathcal{Z} = F,$$

from eqn (1.126), then the mean value of the variable X follows at once as:

$$\langle X \rangle = -\left[\frac{\partial F}{\partial Y}\right]_{Y=0}. \tag{7.43}$$

Then, differentiating again with respect to Y gives the fluctuations in X:

$$\langle X^2 \rangle - \langle X \rangle^2 = \frac{1}{\beta}\frac{\partial \langle X \rangle}{\partial Y} \equiv \frac{\chi}{\beta} \tag{7.44}$$

where χ is the generalized susceptibility. This is known as the linear response equation.

And the physical interpretation? That is quite simple. If we apply a field or a current in order to generate the new term in the Hamiltonian, then this applied field breaks the symmetry of the system. In this way, the response of the system to a symmetry-breaking perturbation is revealed.

7.4.1 Example: spins on a lattice

As a specific example, we shall again consider the Ising microscopic model of a ferromagnet, which consists of a lattice of N spins, with a spin vector S_i at each lattice site i.

7.4.1.1 The mean magnetization. We begin with the mean magnetization, which we denote by

$$M = \langle S \rangle$$

where

$$S = \sum_{i=1}^{N} S_i \equiv \text{total spin.}$$

In line with the procedure of the previous section, we add a term

$$\Delta H = \left(-\frac{1}{\beta} \right) J S$$

to the Hamiltonian or energy. (Actually $J = \beta B$, where B is the magnetic field). Then

$$M = \left. \frac{\partial \ln \mathcal{Z}}{\partial J} \right|_{J=0}. \tag{7.45}$$

7.4.1.2 Correlation functions. Allow J (or B) to be different at each lattice site: add the term

$$\Delta H' = -\sum_i \frac{J_i S_i}{\beta},$$

to the Hamiltonian. Hence we may obtain the mean and correlations as follows:

$$\langle S_i \rangle = \frac{1}{\mathcal{Z}} \frac{\partial \mathcal{Z}}{\partial J_i}, \tag{7.46}$$

$$\langle S_i S_j \rangle = \frac{1}{\mathcal{Z}} \frac{\partial^2 \mathcal{Z}}{\partial J_i \partial J_j}, \tag{7.47}$$

$$\langle S_i S_j S_k \rangle = \frac{1}{\mathcal{Z}} \frac{\partial^3 \mathcal{Z}}{\partial J_i \partial J_j \partial J_k}, \tag{7.48}$$

where $\langle S_i S_j \ldots \rangle$ are the *correlations*.

7.4.1.3 Connected correlations. We can also treat connected correlations, which (as defined by eqn (7.9)) involve fluctuations about the mean value, in this way and these are generated by diferentiating $\ln \mathcal{Z}$, thus:

$$G_c(i, j) = \frac{\partial^2 \ln \mathcal{Z}}{\partial J_i \partial J_j} \tag{7.49}$$

The above is the *pair correlation*: a generalization to the n-point connected correlation is possible:

$$G_c(i_1, \ldots, i_N) = \frac{\partial}{\partial J_{i_1}}, \cdots, \frac{\partial}{\partial J_{i_N}} \ln \mathcal{Z} \equiv \langle S_{i_1} \ldots S_{i_N} \rangle_c. \tag{7.50}$$

7.5   Serious mean-field theory

Basically the idea is just the same as the Weiss theory of Chapter 3. But now we do it more formally and use a variational principle. By combining this with linear response theory we are able to obtain information about correlations. We begin by introducing and then proving the Bogoliubov variational principle.

7.5.1   The Bogoliubov variational theorem

The exact Hamiltonian (or total energy) for an interacting system is often found to be of the form:

$$H = \sum_{i=1}^{N} H_i + \sum_{i,j} H_{i,j}. \tag{7.51}$$

This may be readily understood as a generalization of the forms for the system energy discussed in Section 1.2.4. For example, eqn (1.20) is just a special case of (7.51). Let us *choose* a model Hamiltonian of the form

$$H(\lambda) = H_0 + \lambda H_I, \tag{7.52}$$

such that

$$0 \leq \lambda \leq 1,$$

where H is exact, H_0 is soluble, H_I is the correction term and λ is a variable control parameter. The *Bogoliubov theorem* can be stated in terms of the Helmholtz free energy F as:

$$F \leq F_0 + \langle H_I \rangle_0, \tag{7.53}$$

where F_0 is the free energy of the soluble system with Hamiltonian H_0, and the so-called ground-state expectation value of the correction term is given by

$$\langle H_I \rangle_0 = \frac{tr \; H_I e^{-\beta H_0}}{tr \; e^{-\beta H_0}}. \tag{7.54}$$

This procedure may be interpreted as follows:

1. We are evaluating our estimate of the exact free energy F using the full Hamiltonian $H = H_0 + \lambda H_I$, but only the ground-state (i.e. noninteracting) probability distribution associated with the soluble model Hamiltonian H_0.

2. Then eqn (7.53) gives us a rigorous upper bound on our *estimate* of the free energy corresponding to the exact Hamiltonian.

Our strategy now involves the following steps:

- Choose a trial Hamiltonian H_0 which is soluble.

- Use our freedom to vary the control parameter λ in order to minimize the quantity on the right hand side of the Bogoliubov inequality, as given in (7.53).

Then, in this way, we obtain our best estimate of the exact free energy F for a given choice of soluble model Hamiltonian H_0.

7.5.2   Proof of the Bogoliubov inequality

We shall work in the canonical ensemble; and, in the energy representation, in which case we have the partition function as:[29]

$$Z = tr \ e^{-\beta H}. \tag{7.55}$$

We also have the bridge equation, $F = -kT \ln Z$, which can be rewritten for our present purposes as:

$$-\beta F(\lambda) = \ln tr \ e^{\beta H(\lambda)}. \tag{7.56}$$

Next, we differentiate both sides of this expression with respect to λ, and cancelling factors of β as appropriate, we obtain

$$\frac{\partial F}{\partial \lambda} = \frac{tr \ H_I e^{-\beta(H_0 + \lambda H_I)}}{tr \ e^{-\beta(H_0 + \lambda H_I)}} = \langle H_I \rangle. \tag{7.57}$$

Then we differentiate again with respect to λ (above and below!) and this gives us

$$\frac{\partial^2 F}{\partial \lambda^2} = -\beta[\langle H_I^2 \rangle - \langle H_I \rangle^2] = -\beta \langle (H_I - \langle H_I \rangle)^2 \rangle. \tag{7.58}$$

Now the quantity $\langle (H_I - \langle H_I \rangle)^2 \rangle$ is positive definite. From this it follows that

$$\frac{d^2 F}{d\lambda^2} \leq 0, \tag{7.59}$$

for all λ, and so F is a concave function[30] of λ. This implies that the function $F(\lambda)$ lies below the tangent to $F(\lambda)$ at $\lambda = 0$, or:

$$F(\lambda) \leq F_0 + \lambda \left(\frac{\partial F}{\partial \lambda} \right)_{\lambda=0} + \cdots, \tag{7.60}$$

where we have written $F(0) = F_0$, for convenience.

Our proof is completed as follows. We substitute from (7.57) for $\partial F / \partial \lambda$, evaluated at $\lambda = 0$. Then we set $\lambda = 1$, in order to recover the exact problem, and the Bogoliubov theorem (as given in (7.53)) follows.

---

[29] Remember! The trace notation *tr* just stands for "sum over states."

[30] This is a standard result from elementary calculus.

7.5.3   Mean-field theory of the Ising model

We consider the Ising model with external magnetic field B. From eqn (7.14) the Hamiltonian may be written in the slightly different form:

$$H = - \sum_{i,j} J_{ij} S_i S_j - B \sum_i S_i, \qquad (7.61)$$

where

- $J_{ij} = J$ if i, j are nearest neighbors
- $J_{ij} = 0$ if i, j are NOT nearest neighbors.

Note that this is yet another way of specifying the sum over nearest-neighbor pairs of spins!

In order to reduce the Hamiltonian to a diagonal form,[31] we choose the unperturbed model for H to be

$$H_0 = - \sum_i B' S_i - B \sum_i S_i, \qquad (7.62)$$

where B' is the "collective field" representing the effect of all the other spins with labels $j \neq i$ on the spin at the lattice site i. Sometimes it is convenient to lump the two magnetic fields together as an "effective field" B_E:

$$B_E = B' + B. \qquad (7.63)$$

Now we can work out our upper bound for the system free energy F, using the statistics of the model system. First we obtain the partition function Z_0 by adapting eqn (7.55) to the ground-state case, thus:

$$Z_0 = tr e^{-\beta H_0} = [e^{\beta B_E} + e^{-\beta B_E}]^N, \qquad (7.64)$$

where we have summed over the two spin states of $S = \pm 1$. This may be further written as

$$Z_0 = [2 \cosh(\beta B_E)]^N. \qquad (7.65)$$

The free energy of the model system F_0 follows immediately from the bridge equation, as

$$F_0 = -\frac{N}{\beta} \ln[2 \cosh(\beta B_E)]. \qquad (7.66)$$

Now from (7.53) the Bogoliubov inequality may be written in the form

$$F \leq F_0 + \langle H_I \rangle_0 \leq F_0 + \langle H - H_0 \rangle_0, \qquad (7.67)$$

where we have simply reexpressed the correction term as the difference between the exact and model system Hamiltonians. Then, substituting from eqn (7.61) we may further rewrite

---

[31] For background discussion of this point, see Section 1.2.5.

this condition on the free energy as:

$$F \leq F_0 - \sum_{i,j} J_{ij} \langle S_i S_j \rangle_0 + B' \sum_j \langle S_j \rangle_0. \tag{7.68}$$

We now work out averages over the *model* assembly, thus:

$$\sum_j \langle S_j \rangle_0 = N \langle S \rangle_0, \tag{7.69}$$

and

$$\sum_{ij} J_{ij} \langle S_i S_j \rangle_0 = \sum_{ij} J_{ij} \langle S_i \rangle_0 \langle S_j \rangle_0 = \frac{JNz}{2} \langle S \rangle_0^2, \tag{7.70}$$

where we have made use of the statistical independence of S_i and S_j, which is consistent with the statistics of the zero-order (noninteracting) model, z is the number of nearest neighbors and $Nz/2$ is the number of nearest-neighbor pairs. Then, substituting these results into eqn (7.68), we have

$$F \leq F_0 - \frac{JNz}{2} \langle S \rangle_0^2 + B'N \langle S \rangle_0, \tag{7.71}$$

and, taking the equality rather than the bound,

$$F = F_0 - \frac{JNz}{2} \langle S \rangle_0^2 + (B_E - B) N \langle S \rangle_0, \tag{7.72}$$

where we have also substituted for B' using (7.63). This result will be the basis of our variational method in the next section.

We already know F_0 from eqn (7.66), while $\langle S \rangle_0$ is easily worked out as:

$$\langle S \rangle_0 = \frac{tr \, S \exp(-\beta H_0)}{tr \, \exp(-\beta H_0)} = \frac{tr \, S \exp(\beta B_E S)}{tr \, \exp(\beta B_E S)} = \frac{\exp(\beta B_E) - \exp(-\beta B_E)}{\exp(\beta B_E) + \exp(-\beta B_E)}.$$

The permissible spin states of the Ising model are $S = \pm 1$, hence:

$$\langle S \rangle_0 = \tanh(\beta B_E). \tag{7.73}$$

Our next step is to use the variational method to obtain an effective field B_E such that the free energy is a minimum.

7.5.4 The variational method

The control parameter λ has now been replaced (in effect) by the molecular field B', so we differentiate F as given by (7.72) with respect to B' and set the result equal to zero. Noting that B' always occurs as part of B_E, the condition for an extremum can be written more conveniently as

$$\frac{\partial F}{\partial B_E} = 0, \tag{7.74}$$

and from (7.72), this becomes

$$\frac{\partial F}{\partial B_E} = \frac{\partial F_0}{\partial B_E} - NzJ\langle S\rangle_0 \frac{\partial \langle S\rangle_0}{\partial B_E} + (B_E - B)N\frac{\partial \langle S\rangle_0}{\partial B_E} + N\langle S\rangle_0. \tag{7.75}$$

From eqn (7.66) for F_0, we have:

$$\frac{\partial F_0}{\partial B_E} = -N\langle S\rangle_0, \tag{7.76}$$

which cancels the last term on the right hand side of eqn (7.75), hence

$$\frac{\partial F}{\partial B_E} = (B_E - B)N\frac{\partial \langle S\rangle_0}{\partial B_E} - NzJ\langle S\rangle_0 \frac{\partial \langle S\rangle_0}{\partial B_E} = 0, \tag{7.77}$$

and, equating coefficients of $\partial \langle S\rangle_0/\partial B_E$, we have the condition for an extremum as:

$$B_E - B = zJ\langle S\rangle_0; \tag{7.78}$$

or, rearranging,

$$B_E = B + zJ\langle S\rangle_0, \tag{7.79}$$

is the condition for minimum free energy. In this model, the magnetization is just the mean value of the spin, thus from (7.73) and (7.79):

$$\langle S\rangle_0 = \tanh(\beta B + zJ\beta\langle S\rangle_0). \tag{7.80}$$

In order to identify a phase transition, we put $B = 0$, and (7.80) becomes

$$\langle S\rangle_0 = \tanh(zJ\beta\langle S\rangle_0), \tag{7.81}$$

which is the same as our previous mean-field result as given by eqn (3.14), with the replacement of M/M_∞ by $\langle S\rangle_0$. We have therefore shown that the optimum value of the free energy with an F_0 corresponding to independent spins is exactly that of mean-field theory.

7.5.5 The optimal free energy

From eqns (7.71) and (7.63) we can write the optimal free energy as

$$F = F_0 - \frac{NzJ}{2} \langle S \rangle_0^2 + N(B_E - B)\langle S \rangle_0.$$

Then substituting from (7.66) for the unperturbed free energy F_0, we have

$$F = -\frac{N}{\beta} \ln [2 \cosh (\beta B_E)] - \frac{NzJ}{2} \langle S \rangle_0^2 + N (B_E - B) \langle S \rangle_0,$$

and, lastly, substituting from (7.78) for $\langle S \rangle_0$, yields:

$$F = -\frac{N}{\beta} \ln [2 \cosh (\beta B_E)] - \frac{NzJ}{2} \frac{(B_E - B)^2}{z^2 J^2} + N \frac{(B_E - B)^2}{zJ},$$

and, with cancellations as appropriate, we arrive at the neat form:

$$F = -\frac{N}{\beta} \ln [2 \cosh (\beta B_E)] + \frac{N}{2zJ} (B_E - B)^2 . \qquad (7.82)$$

Bearing in mind that the free energy is always our main objective in statistical physics, as all the macroscopic physics can then be obtained by differentiation using a variety of thermodynamical relationships, this equation may be seen as the overall result of the theory.

Now we can use this result to obtain the exact value of the mean spin by the procedures of linear response theory, as discussed in Section 7.4. That is, we write

$$\langle S \rangle = -\frac{1}{N} \frac{\partial F}{\partial B},$$

and simply differentiate the free energy as given by eqn (7.82) with respect to the external field. The one tricky point is the implicit dependence of B_E on B, but writing

$$\langle S \rangle = -\frac{1}{N} \frac{\partial F}{\partial B_E} \cdot \frac{\partial B_E}{\partial B} - \frac{1}{N} \frac{\partial F}{\partial B} = 0 - \frac{1}{N} \frac{\partial F}{\partial B},$$

we see that there is no explicit dependence on B_E because $\partial F / \partial B_E = 0$ is the condition for the extremum. Then differentiating both sides of (7.82), gives us:

$$\langle S \rangle = \frac{1}{zJ} (B_E - B) = \langle S \rangle_0$$

where the last step follows from (7.79).

Thus we have established the fact that the exact solution (within mean-field theory) is the same as the zero-order result. This implies that we can also write eqn (7.80) in terms of exact values of the mean spin, thus:

$$\langle S \rangle = \tanh(\beta B + zJ\beta \langle S \rangle), \qquad (7.83)$$

and we shall make use of this result in Section 7.7.

7.6   Mean-field critical exponents α, β, γ, and δ for the Ising model

The exponents for the thermodynamic quantities can be obtained quite easily from our present results. However, to get the exponents associated with the correlation function η and the correlation length ν we need to obtain expressions for these quantities. We begin with the easier ones!

7.6.1   Exponent α

For $B = 0$, we have $H = -J \sum_{\langle i,j \rangle} S_i S_j$ where $\sum_{\langle i,j \rangle}$ is the sum over nearest neighbors. The mean energy of the system is given by

$$\overline{E} = \langle H \rangle = -J \sum_{\langle i,j \rangle} \langle S_i S_j \rangle.$$

In lowest-order mean field approximation, the spins are independent and so we may factorize as:

$$\langle S_i S_j \rangle = \langle S_i \rangle \langle S_j \rangle.$$

Hence

$$\overline{E} = -J \sum_{\langle i,j \rangle} \langle S_i \rangle \langle S_j \rangle = -Jz\frac{N}{2}M^2,$$

where $M = \langle S \rangle \equiv$ the order parameter. From the thermodynamic definition of the heat capacity, C_B at constant magnetic field, we have

$$C_B = \left.\frac{\partial \overline{E}}{\partial T}\right)_B = -2Jz\frac{N}{2}M\frac{dM}{dT} = -JzNM\frac{dM}{dT}.$$

Now, for:

$T > T_c : \quad M = 0 \quad \text{therefore} \quad C_B = 0;$

$T \le T_c : \quad M = (-3\theta_c)^{1/2}.$

Thus

$$\frac{\partial M}{\partial T} = \frac{1}{2}(-3\theta_c)^{-1/2} \times -\frac{d\theta_c}{dT} = \frac{-3}{2}M^{-1}\frac{d\theta_c}{dt} = \frac{-3}{2}M^{-1}T_c^{-1},$$

and so

$$\left.\frac{\partial \overline{E}}{\partial T}\right)_B = \frac{3}{2}JzNMM^{-1}T_c^{-1} = \frac{3}{2}\frac{JzN}{c}, \quad \text{(from eqn (3.17))}$$

$$= \frac{3}{2}Nk, \quad \text{as } Jz = kT_c.$$

Hence C_B is discontinuous at $T = T_c$ and so $\alpha = 0$.

7.6.2 Exponent β

The mean magnetization is $M = \langle S \rangle_0$, and from mean-field theory:

$$\langle S \rangle_0 = \tanh(\beta B + 2z J \beta \langle S \rangle_0).$$

Hence we can write:

$$M = \tanh(\beta z J M + b), \quad \text{where } b = \beta B.$$

Now mean-field theory gives $z \beta_c J = 1$ or $z J = 1/\beta_c$, thus it follows that

$$M = \tanh\left[\frac{\beta M}{\beta_c} + b\right] = \tanh\left[M\frac{T_c}{T} + b\right] = \tanh\left[\frac{M}{(1 + \theta_c)} + b\right].$$

Set $B = 0$ and expand for $T \sim T_c$, in which case θ_c is small:

$$M = \frac{M}{1 + \theta_c} - \frac{1}{3}\frac{M^3}{(1 + \theta_c)^3},$$

and rearranging:

$$M\left(1 - \frac{1}{1 + \theta_c}\right) = -\frac{1}{3}\frac{M^3}{(1 + \theta_c)^3},$$

hence, either:

$$M = 0$$

or

$$M^2 = -3\theta_c\frac{(1 + \theta_c)^3}{(1 + \theta_c)} = -3\theta_c(1 + \theta_c)^2.$$

Taking the nontrivial case,

$$M \sim |-3\theta_c|^{1/2},$$

and by comparison with eqn (3.33) which defines the critical exponent:

$$\beta = 1/2.$$

7.6.3 Exponents γ and δ

From the definition of the isothermal susceptibility χ_T, we have:

$$\chi_T = \frac{\partial M}{\partial B} = \beta \frac{\partial M}{\partial b},$$

and also

$$M = \tanh\left(\frac{M}{1+\theta_c} + b\right) \simeq \frac{M}{1+\theta_c} + b \quad \text{for} \quad T > T_c.$$

Now, with some rearrangement,

$$M - \frac{M}{1+\theta_c} = b,$$

to this order of approximation and, rearranging further, we have:

$$M = \left(\frac{1+\theta_c}{\theta_c}\right) b.$$

Hence

$$\chi_T \sim \frac{\partial M}{\partial b} \sim \frac{1}{\theta_c} \quad \text{as} \quad \theta_c \to 0$$

and so

$$\chi_T \sim \theta_c^{-1}, \quad \gamma = 1,$$

which follows from the definition of γ as given in eqn (3.34). Next, consider the effect of an externally imposed field at $T = T_c$, where $\theta_c = 0$, and so $1 + \theta_c = 1$. We use the identity:

$$M = \tanh(M + b) = (\tanh M + \tanh b)(1 + \tanh M \tanh b),$$

which leads to

$$M \simeq \left(M - \frac{M^3}{3} + b - \frac{b^3}{3}\right)(1 + \tanh M \tanh b).$$

Cancel the factor of M on both sides and rearrange, to obtain:

$$b \sim \frac{M^3}{3} + \frac{b^3}{3} - \left(M - \frac{M^3}{3} + b - \frac{b^3}{3}\right)\left(Mb - \frac{Mb^3}{3} - \frac{Mb^3}{3} + \cdots\right).$$

Therefore $b \sim M^3/b$ for small b, M and, by comparison with eqn (3.35), $\delta = 3$.

If we set $b \sim M^3$ on the right hand side, we can verify that all terms of order higher than $\mathcal{O}(M^3)$ have been neglected.

It should be noted that these values for the critical exponents are the same as those found using the Landau theory in Section 3.3.

7.7 The remaining mean-field critical exponents for the Ising model

We now turn our attention to the problem of calculating the connected correlation function, as this leads us to the remaining exponents, ν and η.

7.7.1 The connected correlation function G_{ij}^c

In mean field theory, we have approximated the exact equilibrium probability distribution P by a "ground state" value P_0 which is based on a model of independent spins. It follows that if we evaluate correlations on this basis we must find (e.g.) that

$$\langle S_i S_j \rangle = \sum_S P_0 S_i S_j = \langle S_i \rangle \langle S_j \rangle,$$

as the spins are independent variables on this model. Hence connected correlations like G_{ij}^c are zero, at the level of the zero-order approximation.

If we have

$$P - P_0 = \delta P$$

(assumed small) then it is easily seen that the error in working out a correlation against P_0 is also $\mathcal{O}(\delta P)$.

However, let us consider the free energy F and expand about the ground-state value F_0, thus

$$F = F_0 + \left(\frac{\delta F}{\delta P} \right) \delta P + O(\delta P)^2.$$

But $\delta F / \delta P = 0$ for a minimum and so it follows that $F = F_0 + O(\delta P)^2$. Hence, we can expect to obtain a better answer by resorting to linear response theory, as discussed in Section 7.5, and by using F to work out correlations.

Specifically, we may rewrite eqn (7.46), using eqn (1.125), for the mean value of the spin as

$$\langle S_i \rangle = -\partial F / \partial B_i,$$

and, from eqn (7.49), an equation for the connected pair correlation as

$$G^c(i, j) = -\frac{1}{\beta} \frac{\partial^2 F}{\partial B_i \partial B_j} = \frac{1}{\beta} \frac{\partial \langle S_i \rangle}{\partial B_j}, \tag{7.84}$$

where B_i is the inhomogeneous external field.

From Appendix E we can can write F for this case as:

$$F = -\frac{1}{\beta} \sum_i \ln \left[2 \cosh \left(\beta B_E^{(i)} \right) \right] + \frac{1}{2} \sum_i \left(B_E^{(i)} - B_i \right) \tanh \left(\beta B_E^{(i)} \right). \tag{7.85}$$

The implications of the generalization to a non-uniform B should be noted. First, we make the replacement: $B_E \rightarrow B_E^{(i)}$. Second, the factor N is replaced by the summation over the index i. We also have the condition (E.9) that:

$$B_{\mathrm{E}}^{(i)} - B_i = J \sum_{\langle j \rangle} \tanh\left(\beta B_{\mathrm{E}}^{(j)}\right), \tag{7.86}$$

where $\sum_{\langle j \rangle}$ is the sum over nearest neighbors of i.

Now, in principle we differentiate both sides of (7.85) with respect to B_i and obtain the mean spin $\langle S_i \rangle$, just as we did in Section 7.5.5 for the homogeneous case. However, this can be tricky due to the need to invert (7.86), and it is easier to generalize (7.83). For the inhomogeneous case we may write this as:

$$\langle S_i \rangle = \tanh\left[\beta\left(B_i + J \sum_{\langle j \rangle}\langle S_j \rangle\right)\right]. \tag{7.87}$$

We may simplify this result by considering the case where $T > T_{\mathrm{c}}$ and the external field B_i is weak. Under these circumstances $\langle S_i \rangle$ will be small and this allows us to expand tanh to first order in Taylor series, thus:

$$\langle S_i \rangle = \beta\left(B_i + J \sum_{\langle j \rangle}\langle S_j \rangle\right), \tag{7.88}$$

which still leaves us with N linear simultaneous equations to be solved for the $\langle S_i \rangle$.

We can use Fourier transformation to produce N independent equations, but in order to avoid confusion with $i = \sqrt{-1}$, we shall first change the index i to n, and rearrange eqn (7.88) as:

$$\langle S_n \rangle - J \sum_{\langle j \rangle}\langle S_j \rangle = \beta B_n. \tag{7.89}$$

In the next section we shall make a digression and introduce the discrete form of the Fourier transform.[32]

7.7.2 The discrete Fourier transform of eqn (7.89)

Let us consider a set of N numbers $\{x_n\}$, such that $(n = 0, \ldots, N - 1)$. We may define the Fourier transform of x_n, which we shall call $\tilde{x}_q$, by

$$\tilde{x}_q = \sum_{n=0}^{N-1} x_n e^{-2\pi i n q/N} \quad \text{where } (q = 0, \ldots, N - 1), \tag{7.90}$$

---

[32] You might think that the discrete form of the Fourier transform is the Fourier series! But in Fourier series, one of the Fourier pair is still a continuous function.

and correspondingly we also have

$$x_n = \frac{1}{N} \sum_{q=0}^{N-1} \tilde{x}_q e^{2\pi i n q/N}. \tag{7.91}$$

It is trivial to extend this definition to d-dimensional vectors $\mathbf{n}$ and $\mathbf{q}$, thus:

$$2\pi i n q/N \rightarrow 2\pi i \mathbf{n} \cdot \mathbf{q}/L,$$

where $L = N^{1/d}$ is the number of points in each coordinate direction and $\mathbf{n} \cdot \mathbf{q} = n_\alpha q_\alpha$ with $\alpha = 1, \ldots, d$.

Equations (7.90) and (7.91) define the Fourier transform pair x_n and $\tilde{x}_q$. We shall drop the tilde on the latter quantity as the label (n or q) makes it clear which space is being referred to.

Now we substitute for $\langle S_n \rangle$ in terms of its Fourier transform in eqn (7.89), thus

$$\frac{1}{N} \sum_{q=0}^{N-1} \langle S_q \rangle e^{2\pi i \mathbf{n} \cdot \mathbf{q}/L} - \frac{\beta J}{N} \sum_{\langle j \rangle} \sum_{q-0}^{N-1} \langle S_q \rangle e^{2\pi i \mathbf{j} \cdot \mathbf{q}/L} = \frac{1}{N} \sum_{q=0}^{N-1} \beta B_q e^{2\pi i \mathbf{n} \cdot \mathbf{q}/L}, \tag{7.92}$$

where $\sum_{\langle j \rangle}$ indicates the sum over j restricted to nearest neighbors of n.

This result holds for arbitrary N and for arbitrary exponential factors, but in order to make use of that fact we need to have some way of handling the second term on the left hand side which involves the inter-spin coupling. We consider this term on its own for the moment:

$$-\frac{\beta J}{N} \sum_{\langle j \rangle} \sum_{q=0}^{N-1} \langle S_q \rangle e^{2\pi i \mathbf{j} \cdot \mathbf{q}/L}.$$

The operation $\sum_{\langle j \rangle}$ is the sum over nearest neighbors and depends on the dimension d and type of lattice. We shall, as usual, assume that the lattice is cubic so that the number of nearest neighbors is given by $2d$. Accordingly, we may write

$$\mathbf{j} \cdot \mathbf{q} = \sum_{\alpha=1}^{d} j_\alpha q_\alpha.$$

and for each α, we set $j_\alpha = n_\alpha \pm 1$, whereupon the sum over nearest neighbors becomes the sum over coordinate directions.

The number of points in each coordinate direction α is $L = N^{1/d}$. Therefore

$$\sum_{\langle j \rangle} \exp\{2\pi i j_\alpha q_\alpha/L\} = \sum_{\alpha=1}^{d} [\exp\{2\pi i (n_\alpha q_\alpha + q_\alpha)/L\} + \exp\{2\pi i (n_\alpha q_\alpha - q_\alpha)/L\}]$$

$$= \exp\{2\pi i \mathbf{n} \cdot \mathbf{q}/L\} \sum_{\alpha=1}^{d} 2 \cos(2\pi q_\alpha L).$$

Substituting, this back into eqn (7.92), and cancelling factors as appropriate, we obtain

$$\left[1 - 2\beta J \sum_{\alpha=1}^{d} \cos(2\pi q_\alpha/L)\right]\langle S_q \rangle = \beta B_q. \tag{7.93}$$

7.7.3   The connected correlation function in Fourier space

Now we work out the spin–spin correlation but we do it in Fourier space. Fourier transforming both sides of eqn (7.84) we find

$$G^c(q) = \frac{1}{\beta} \frac{\partial \langle S_q \rangle}{\partial B_q}. \tag{7.94}$$

Then, differentiating both sides of eqn (7.93), and rearranging, we have

$$G^c(q) = \frac{1}{1 - 2\beta J \sum_{\alpha=1}^{d} \cos(2\pi q_\alpha/L)}. \tag{7.95}$$

For critical phenomena, we are interested in long-range correlations in real space and so we consider the behavior of $G^c(q)$ at small q. To do this, we set $k_\alpha = 2\pi q_\alpha/L$, and expand for small k_α up to k_α^2:

$$G^c(k) = \frac{1}{1 - 2\beta J \sum_{\alpha=1}^{d} \cos k_\alpha} = \frac{1}{1 - 2\beta J \sum_{\alpha=1}^{d}(1 - k_\alpha^2/2)}$$

$$= \frac{1}{1 - z\beta J + \beta J k^2} \tag{7.96}$$

where, as usual, $z = 2d$ is the coordination number of the lattice.[33] If we introduce the quantity ξ such that

$$\xi^2 = \frac{\beta J}{1 - z\beta J}, \tag{7.97}$$

then we can rearrange our expression, as given by eqn (7.96), for the spin–spin correlation as

$$G^c(k\xi) = \left(\frac{\xi^2}{\beta J}\right)\left(\frac{1}{1 + k^2\xi^2}\right). \tag{7.98}$$

We may note that as G^c in Fourier space is a function of $k\xi$, it follows that in real space we must have $G^c(i, j)$ a function of r/ξ, where r is the distance between lattice sites i and j. Hence we interpret ξ as the *correlation length*.

---

[33] In this book we only consider the simple cubic lattice.

7.7.4   Critical exponents ν and η

First we have the correlation length, with the assumption $\xi = |\theta_c|^{-\nu}$ defining the exponent ν. From mean-field theory the correlation length is given by (7.97) and from eqn (3.17) we have $zJ = 1/\beta_c$, so it follows that

$$\xi^2 = \frac{\beta J}{1 - \beta/\beta_c} \to \infty,$$

as $\beta/\beta_c \to 1$, which is the correct qualitative behavior. Now,

$$\frac{\beta_c}{\beta} = 1 + \theta_c \quad \text{therefore } 1 - \beta/\beta_c = \theta_c,$$

and so it also follows that

$$\xi^{-2} \sim \theta_c, \quad \text{or} \quad \xi \sim \theta_c^{-1/2}.$$

Hence, comparison of this result with the defining relationship indicates that the mean-field value of ν is

$$\nu = 1/2. \tag{7.99}$$

For the correlation itself, when the system is near the critical point, we have

$$G^c(r) \sim r^{2-d-\eta},$$

which defines the exponent η. Fourier transforming this gives us

$$G^c(k) \sim k^{-2+\eta},$$

for small k. From mean-field theory, eqn (7.96) gives

$$G^c(k) = \frac{1}{1 - 2\beta J \sum_{\alpha=1}^{d} \cos k_\alpha},$$

and expanding out the cosine for small values of k we obtain the denominator as:

$$\text{denominator} = 1 - 2d\beta J - 2\beta J \frac{k^2}{2} + \cdots.$$

Now for $T = T_c$, we have $\beta = \beta_c$ and $2dJ = zJ = \beta_c^{-1}$, hence the above expression for the denominator reduces to

$$\text{denominator} \approx -\beta J k^2 + \cdots$$

therefore $G^c(k) \sim k^{-2}$ and comparison with the defining relationship for η yields the mean-field value of the critical exponent as:

$$\eta = 0. \tag{7.100}$$

Mean-field theory exponents do not depend on dimension d whereas the Ising model exponents do depend on d. Mean field theory gives quite good values for $d = 3$ and, as we shall see, exact results for $d = 4$, but it consistently overestimates T_c.

7.8   Validity of mean-field theory

Mean-field theory can be shown to be equivalent to an assumption that each spin interacts equally with every other spin in the lattice: this implies infinite interaction range in the limit $N \to \infty$.

7.8.1   The model

We wish to set up a model in which all spins interact with each other. For N spins, we have effectively $N^2/2$ pairs of spins, so in order to have the same overall energy as the Ising model the interaction of each pair must be proportional to $1/N$. Then the total energy behaves as

$$E \sim \frac{N^2}{2} \times \frac{1}{N} \sim N,$$

which is correct.

In view of this, let us suppose that the N spins have an interaction energy $-J/N$ between each pair and, assuming zero external field, write the Hamiltonian as

$$H = -\frac{2J}{N} \sum_{i<j}^{N} S_i S_j = -\frac{J}{N} \sum_{i=1}^{N}\sum_{j=1}^{N} S_i S_j + \frac{J}{N} \sum_{i=1}^{N} S_i^2. \tag{7.101}$$

Then the following points should be noted:

1. In the second equality we have abandoned the condition $i < j$ in the double sum therefore each off-diagonal term is now counted twice. Accordingly we drop the factor of two.

2. The double sum now includes (erroneously) the diagonal terms, for which $i = j$, and so we cancel these by adding the last term.

Now each $S_i^2 = 1$, hence the last term is equal to $N \times J/N$ and so the Hamiltonian becomes:

$$H = J - \frac{J}{N}\left(\sum_{i=1}^{N} S_i\right)\left(\sum_{i=1}^{N} S_j\right) = J - \frac{J}{N}\left(\sum_{i=1}^{N} S_i\right)^2. \tag{7.102}$$

Thus the partition function (in the canonical ensemble) may in turn be written as

$$Z_N = \sum_{\{S\}} \exp\left[-K + \frac{K}{N}\left(\sum_{i=1}^{N} S_i\right)^2\right] = \exp[-K] \sum_{\{S\}} \exp\left[K\left(\sum_{i=1}^{N} -S_i N^{-1/2}\right)\right]^2, \tag{7.103}$$

where $K \equiv \beta J$.

7.8.2 A trick to evaluate the partition sum

Our next step is to use the trick that we used in Section 1.3.4, and "complete the square" in the integrand.

This time we begin with the standard form

$$\int_{-\infty}^{\infty} dX e^{-X^2} = \sqrt{\pi}.$$

If we set

$$X = \left(\frac{x}{\sqrt{2}} - a\right), \quad dX = \frac{1}{\sqrt{2}} dx,$$

then the standard form can be written as the identity:

$$\exp\left[a^2\right] = (2\pi)^{-1/2} \int_{-\infty}^{\infty} \exp\left[(-x^2/2 + 2^{1/2}ax)\right] dx.$$

We may then use this identity by setting

$$a = \left(\frac{K}{N}\right)^{1/2} \sum_{i=1}^{N} S_i,$$

which allows us to rewrite the partition function as

$$Z_N = \exp[-K](2\pi)^{-1/2} \sum_{\{S\}} \int_{-\infty}^{\infty} dx \exp\left[-\frac{x^2}{2}\right] \exp\left[x \left(\frac{2K}{N}\right)^{1/2} \sum_{i=1}^{N} S_i\right]$$

$$= \exp[-K](2\pi)^{-1/2} \int_{-\infty}^{\infty} dx \exp\left[-x^2/2\right] \sum_{S_1=-1}^{1} \cdots \sum_{S_N=-1}^{1} \exp\left[x \left(\frac{2K}{N}\right)^{1/2} \sum_{i=1}^{N} S_i\right].$$

The summations can be done before the integration, thus:

$$Z_N = (2\pi)^{-1/2} e^{-K} \int_{-\infty}^{\infty} \exp\left[-x^2/2\right] \left[2 \cosh\left(x\sqrt{2K/N}\right)\right]^N dx,$$

and substituting $x = yN^{1/2}$

$$Z_N = \left(\frac{N}{2\pi}\right)^{1/2} e^{-K} \quad 2^N \int_{-\infty}^{\infty} \left[e^{-y^2/2} \cosh\left(y\sqrt{2K}\right)\right]^N dy.$$

The integral may be approximated by taking only its largest term[34] to give

$$Z_N \propto N^{1/2} e^{-K} 2^N \max_{-\infty \le y \le \infty} \left[e^{-y^2/2} \cosh\left(y\sqrt{2K}\right)\right]^N.$$

---

[34] This is known as the *method of steepest descents* or a *saddle-point approximation*.

7.8.3   The thermodynamic limit

Consider the free energy F per spin in the thermodynamic limit: we have

$$F = -kT \lim_{N \to \infty} (N^{-1} \ln Z_N),$$

and so

$$F = -kT \left\{ \ln 2 + \ln \max_{-\infty \le y \le \infty} \left[\exp\left[-\frac{y^2}{2} \right] \cosh\left(y\sqrt{2K} \right) \right] \right\}.$$

We maximize the second term on the right hand side, as follows: Let

$$f = \ln[e^{-y^2/2} \cosh\left(y\sqrt{2K} \right)] = -\frac{y^2}{2} + \ln \cosh\left(y\sqrt{2K} \right),$$

hence

$$\frac{\partial f}{\partial y} = -y + \frac{1}{\cosh\left(y\sqrt{2K} \right)} 2K \sinh\left(y\sqrt{2K} \right),$$

and so

$$y = \sqrt{2K} \tanh\left(y\sqrt{2K} \right),$$

is the condition for a maximum. This is just the molecular field equation for $B = 0$. This limit, in which each spin is coupled to infinitely many others, can also be realized by having an infinity of nearest neighbors. We may achieve this by increasing the number of dimensions. In principle $d \to \infty$ should give an exact result. In fact, as we shall see in Chapter 9, $d = 4$ is good enough for this model!

7.9   Upper critical dimension

We can give a heuristic argument for the value of d, at which mean-field theory becomes valid. Essentially the limitation of the theory is that it ignores fluctuations compared to the mean. In the real world there are fluctuations in which spins are correlated over a distance ξ. Each fluctuation has energy of order kT and so the energy of fluctuations per unit volume in a d-dimensional space may be estimated as

$$\Delta F_{fluct} \sim \frac{kT}{\xi^d}.$$

For T near T_c, we can write the correlation length as $\xi \sim |\theta_c|^{-\nu}$, and so the above expression becomes

$$\Delta F_{fluct} \sim kT_c |\theta_c|^{d\nu}. \tag{7.104}$$

How does this compare to the *total* free energy per unit volume F near T_c? We may estimate the latter quantity from a consideration of the thermodynamic relationship:

$$C_B \sim -T \left(\frac{\partial^2 F}{\partial T^2} \right)_B \sim -\frac{T}{T_c^2} \left(\frac{\partial^2 F}{\partial \theta_c^2} \right)_B. \tag{7.105}$$

Now, for $T \to T_c$, we may write $C_B \sim |\theta_c|^{-\alpha}$. So, substituting this on the left hand side of eqn (7.105), and integrating twice with respect to the temperature, we obtain an expression for the total free energy in the form

$$F \sim T_c |\theta_c|^{2-\alpha}. \tag{7.106}$$

For a consistent theory, we want to be able to justify our neglect of the free energy of fluctuations:

$$\Delta F_{fluct} \ll F,$$

and so, a comparison of eqns (7.104) and (7.106) indicates that, as $\theta_c \to 0$, what we really require is that

$$\nu d > 2 - \alpha. \tag{7.107}$$

In mean-field theory we have $\nu = 1/2$ and $\alpha = 0$. Accordingly, substituting these values into the above equation gives us the requirement for mean-field theory to be valid that the dimension d should satisfy the relationship

$$d > d_{\mathrm{mf}} = 4, \tag{7.108}$$

where d_{mf} is known as the *upper critical dimension*.

Further reading

General introductory discussions of phase transitions and critical phenomena can be found in the books [6], [16], [12], and [26]. However, for a remarkably comprehensive, yet concise, introduction to the subject, the book by Stanley [29] would be hard to beat.

7.10   Exercises

Solutions to these exercises, along with further exercises and their solutions, will be found at: www.oup.com

1. Prove the critical exponent inequality $\alpha + 2\beta + \gamma \geq 2$
   [Hint: you may assume the relationship

$$\chi_T (C_B - C_M) = T (\partial M / \partial T)_B^2,$$

   where C_B and C_M are the specific heats at constant field and magnetization, respectively.]

2. Additional critical exponents can be obtained if we differentiate the Gibbs free energy repeatedly with respect to the external magnetic field B, thus:

$$(\partial^\ell G/\partial B^\ell)_T = G^{(\ell)} \sim \theta_c^{\Delta_\ell} G^{(\ell-1)},$$

where the Δ_ℓ are known as the "gap exponents" and θ_c is the reduced temperature. On the basis of the Widom scaling hypothesis, show that the gap exponents are all equal and give their value in terms of the parameters of the Widom scaling transformation.
3. If we denote the order parameter (or specific magnetization) by M, show that the mean-field solution of the Ising model can be written as

$$M = \tanh\left[\frac{M}{1+\theta_c} + b\right],$$

where θ_c is the reduced temperature and $b = \beta B$ is the reduced external magnetic field.
   By considering m for temperature close to T_c and for zero external field, show that the associated critical exponent takes the value $\beta = \frac{1}{2}$.
   [Hint: the following expansion

$$\tanh x = x - \tfrac{1}{3}x^3 + 0(x^5),$$

for small values of x, should be helpful.]
4. Obtain an expression for the mean energy $\overline{E}$ of the Ising model when the applied field is zero, using the simplest mean-field approximation. Hence show that the specific heat C_B has the behavior:

$$C_B = 0, \qquad \text{for } T > T_c;$$
$$= 3Nk/2, \quad \text{for } T < T_c.$$

What is the value of the associated critical exponent α ?
5. By considering the behavior of the order parameters at temperatures just above the critical point, show that the critical exponent γ, which is associated with the isothermal susceptibility, takes the value $\gamma = -1$ according to mean-field theory.
   Also, by considering the effect of an externally applied magnetic field at $T = T_c$, show that the exponent associated with the critical isotherm takes the value $\delta = 3$. [In the latter case, the identity

$$\tanh(x+y) = \frac{\tanh x + \tanh y}{1 + \tanh x \tanh y},$$

may be helpful.]
6. Show that the optimal free energy of an Ising model, which corresponds to the mean-field theory, may be written in the form:

$$F = -\frac{N}{\beta}\ln[2\cosh(\beta B_E)] + \frac{N}{2zJ}(B_E - B)^2,$$

where $\beta = 1/kT$, B is the externally applied magnetic field, B_E is the effective field experienced by each spin, z is the coordination number and J is the interaction strength.

7. Consider an Ising model where the external field B_i depends on the position of the lattice site i. Show that the condition for the free energy to be a minimum takes the form:

$$B_{\mathrm{E}}^{(i)} - B_i = J \sum_{\langle j \rangle} \langle S_j \rangle_o,$$

where all the symbols have their usual meaning and the notation indicates that the sum over j is restricted to nearest neighbors of i. Also show that the optimal free energy takes the form:

$$F = \frac{-N}{\beta} \ln \left[2 \cosh \left(\beta B_{\mathrm{E}}^{(i)} \right) \right] + \frac{1}{2} \sum_i \left(B_{\mathrm{E}}^{(i)} - B_i \right) \langle S_i \rangle_o.$$

8. A generalized Ising model has the usual Hamiltonian but each spin variable takes the values:

$$S_i = -t, -t+1, \ldots, t-1, t,$$

where t may be either an integer or a half-odd integer. Using mean-field theory find the critical temperature of this system, and then use this result to recover the critical temperature corresponding to the standard two-state Ising model.
[Note: you may assume the relationship:

$$\sum_{S=-t}^{t} e^{xS} = \frac{\sinh \left[(t+1/2)x \right]}{\sinh \left[x/2 \right]}.]$$

9. The Heisenberg model for ferromagnetism is given by the Hamiltonian:

$$H = -J \sum_{<i,j>} \mathbf{S}_i . \mathbf{S}_j - \sum_i \mathbf{B} . \mathbf{S}_i,$$

where $\mathbf{S}_i$ is a three-dimensional unit vector and $\mathbf{B}$ is a uniform external magnetic field. Use mean-field theory to obtain an expression for the critical temperature T_c.
[Hint: Take the external field to be in the $\hat{z}$ direction.]

10. The Hamiltonian of a certain model system is given by

$$H = \frac{-J}{N} \sum_{ij} S_i S_j - B \sum_i S_i,$$

with $S_i = \pm 1$. Show that the system undergoes a phase transition in all dimensions and find the critical temperature in the thermodynamic limit $N \to \infty$.
Comment on the implications of this result for the Ising model.

8

REAL-SPACE RG

We have already met the basic idea of renormalization group (or RG) in Section 1.5.2. It involved a two-step transformation:

(a) Coarse-grain our theoretical description by organizing spins on a lattice into blocks and then carry out some form of averaging which assigns an effective or average value of the spin variable to each block.

(b) Rescale the new coarse-grained lattice to reduce its lattice constant to the original value for the lattice of spin sites.

These two steps make up the renormalization group transformation (or RGT, for short).

As we saw in Chapter 2, the effect of the RGT is to change the strength of the coupling, and successive applications of the RGT can cause the value of a particular coupling constant to *flow* in a space of coupling constants. We use these ideas now to provide a general, more quantitative, formulation of RG, and in particular to identify the fixed point which corresponds to the critical point.

In this chapter, after a more mathematical statement of the RGT, we present a linearized analysis in the neighborhood of the fixed point. We then make use of this analysis to carry out two tasks:

1. Provide a justification of the Widom scaling hypothesis as discussed in Section 7.3 and derive relationships between critical exponents.

2. Calculate critical exponents for simple systems.

8.1 A general statement of the RG transformation

We shall consider the Ising model for definiteness, but the procedure itself is general. Although, for it to be possible, the lattice must have a discrete scaling symmetry, so that blocking and re-scaling leads to a lattice exactly like the one we started with. Near the critical point, the number of degrees of freedom effectively interacting with one another varies as $\sim \xi^d$ where ξ is, as usual, the correlation length and we take the lattice to have d spatial dimensions. Thus, as $T \to T_c$, the correlation length ξ tends to infinity and consequently the number of degrees of freedom also tends to infinity.

We seek to reduce the number of degrees of freedom from N to N' (say) by integrating out fluctuations with wavelengths λ such that:

$$a \leq \lambda \leq ba, \tag{8.1}$$

where $b > 1$ is the spatial re-scaling factor or dilatation, and in d-dimensions:

$$b^d = N/N'. \tag{8.2}$$

In order to achieve this, spins on the lattice are grouped into blocks of length ba, and essentially we now follow the procedure of Kadanoff, as described in Section 7.3.5, albeit with some differences in notation. As before we label spin sites on the lattice by i and blocks by α. If we denote "block spins" by S'_α, then:

$$S'_\alpha = f(S_i) \quad \text{for } i \in \text{block } \alpha.$$

where f stands for "majority rule"; or decimation; or average, as discussed in Section 1.5.2.

Now, there must exist a Hamiltonian $H'[S'_\alpha]$ such that the probability of observing a configuration $[S'_\alpha]$ is given by:

$$P[S'_\alpha] \sim \exp\{-\beta H'[S'_\alpha]\}. \tag{8.3}$$

Formally we have for H':

$$\exp\{-\beta H'[S'_\alpha]\} = \sum_{[S_i]} \prod_\alpha \delta(S'_\alpha - f(S_i)|_{i \in \alpha}) \exp\{-\beta H[S_i]\}. \tag{8.4}$$

Given a configuration $[S_i]$ there exists just one configuration $[S'_\alpha]$ for which the delta function is nonzero, hence we have:

$$\sum_{[S_i]} \prod_\alpha \delta(S'_\alpha - f(S_i)|_{i \in \alpha}) = 1, \tag{8.5}$$

and so

$$\mathcal{Z} = \sum_{[S_i]} \exp\{-\beta H[S_i]\} = \sum_{[S'_\alpha]} \exp\{-\beta H'[S'_\alpha]\} = \mathcal{Z}'. \tag{8.6}$$

Thus the partition function $\mathcal{Z}$ must be unchanged or *invariant under RGT*.

8.1.1 Invariance of expectation values

The invariance of the partition function under RGT implies, through the bridge equation, the invariance of the free energy. This, as we shall see, is a key feature of the RGT and is the basis of the scaling laws. However, the above arguments also imply the invariance of expectation values in general.

If $G[S]$ is any function of the lattice spins, then the usual definition of the expectation value is

$$\langle G \rangle = \frac{1}{\mathcal{Z}} \sum_{\{S\}} G[S_i] \exp\{-\beta H[S_i]\}. \tag{8.7}$$

It follows from the preceeding analysis that we should obtain the same value for the mean of G by evaluating the average over the coarse-grained lattice:

$$\langle G \rangle = \frac{1}{\mathcal{Z}'} \sum_{\{S'_\alpha\}} G[S'_\alpha] \exp\{-\beta H[S']_\alpha\}. \tag{8.8}$$

Of course, this result would be trivially obvious, if the Hamiltonian itself were invariant under RGT. In fact it is not and this is the next topic to be considered.

8.2   RG transformation of the Hamiltonian and its fixed points

In general there is no reason why the Hamiltonian itself should be unchanged under the RGT. We should expect that

$$-H = K \sum_{\langle i,j \rangle} S_i S_j, \tag{8.9}$$

where we have temporarily moved the minus sign to the left hand side of the equation, is changed to:

$$-H = K_1 \sum_{\langle i,j \rangle} S_i S_j + K_2 \sum_{\ll i,j \gg} S_i S_j + K_3 \sum_{\langle ijkl \rangle} S_i S_j S_k S_l + \cdots , \tag{8.10}$$

and so on. This may easily be seen for a two-dimensional lattice, as discussed in Section 2.3 and illustrated here in Fig. 8.1. The coefficients $K_1, K_2, \ldots, K_n, \ldots,$ are new coupling constants.

The state of a physical system at a given temperature can be conveniently represented by a point in an abstract parameter space. We can represent this by a vector μ defined as

$$\mu = \{K_1, K_2, \ldots, K_n, \ldots\}. \tag{8.11}$$

The RGT maps the point μ into another point μ', in the parameter space,

$$\mu' = \{K'_1, K'_2, \ldots K'_n, \ldots\}, \tag{8.12}$$

where

$$\mu' = R_b \mu. \tag{8.13}$$

The RG strategy consists of iterating R_b many times. We can only hope to carry out the RGT for $b \sim 1$ with just a few degrees of freedom to be integrated out at each stage. To get $b \gg 1$, requires many iterations of the RGT.

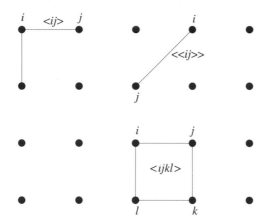

FIG. 8.1 Interactions with nearest neighbors, next nearest neighbors and so on.

Note that the RGT must possess the group closure or transitive property:

$$R_{b_1} R_{b_2} = R_{b_1 b_2}. \qquad (8.14)$$

That is, successive transformations with re-scaling factors b_1 and b_2 must have the same effect as one such transformation with a re-scaling factor $b_1 \times b_2$. From this we see that the set of RGTs forms a simple group and this is why we refer to *renormalization group*. We can also represent the mapping operation in parameter space as a renormalization group transformation of the Hamiltonian:

$$H_{n+1} = R_b H_n, \qquad (8.15)$$

where each Hamiltonian depends only on the immediately preceeding one. If successive Hamiltonians are equal then we have a fixed point, thus:

$$H_{n+1} = R_b H_n = H_n = H_N = H^*. \qquad (8.16)$$

(showing different notations H_N and H^* for the fixed point where $n = N$). As we iterate this map, a representative point moves about in parameter space in discrete steps. We can join up the points to make the motion clearer. Types of fixed point are shown in Fig. 8.2 for the case of a two-dimensional parameter space. We have already met the idea of attractive and repelling fixed points in Section 1.6, when we discussed discrete dynamical systems. However, in the case of point C, we see that in two dimensions there is a new possibility. That is, there can be a fixed point which is attractive in one direction but repelling in another. We call such a point a **mixed** fixed point.

8.2.1   Linearization of the RGT about the fixed point: critical indices

Now, if the RGT reaches a fixed point,

$$\mu' = \mu \equiv \mu^*, \qquad (8.17)$$

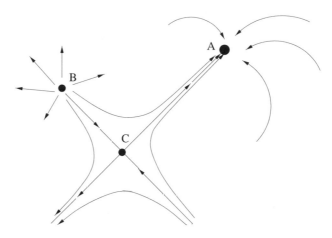

FIG. 8.2 Types of fixed point: A attractive; B repelling; C mixed.

then we can linearize the transformation about this fixed point by considering only very small, and in fact infinitesimal, variations. To do this, we set:

$$\mu = \mu^* + \delta\mu, \tag{8.18}$$

$$\mu' = \mu^* + \delta\mu', \tag{8.19}$$

and "Taylor expand" the RGT about the fixed point, thus:

$$\delta\mu' = A_b(\mu^*)\delta\mu, \tag{8.20}$$

where the matrix $A_b(\mu^*)$ is given by

$$A_b(\mu^*) \equiv \left[\frac{\mathrm{d}R_b}{\mathrm{d}\mu}\right]_{\mu^*}. \tag{8.21}$$

Now let $\mathbf{X_i}$ be the eigenvectors and λ_i be eigenvalues of $A_b(\mu^*)$. Then by definition these satisfy the usual type of equation:

$$A_b(\mu^*)\mathbf{X}_i = \lambda_i\mathbf{X}_i \tag{8.22}$$

Successive transformations with scale factors b_1, b_2 must be equivalent to one transformation with $b_1 \times b_2$. This is the group closure property as given by (8.14) and implies

$$A_{b_1}A_{b_2} = A_{b_1b_2}, \tag{8.23}$$

where we have stopped showing the dependence on $\boldsymbol{\mu^*}$, in order to lighten the notation.

We should note that, as the system comes infinitesimally close to the fixed point, the trajectory is specified completely by the eigenvalues and eigenvectors of A_b.

As we shall see shortly, this information will allow us to evaluate critical exponents and to establish relations between the exponents. In order to do this, we first show that the group closure property implies that the eigenvalues must satisfy the following equation:

$$\lambda_i(b_1)\lambda_i(b_2) = \lambda_i(b_1b_2). \tag{8.24}$$

We do this as follows: from the eigenvalue eqn (8.22), we have

$$A_{b_1b_2}\mathbf{X}_i = \lambda_i(b_1b_2)\mathbf{X}_i, \tag{8.25}$$

for a transformation with scaling parameter $b = b_1b_2$. Now if we transform first with $b = b_2$ and second with $b = b_1$, we obtain

$$A_{b_1}\left[A_{b_2}\mathbf{X}_i\right] = A_{b_1}\left[\lambda_i(b_2)\mathbf{X}_i\right],$$
$$= \lambda_i(b_2)A_{b_1}\mathbf{X}_i,$$
$$= \lambda_i(b_2)\lambda_i(b_1)\mathbf{X}_i. \tag{8.26}$$

Then, equating (8.25) to (8.26), and with some rearrangement, eqn (8.24) follows. Hence the eigenvalues of the transformation matrix must take the form

$$\lambda_i(b) = b^{y_i}. \tag{8.27}$$

This is a key result. The exponents y_i are known as *critical indices*.

8.2.2   System-point flows in parameter space

From our discussion of a ferromagnet in Section 1.5.6, we can expect that there will be two trivial fixed points:

- The high-temperature fixed point: $T \to \infty$;
- The low-temperature fixed point: $T \to 0$.

and we now present arguments to show that both of these are attractive.
    Under the RGT, we have the spatial rescaling process such that

$$x' = x/b; \quad b > 1. \tag{8.28}$$

This applies to all lengths, so that under each RGT, the correlation length ξ is also rescaled as:

$$\xi' = \xi/b. \tag{8.29}$$

At a fixed point, this equation becomes

$$\xi' = \xi'/b,$$

and this has only two possible solutions:

- $\xi = 0;$ (High temperatures)
- $\xi = \infty.$ (Low temperatures)

We now discuss these two cases in turn.

8.2.2.1 Attractive high-temperature fixed point.   If we consider the case where the system temperature is above the critical value, then we may expect to observe fluctuations with a finite
correlation length. This situation is illustrated in Fig. 1.14. Now apply the RGT. First the
coarse-graining operation, or partial average, will tend to fold in random effects and contam-
inate the "island" of aligned spins. Thus coarse-graining will tend to make the fluctuation less
ordered. Second, the rescaling according to eqn (8.29) will decrease the correlation length,
making the coarse-grained fluctuation even less significant. Hence operation of the RGT will
make the system look more random and so the flow must be such that the coupling constants
decrease, which is equivalent to the temperature increasing.[35] It follows that the system point
must flow to the high-temperature fixed point.

8.2.2.2 Attractive low-temperature fixed point.   Below the critical temperature the situation is
quite different, as the entire macroscopic system can have a mean magnetization. Fluctuations
about this mean value can be positive or negative with equal probability. Accordingly any
coarse-graining or averaging procedure will tend to average out fluctuations and make the
system appear more ordered. Thus the operation of the RGT will make the system seem more
ordered and this can be interpreted as an increase in the coupling strength or equally as a
decrease in the temperature. Thus, if we start at temperatures under T_c, the low-temperature
critical point must be attractive.

8.2.2.3 The idea of a "watershed" between the trivial fixed points.   From this discussion of the
trivial high-temperature and low-temperature fixed points, it follows that coarse-graining
operations carried out above the critical temperature, will lead to the high-temperature fixed
point. Conversely, if we begin coarse-graining operations below the critical temperature, such
operations will lead to the low-temperature fixed point. We may sum this up as:

- For $T > T_c$, system points flow under RGT to the high-temperature fixed point
- For $T < T_c$, system points flow under RGT to the low-temperature fixed point

In our present two-dimensional illustration, effective Hamiltonians which flow to $T = 0$
under successive RG transformations must be separated by a line across parameter space
from those which flow to $T = \infty$. This is basically just the same idea as a "watershed."

   In general, in d-dimensions, this line will be a surface, and is called the *critical surface*.
Each point on this surface corresponds to an effective Hamiltonian which is at the critical
temperature appropriate to the values of the other parameters at that point.

8.2.3 Scaling fields

If we choose a point on the critical surface, and iterate the transformation

$$\mu' = R_b \mu,$$

then the system point will move on the critical surface and may ultimately reach a fixed point.
This is the critical fixed point and within the critical surface it is attractive (see Fig. 8.3). Off
the critical surface, points will move to $T = 0$ or $T = \infty$. Hence, the critical point is a

---

[35] Remember: the coupling constant is $K = J/kT$.

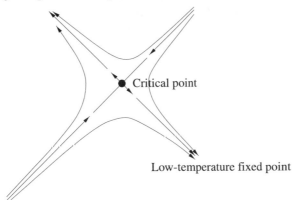

High-temperature fixed point

Critical point

Low-temperature fixed point

FIG. 8.3 Critical fixed point.

mixed, fixed point. The deviation of the Hamiltonian from its value at the fixed point may be expanded in terms of the eigenvectors $\mathbf{X_i}$ of $A_b(\mu^*)$ by writing it as

$$\delta\mu = \sum_i g_i \mathbf{X_i}, \tag{8.30}$$

for small values of $|\delta\mu| = |\mu - \mu^*|$. In other words, the $\mathbf{X}_i$ are the basis vectors for infinitesimal displacements in the parameter space of the coupling constants. The coefficients g_i are the components of the vector $\delta\mu$ in that space. The g_i are called *scaling fields*.

In order to assess the effect of the RGT, we can also expand

$$\delta\mu' = \sum_i g_i' \mathbf{X_i}. \tag{8.31}$$

However, operating on both sides of eqn (8.30) with the RGT gives

$$\delta\mu' = \sum_i R_b g_i \mathbf{X_i},$$

$$= \sum_i g_i R_b \mathbf{X_i},$$

$$= \sum_i g_i \lambda_i \mathbf{X_i},$$

$$= \sum_i g_i b^{y_i} \mathbf{X_i}. \tag{8.32}$$

where we made use of equations (8.20), (8.22), and (8.27). Hence, from the equivalence of equations (8.30) and (8.31), we have for the behavior of the scaling fields under a RGT:

$$g_i' = b^{y_i} g_i. \tag{8.33}$$

Note that if y_i is positive, as we have $b > 1$, RGTs increase g_i', and the system moves away from the fixed point (incidentally, destroying the linearization, hence our analysis does not tell us about the manner of its going!). We may distinguish 3 cases:

$y_i > 0$: g_i increases under RGT, and the system moves away from the fixed point. We say that g_i is a *relevant* variable.

$y_i < 0$: g_i decreases under RGT, and the system moves toward the fixed point. In this case, g_i is an *irrelevant* variable.

$y_i = 0$: g_i is a *marginal* variable. In this case higher-order terms become important.

In terms of the usual analysis of stability of systems, we may also state this as:

- If A_b has an eigenvalue $\lambda_R > 1$, then this corresponds to a repelling fixed point.

- If A_b has an eigenvalue $\lambda_A < 1$, then this corresponds to an attractive fixed point.

8.3   Relations between critical exponents from RG

In Chapter 7, we used the Widom scaling hypothesis to obtain critical exponents in terms of two parameters r and p and this led to relationships among the exponents. Now we do the same again; but this time we express critical exponents in terms of the y_i.

The partition function remains invariant under RGT, thus

$$Z_{N'}(H') = Z_N(H), \tag{8.34}$$

and hence the reduced free energy $\bar{F} = \beta F$ is, by the bridge equation, also invariant. Let us introduce

$$\bar{f} = \text{ reduced free energy per unit volume.} \tag{8.35}$$

Then we may express the invariance of the reduced free energy as:

$$\bar{f}(H')\mathrm{d}^d x' = f(H)\mathrm{d}^d x, \tag{8.36}$$

or, using (8.28),

$$f(H')b^{-d}\mathrm{d}^d x = f(H)\mathrm{d}^d x, \tag{8.37}$$

and so

$$\bar{f}(H) = b^{-d}\bar{f}(H'). \tag{8.38}$$

Now instead of working with the Hamiltonians H and H' we can work equally well with the sets of coupling constants μ and μ', as these parameterize the respective Hamiltonians, so we may also write

$$\bar{f}(\mu) = b^{-d}\bar{f}(\mu'), \tag{8.39}$$

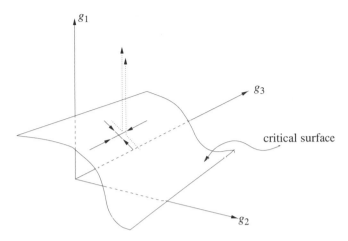

FIG. 8.4 Relevant and irrelevant fixed point.

and near the fixed point (Fig. 8.4)

$$\bar{f}(g_1, g_2, g_3, \ldots) \sim b^{-d} \bar{f}(b^{y_1} g_1, b^{y_2} g_2, b^{y_3} g_3, \ldots). \qquad (8.40)$$

This holds for arbitrary values of the spatial rescaling factor b as, by repeated iterations of the RGT, we can make the total change of scale as large as we please. Hence, this is a generalized homogeneous function, as discussed in Section 7.3.

8.3.1 Application to magnetic systems

This is a general result and in order to apply it to a specific situation we must think about the nature of the scaling fields. Evidently we are concerned with relevant fields and, from the discussions given in Section 8.2, we note that the RGT can only reach a fixed point if all the relevant fields are zero. That is, if they take values appropriate to the critical surface.

In the case of the magnetic systems that we have been discussing at various stages, the obvious relevant field is the temperature; or, more specifically, the reduced temperature θ_c. However, it is often convenient to include the external field B in the analysis and, introducing the reduced external magnetic field $\bar{B} = B/kT$, we set $\bar{B} = 0$ as part of the definition of a critical surface. So, formally we shall work with θ_c and $\bar{B}$ and define the critical point as $\theta_c = 0$, $\bar{B} = 0$. We choose

$$g_1 = \theta_c, \qquad g_2 = \bar{B};$$

and assume all other scaling fields are irrelevant. Then we have

$$\bar{f}(\theta_c, \bar{B}) \sim b^{-d} \bar{f}(b^{y_1} \theta_c, b^{y_2} \bar{B}), \qquad (8.41)$$

as $\theta_c, \bar{B} \to 0$.

8.3.2 The critical exponent α

We can express the critical exponent α in terms of y_1 as follows. Using a standard result from thermodynamics, an expression for the heat capacity C_B can be written as

$$C_B \sim \left(\frac{\partial^2 \bar{f}}{\partial \theta_c^2} \right)_{B=0} \sim |\theta_c|^{-\alpha}, \qquad (8.42)$$

where the second step comes from eqn (3.32) and is just the definition of the exponent α. Differentiate the expression for $\bar{f}$ twice with respect to θ_c and set $\bar{B} = 0$, thus:

$$\frac{d\bar{f}}{d\theta_c} \sim b^{-d+y_1} \bar{f}'(b^{y_1}\theta_c, b^{y_2}\bar{B}); \qquad (8.43)$$

and

$$\frac{d^2 \bar{f}}{d\theta_c^2} \sim b^{-d+2y_1} \bar{f}''(b^{y_1}\theta_c, 0), \qquad (8.44)$$

for $\bar{B} = 0$.

We now use the same argument as in Widom scaling. As b is arbitrary, we may choose

$$b = |\theta_c|^{-1/y_1}, \qquad (8.45)$$

and so

$$C_B \sim |\theta_c|^{(d-2y_1)/y_1} f''(1, 0). \qquad (8.46)$$

For the case $\theta_c \to 0$,

$$C_B \sim |\theta_c|^{-\alpha},$$

defines the critical exponent α and comparison with (8.46) yields

$$\alpha = 2 - d/y_1. \qquad (8.47)$$

8.3.3 The critical exponent ν

Near the critical point, we have:

$$\xi \sim |\theta_c|^{-\nu},$$

for the divergent correlation length. But from (8.45) lengths scale as

$$b \sim |\theta_c|^{-1/y_1}.$$

Comparison yields:

$$\nu = 1/y_1. \qquad (8.48)$$

Now we can eliminate y_1 from the expression for α, thus from (8.47) and (8.48):

$$\alpha = 2 - \nu d, \qquad (8.49)$$

which is known as the "Josephson relation", or the "hyperscaling relation".

8.4 Applications of the linearized RGT

In Chapter 2 we applied RG to both percolation and the Ising model, and obtained RG equations and fixed points. Now we use the linearized procedure in order to establish the nature of the fixed points and to calculate numerical values for the critical exponent ν.

8.4.1 Example: two-dimensional percolation

In Section 2.2 we applied RG to the problem of bond percolation in two dimensions and obtained a fixed-point value for the critical probability

$$p_c = p^* = 0.62,$$

as given in eqn (2.54).

Now we consider how the linearized analysis presented in this chapter will allow us to work out a value for the critical exponent ν, as defined by eqn (7.12). We begin with eqn (8.48) which expresses ν in terms of a critical index y_1, where equation (8.27) defines the critical index y_1, thus:

$$\lambda_1 = b^{y_1}, \tag{8.50}$$

where λ_1 is an eigenvalue of the transformation matrix (8.21) and b is the spatial rescaling factor of the RGT.

The transformation matrix is obtained by linearizing the RG equations about the fixed point. For this problem the RG equation is given by (2.46) and we reproduce this here for convenience as:

$$p' = 2p^2 - p^4. \tag{8.51}$$

Following the procedure in Section 8.2.1, we readily obtain the linearized transformation as follows.

At the fixed point, we have

$$p' = p = p^*,$$

and the RG eqn (8.51) takes the form

$$p^* = 2p^{*2} - p^{*4}. \tag{8.52}$$

Then set:

$$p' = p^* + \delta p'; \qquad p = p^* + \delta p, \tag{8.53}$$

and substitute into (8.50), which becomes:

$$\delta p' + p^* = 2(p^* + \delta p)^2 - (p^* + \delta p)^4 = 2p^{*2}\left(1 + \frac{\delta p}{p^*}\right)^2 - p^{*4}\left(1 + \frac{\delta p}{p^*}\right)^4. \tag{8.54}$$

Now, on the right hand side we use the binomial theorem to expand out both terms in brackets to first order in δp, thus:

$$\delta p' + p^* = 2p^{*2} - p^{*4} + 4p^*\delta p - 4p^{*3}\delta p = p^* + 4p^*(1 - p^{*2})\delta p, \tag{8.55}$$

where we used eqn (8.51) to obtain the second equality on the right hand side. Then, cancelling p^* on both sides of the equation we are left with:

$$\delta p' = 4p^*(1 - p^{*2})\delta p. \tag{8.56}$$

Or,

$$\delta p' = A_b(p^*)\delta p, \tag{8.57}$$

where

$$A_b(p^*) = 4p^*(1 - p^{*2}). \tag{8.58}$$

This is rather an easy problem. Because this is a single-parameter space, the transformation matrix is a scalar, and there is only one eigenvalue, λ_1, which is:

$$\lambda_1 = 4p^*(1 - p^{*2}), \tag{8.59}$$

and, as $p^* = 0.62$, gives $\lambda_1 = 1.53$.

As we saw in Section 2.2, the rescaling factor $b = \sqrt{2}$ in this case. Thus taking logs of both sides of eqn (8.50), we have the critical index y_1, as:

$$y_1 = \frac{\ln \lambda_1}{\ln b} = \frac{\ln 1.53}{\ln \sqrt{2}}, \tag{8.60}$$

and from (8.48), the critical exponent is given by:

$$\nu = \frac{1}{y_1} = \frac{\ln \sqrt{2}}{\ln 1.53} \simeq 0.82. \tag{8.61}$$

The correct value is believed to be about 1.34. Modified RG approaches are capable of giving better answers: see Stauffer and Aharony (1994), pp. 75–81.

8.4.2   Example: two-dimensional magnet

Given the RG equations (see Section 2.3 for a discussion) for the coupling constants K and L as:

$$K' = 2K^2 + L; \quad L' = K^2, \tag{8.62}$$

at the fixed points we have

$$K^* = 2K^{*2} + L^*; \quad L^* = K^{*2}. \tag{8.63}$$

It is easily verified by direct substitution that the fixed points are $(K^*, L^*) = (0, 0)$, (∞, ∞) and $(1/3, 1/9)$. The first two are the high-temperature and low-temperature points and are trivial. The nontrivial fixed point is $(K^*, L^*) = (1/3, 1/9)$ and we linearize about this. Set:

$$K' = K^* + \delta K'; \quad L' = L^* + \delta L',$$
$$K = K^* + \delta K; \quad L = L^* + \delta L. \tag{8.64}$$

Eqns (8.62) become:

$$\delta K' = 4K^* \delta K + \delta L + (2K^{*2} - K^* + L^*),$$
$$\delta L' = 2K^* \delta K + (K^{*2} - L^{*2}).$$

Substituting $K^* = 1/3$, $L^* = 1/9$, these may be written as:

$$\begin{pmatrix} \delta K' \\ \delta L' \end{pmatrix} = \begin{pmatrix} 4K^* & 1 \\ 2K^* & 0 \end{pmatrix} \begin{pmatrix} \delta K \\ \delta L \end{pmatrix} = \begin{pmatrix} 4/3 & 1 \\ 2/3 & 0 \end{pmatrix} \begin{pmatrix} \delta K \\ \delta L \end{pmatrix}, \tag{8.65}$$

and by comparison with eqns (8.20) and (8.21) we deduce that the transformation matrix for this problem takes the form

$$A_b = \begin{pmatrix} 4/3 & 1 \\ 2/3 & 0 \end{pmatrix}. \tag{8.66}$$

The eigenvalues may be found in the usual way from the condition:

$$\begin{vmatrix} 4/3 - \lambda & 1 \\ 2/3 & -\lambda \end{vmatrix} = 0.$$

Expanding out the determinant gives

$$\lambda^2 - 4/3\lambda - 2/3 = 0, \tag{8.67}$$

with roots:

$$\lambda = \tfrac{1}{3}(2 \pm \sqrt{10}),$$

and so

$$\lambda_1 = 1.722; \quad \lambda_2 = -0.390.$$

For the critical exponent ν we have $\nu = 1/y_1$, where $\lambda_1 = b^{y_1}$, $b = \sqrt{2}$, and hence

$$\nu = \frac{\ln b}{\ln \lambda_1} = \frac{\ln(\sqrt{2})}{\ln(1.722)} = 0.652.$$

The exact value for a two-dimensional Ising model is $\nu = 1$, so this is not a bad result for such a simple approximation.

Further reading

Most modern books on critical phenomena or advanced statistical physics give an account of the linearized analysis presented in this chapter. The reader who wishes to take the topic further should find the books [32], [5], [7], and [4] helpful. Background material on discrete dynamical systems can be found in [28].

8.5   Exercises

Solutions to these exercises, along with further exercises and their solutions, will be found at:
www.oup.com

1. The RG recursion relations for a one-dimensional Ising model in an external field B may be written as;

$$x' = \frac{x(1+y)^2}{(x+y)(1+xy)}; \qquad y' = \frac{y(x+y)}{(1+xy)},$$

where $x = e^{-4J/kT}$ and $y = e^{-B/kT}$. Verify the existence of fixed points as follows: $(x^*, y^*) = (0, 0)$; $(x^*, y^*) = (0, 1)$ and a line of fixed points $x^* = 1$ for $0 \le y^* \le 1$. Discuss the physical significance of these points and sketch the system point flows in the two-dimensional parameter space bounded by $0 \le x \le 1$ and $0 \le y \le 1$. By linearizing about the ferromagnetic fixed point, obtain the matrix of the RGT and show that the associated critical indices are $y_1 = 2$, $y_2 = 1$ for scaling factor $b = 2$.

2. The Hamiltonian for the one-dimensional Ising model in an external field B may be written as

$$H = -K \sum_{<ij>} S_i S_j - \beta B \sum_i S_i - \sum_i C,$$

where C is a constant, background term. Show that the partition function for the system may be written as a product of terms, each of which depends on only one of the even-numbered spins. Hence, by partial summation over the even-numbered spins, obtain the recursion relations for K', B' and C'.

3. Consider the two-dimensional Ising model under decimation on a square lattice. If we only take into account the coupling constants K (nearest-neighbor interactions) and L (next-nearest-neighbor interactions), the recursion relations are given by

$$K' = 2K^2 + L; \qquad L' = K^2.$$

Find the fixed points for these renormalization group equations and identify the critical one. Linearizing the RGT about this point, obtain a value for the critical exponent ν.

4. Discuss bond percolation on a two-dimensional square lattice and, drawing analogies with the Ising model where appropriate, introduce the concepts of *critical probability, correlation length,* and *critical exponent*.

   Show that the critical probability for this problem is $p_c = 1/2$. Apply the renormalization group to bond percolation, using a scheme in which alternate sites are removed from the lattice. Show that the RG equation takes the form

$$p' = 2p^2 - p^4,$$

where p' is the probability of two sites being connected on the new lattice and p is the analogous quantity for the old lattice. Verify that the fixed points are $p^* = 0$, $p^* = 1$, and $p^* = (\sqrt{5}-1)/2 \simeq 0.62$, and discuss the nature of these fixed points.

   By linearizing the RG transformation about a fixed point, show that the relevant eigenvalue of the transformation matrix is given by

$$\lambda_1 = 4p^*(1 - p^{*2}).$$

Using this result, obtain a numerical value for the critical exponent ν.

9

MOMENTUM-SPACE RENORMALIZATION GROUP

In this chapter we again use the linearized analysis of Chapter 8 in order to investigate the fixed points associated with renormalization group (RG) equations. But this time, instead of performing simple coarse-graining operations on a discrete set of lattice spin sites, we integrate out high-wave number modes from a continuous spin field. We then rescale the coarse-grained field in order to maintain the invariance of the Hamiltonian and this results in the RG equations.

A new feature here is the role of the dimensionality of the field as a control parameter. Previously we found that varying the temperature or density could weaken the interaction strength and this allowed one to carry out a perturbation expansion. In this chapter we shall find that the zeroorder state (the Gaussian model) is exact for $d > 4$, and that a nontrivial fixed point can be calculated perturbatively for $d = 4 - \epsilon$, where ϵ is taken to be small.

This is only a simple form of perturbation theory but it will allow us to calculate corrections to the mean-field values of the critical exponents to order ϵ.

9.1 Overview of this chapter

As this is quite a long chapter with some substantial algebraic manipulations, it may be helpful to begin with a detailed overview.

- In Section 9.2 we formulate the Ginsburg–Landau microscopic model and show that it is consistent with the macroscopic Landau model as discussed in Chapter 7.

- In Section 9.3 we re-formulate the RG transformation in wave number space (as opposed to real space) and discuss the need to carry out the RGT perturbatively about the soluble Gaussian case. We then introduce the idea of *scaling dimension*, as defined by the renormalization group transformation (RGT), as this is a key concept in the later parts of the chapter.

- Section 9.5 is self-explanatory: we restate our objective, which is to obtain numerical values of the critical exponents. For sake of convenience, we write down the four scaling relationships which connect the six exponents, and note that this means that we only have to calculate two independent exponents.

- Sections 9.6 to 9.8 deal with the exactly soluble Gaussian model, which is the "zero-order" for our perturbation theory. Paradoxically this is the most demanding part of the chapter in mathematical terms. The emphasis here is on formulating the functional integral, showing that it can be evaluated as a standard form, and then obtaining the partition functional for the Gaussian model. After that, application of the RGT is relatively easy, and we obtain the fixed points.

- The implementation of the RGT for the Ginsburg–Landau theory requires perturbation theory in order to be able to handle the ϕ^4 term. This is quite straightforward but we end up with the interesting result that the nature of the fixed points depends on the dimensionality d of the lattice (or the space). For $d > 4$, the fixed point is the same as in the Gaussian (or mean-field case), while for $d < 4$ we can calculate corrections of order ϵ to mean-field values, where $\epsilon = 4 - d$.

9.2 Statistical field theory

In both classical and quantum mechanics it is usual, in appropriate circumstances, to make the transition from a description based on discrete particles to one based on continuous fields. Invariably some limiting process is involved in so doing and the appropriate generic term is *the continuum limit*. For example, continuous transverse waves on a stretched string can be modeled as the limiting case of many discrete masses joined by a massless elastic string. Or, to take another example, in statistical mechanics we begin with an assembly consisting of a large number of individual molecules and end up with the derivation of macroscopic kinetic equations for the continuous mass density, pressure, and velocity fields, in the so-called hydrodynamic limit.

In this chapter we shall still be concerned with "spins on a lattice." But now we shall take the view that, when we are also faced with fluctuations of large spatial extent (as the system approaches the critical point), the fine structure of the lattice may, in comparison, be smeared out into a continuum.

In essence, this means that instead of measuring distance along the lattice in terms of a discrete number of unit cells we measure the distance from some origin as a continuously varying coordinate.

We shall also make the change into momentum space which is conjugate to real $(\mathbf{x})$ space. This takes us into the territory of quantum field theory but we shall invariably work in units where Planck's constant is unity and hence there will be no distinction drawn between momentum space and wave number $(\mathbf{k})$ space.

In this section we shall formulate a continuum model in such a way that it should have some of the features of both the Ising model and the Landau mean-field theory. The resulting model is known as the Ginsburg–Landau model. We conclude the section by verifying the consistency of the model with Landau theory, as previously discussed in Section 3.3.

9.2.1 The continuum limit

We begin by changing from a discrete lattice to a continuous field. As the first step, we write the discrete spins as a discretization (or sampling) of a continuous function $S(x)$, thus:

$$S_i \equiv S(x_i), \tag{9.1}$$

where S_i is the spin at ith site on the lattice and

$$x_i = i \times a \quad \text{(integer} \times \text{lattice constant).} \tag{9.2}$$

Next we introduce a *coarse-graining* transformation such that

$$S^{(D)}(x_i) \rightarrow \phi(\mathbf{x}) \tag{9.3}$$

where the D-component spin vector on a d-dimensional lattice becomes the D-component field on a d-dimensional space. This means we average $S^D(x_i)$ over a volume $(ba)^d$ where b is of the order of a few lattice steps and $ba \ll \xi$ where, as usual, ξ is the correlation length. In principle this is the same idea as used in statistical mechanics to derive kinetic equations.

In order to do this, we require the system to be near to the critical point where ξ is large. Under these circumstances we have the equivalence:

continuum version of the theory $\rightleftharpoons$ lattice version of the theory.

We shall need to work in wave number space (momentum space) and we should note that there is an ultraviolet cut-off on the permitted wave number values:

$$|\mathbf{k}| \leq \Lambda \sim \pi/a. \tag{9.4}$$

As a is the lattice constant, this cutoff is due to the lattice: in quantum field theory there is no cut-off and $\Lambda \rightarrow \infty$.

In order to make the transition to wave number space, we introduce the Fourier transform of the field:

$$\phi(\mathbf{x}) = \int_{0 \leq k \leq \Lambda} \frac{d^d k}{(2\pi)^d} \exp\{i\mathbf{k} \cdot \mathbf{x}\}\phi(\mathbf{k}), \tag{9.5}$$

in d dimensions. We shall require this relationship later on, where the Fourier modes will turn out to be the natural "diagonalizing functions" in our evaluation of the Gaussian functional integral.

9.2.2   Densities

In general the transition from microscopic to continuum description involves the introduction of densities. For example, in elementary continuum mechanics the masses of individual particles are taken over into the continuum density $\rho(\mathbf{x}, t)$ by an averaging operation. As the system tends to thermal equilibrium, we have the result:

$$\rho(\mathbf{x}, t) \rightarrow \rho \equiv \lim_{N \to \infty} \frac{N}{V}.$$

We have already introduced the free energy per unit volume in Chapter 8. Now we introduce the Hamiltonian density h, such that

$$H = \int d^d x \, h. \tag{9.6}$$

Both h and H are now *functionals*[36] of the field $\phi(\mathbf{x})$. Thus the probability distribution becomes a functional

$$P[\phi(\mathbf{x})] \sim \exp\{-\beta H[\phi(\mathbf{x})]\} \sim \exp\{-\beta \int d^d x\, h[\phi(\mathbf{x})]\}, \tag{9.7}$$

where, as usual, $\beta = 1/kT$. The associated *partition functional* is

$$Z = \int \mathcal{D}\phi \exp\{-\beta \int d^d x\, h[\phi(\mathbf{x})]\}, \tag{9.8}$$

where the meaning of functional integration will be discussed later. For the moment, we shall make the interpretation:

$$\int \mathcal{D}\phi \equiv \text{``sum over states''}. \tag{9.9}$$

9.2.3 The Ginsburg–Landau model

As we concluded from Sections 3.3 and 7.6, Landau theory gives the same critical exponents as the mean-field theory of the Ising model. This suggests that we formulate a microscopic model guided by Landau theory and Ising characteristics. Let us recall the Landau theory from Section 3.3. We have the free energy F in the form of an expansion in the magnetization M, thus:

$$F(T, M) = F_0(T) + A_{21}(T - T_c)M^2 + A_4 M^4 \cdots,$$

where $A_4 > 0$ for stability. If we take higher terms, we must have $A_6, A_8, \ldots$, all greater than zero, so it is normally argued that their inclusion would make no real difference to the result.

Let us assume that the microscopic equivalent of the above expansion is to postulate a Hamiltonian density h of the form:

$$h[\phi(\mathbf{x})] = h_0 + \frac{1}{2}u|\phi|^2 + \frac{\lambda}{4!}|\phi|^4, \tag{9.10}$$

where the choice of constants $u/2$ and $\lambda/4!$ is conventional. Now we wish to build in the Ising "nearest neighbors" characteristic. In the continuum limit, the local interaction is given by the gradient of the field:

$$\lim_{a \to 0} \frac{S^{(D)}(x_i) - S^{(D)}(x_{i-1})}{a} = \nabla\phi.$$

Accordingly we choose

$$h_0 = \frac{1}{2}\alpha|\nabla\phi|^2, \tag{9.11}$$

where α is a constant.[37] Now when there is an external magnetic field of magnitude $B \neq 0$, ϕ will lie along the direction of the vector field $\mathbf{B}$. With a suitable choice of coordinates

---

[36] In simple terms, a functional is just a "function of a function." More technical aspects will be dealt with as they arise.

[37] There should be no confusion between this constant and the use of the symbol α for one of the critical exponents.

we can work with the scalar field

$$\phi(\mathbf{x}) \equiv |\phi(\mathbf{x})|.$$

Then, from (9.10) and (9.11), we have the functional Hamiltonian density in the form

$$h[\phi(\mathbf{x})] = \frac{1}{2}\alpha|\nabla\phi|^2 + \frac{1}{2}u\phi^2 + \frac{\lambda}{4!}\phi^4. \tag{9.12}$$

This form defines the Ginsburg–Landau model. It may be compared with the analogous form in quantum field theory,

$$h = \frac{1}{2}(\nabla\phi)^2 + \frac{1}{2}m_0^2\phi^2 + \frac{\lambda_0}{4!}\phi^4, \tag{9.13}$$

which is known as the ϕ^4 scalar field theory, with minor variations in notation from one source to another.

Equation (9.12) can be extended to the case of an external field B,

$$h[\phi(\mathbf{x})] = \frac{1}{2}\alpha|\nabla\phi|^2 + \frac{1}{2}u\phi^2 + \frac{\lambda}{4!}\phi^4 - B\phi. \tag{9.14}$$

Or, in terms of a source $J = \beta B$

$$h[\phi(\mathbf{x})] = \frac{1}{2}\alpha|\nabla\phi|^2 + \frac{1}{2}u\phi^2 + \frac{\lambda}{4!}\phi^4 - \frac{J\phi}{\beta}. \tag{9.15}$$

Here:

- α is a dimensionless constant which controls the "nearest neighbor" interaction.

- u is a parameter which depends on T and controls the phase transition. This will be clearer in the next section where we identify u in terms of Landau theory. However, as we shall see in Section 9.7.2, Case 2, we must have $u > 0$.

- λ is the coupling parameter.

- J is the source term which allows us to obtain correlations by functional differentiation. This is an extension of the techniques of linear response theory, as discussed in Section 7.4.

Now we have two objectives:

1. To check the consistency of this model with the earlier macroscopic Landau theory, as discussed in Section 3.3.

2. To take a first look at the technical problem of working out the partition functional.

9.2.4 Consistency with the Landau model

For sake of convenience we shall absorb the β factor into the constants α, u, and λ. Hence (9.8) for the partition functional becomes:

$$Z = \int \mathcal{D}\phi \exp\left\{-\int d^d x\, h[\phi(\mathbf{x})]\right\}, \tag{9.16}$$

with $h[\phi(\mathbf{x})]$ given by (9.14) or (9.15) if using J.

This functional integral cannot be evaluated exactly for nonzero λ. For our first look at the problem, we shall make the approximation of replacing the partition sum by its largest term. To obtain the largest term, we need to minimize the exponent. Clearly, one way of minimizing $h[\phi(\mathbf{x})]$ is to choose $\phi(\mathbf{x}) = $ constant and hence $\nabla\phi = 0$. Thus the exponent in Z becomes

$$\int d^d x h[\phi] = V \left(\frac{1}{2} u\phi^2 + \frac{\lambda}{4!} \phi^4 - B\phi \right), \tag{9.17}$$

where $\int d^d x \equiv V$, the volume of the system. Using $F = -\ln Z$ (the bridge equation in disguise: remember that we are now working in units such that $\beta = 1$), and setting $B = 0$, we obtain

$$F = \text{constant} + \frac{1}{2} u\phi^2 + \frac{\lambda}{4!} \phi^4.$$

But with $\phi = $ constant, increasing the coarse-graining cannot affect it and so we must have $\phi \equiv M$ where M is the mean magnetization. Hence we have recovered Landau theory, if we make the identification

$$u/2 = A_2(T) \quad \text{and} \quad \lambda/4! = A_4(T).$$

9.3   Renormalization group transformation in wave number space

In the continuum limit our RGT, with scale factor b, now becomes:

1. Integrate over modes in the band of wave numbers

$$\Lambda/b \leq k \leq \Lambda$$

   while leaving other modes with wave numbers $k \leq \Lambda/b$ unaffected.

2a. Rescale the system in real space, thus:

$$x \rightarrow x/b = x';$$

$$\int d^d x \rightarrow b^{-d} \int d^d x = \int d^d x'.$$

2b. Rescale the system in wave number space, thus:

$$k \rightarrow bk = k';$$

$$\int d^d k \rightarrow \int b^d d^d k = \int d^d k'. \tag{9.18}$$

This is a quite general approach but we will focus on the much studied Ginsburg–Landau model. Even so we run into problems immediately: it is impossible to do the necessary functional integrals with the Ginsburg–Landau partition function. We are forced to resort to the same idea as with the Bogoliubov mean-field theory of the Ising model (see Section 7.5): and split the Hamiltonian up into a soluble part and a correction:

$$H_{GL} = H_0 + H_I; \tag{9.19}$$

$$H_0 = \frac{1}{2} \int d^d x [\alpha (\nabla \phi)^2 + u\phi^2]; \tag{9.20}$$

$$H_I = \int d^d x \frac{\lambda}{4!} \phi^4. \tag{9.21}$$

The Ginsburg–Landau model with $H_{GL} = H_0$ is called the *Gaussian model*, and is soluble. Thus it provides the necessary soluble zero-order model upon which we can base our perturbation theory.

The rest of this chapter will be given over to the following programme:

A. Solve (i.e. obtain the partition function $\mathcal{Z}_0$) for the case $H_{GL} = H_0$.

B. Solve for $H_{GL} = H_0 + H_I$ perturbatively, in order to carry out the RGT:

$$H'_{GL} = R_b H_{GL} = R_b H_0 + \langle R_b H_I \rangle_0 + \cdots .$$

C. Use functional differentiation to obtain correlation functions.

D. Hence find the critical exponent ν.

In carrying out this programme, we have to do the following.

1. Learn how to do the Gaussian functional integral.

2. Carry out a perturbation analysis, about the Gaussian solution, in powers of λ.

The second step involves two aspects:

2a. The technology of actually doing the perturbation analysis, including renormalization and the use of Feynman diagrams.[38]

2b. As $T \to T_c$, the analytic basis of the perturbation theory is lost. This is indicated by an infrared divergence for $d < 4$. We shall "cure" this by working in d dimensions and expanding in Taylor series about $d = 4$ in powers of ϵ.

This procedure gives a theory in $d = 4 - \epsilon$. We treat ϵ as small, but ultimately we shall set $\epsilon = 1$ in order to recover a theory in three dimensions. Coefficients in the original perturbation expansion now become power series in ϵ, and the method is known as the "ϵ-expansion."

9.4 Scaling dimension: anomalous and normal

The use of the term "dimension" in the theory of critical phenomena can be quite misleading. The only writer who is at all clear on the subject is Ma (see [15]) who defines the *scaling*

---

[38] Although the use of diagrams will actually be postponed to the following chapter, where we give a more "field theoretic" treatment of the subject.

dimension (or *scale dimension*) of a quantity in terms of its behavior under a scaling transformation.

In the context of RG, the only dimension which matters is the length. For instance, under RGT, the length interval Δx transforms as

$$\Delta x \rightarrow \Delta x' = b^{-1}\Delta x.$$

Now we would usually say that Δx has dimension unity but in terms of the power of the rescaling factor b, it has *scale dimension* minus one, as determined by the scaling transformation.

We can introduce a general definition by letting the scale dimension of any quantity A be d_s and defining this by the relationship:

$$A \rightarrow A' = b^{d_s} A, \tag{9.22}$$

under RGT with rescaling factor b.

As an example, we may take the basic volume element $dV = d^d x$ in d-dimensional space. Under RGT we have

$$d^d x \rightarrow b^{-d} d^d x = d^d x',$$

and hence, from eqn (9.22), the counter-intuitive conclusion that

$$d_s = -d.$$

Ma then states that thereafter "scale dimension" is shortened to "dimension." Other writers simply use the term "dimension" without any further qualification, but from the context it may be deduced that they are really referring to "scale dimension."

9.4.1 Anomalous dimension

Now we can carry out the two steps which make up the RG transformation, as outlined in eqn (9.18). We concentrate on doing this in terms of wave numbers, thus:

1. Integrate over wave numbers in the shell $\Lambda/b \le k \le \Lambda$.

2. Rescale the lengths according to: $x = bx'$, $k = k'/b$.

We renormalize fields by the transformation

$$\phi(\mathbf{x}) = b^{-d_\varphi}\phi'(\mathbf{x}'), \tag{9.23}$$

where d_φ is called *anomalous dimension* of the field. Hence, in wave number space,

$$\phi'(\mathbf{k}') = \int d^d x' \exp\{-i\mathbf{k}' \cdot \mathbf{x}'\}\phi'(\mathbf{x}')$$

$$= b^{d_\varphi - d} \int d^d x \exp\{-i\mathbf{k} \cdot \mathbf{x}\}\phi(\mathbf{x})$$

$$= b^{d_\varphi - d}\phi(\mathbf{k}); \tag{9.24}$$

that is,

$$\phi'(\mathbf{k}') = b^{d_\varphi - d}\phi(\mathbf{k}). \tag{9.25}$$

This relationship, when added to eqn (9.18), completes our specification of the RGT in wave number space. We can obtain an expression for the anomalous dimension by considering the effect on correlations and in particular the pair correlation. We have:

$$G'(r') = b^{2d_\varphi}G(r),$$

and at the fixed point:

$$G^*(r/b) \sim b^{2d_\varphi}G^*(r)$$

As the rescaling factor b is arbitrary we may choose $b = r$, thus:

$$G^*(r) \sim r^{-2d_\varphi}G^*(1) \sim \frac{1}{r^{d-2+\eta}}, \tag{9.26}$$

from eqn (7.13), and so the anomalous dimension d_φ is given by

$$2d_\varphi = d - 2 + \eta,$$

or:

$$d_\varphi = \frac{1}{2}(d - 2 + \eta). \tag{9.27}$$

9.4.2 Normal dimension

The *normal* or *canonical* dimension of a field is the (scale) dimensionality as given by simple dimensional analysis. For instance, as the exponent in (9.8) must be dimensionless, it follows that now we have absorbed the β factors into the constants then (e.g.)

$$\left[\frac{1}{2}\int d^d x \alpha(\nabla\phi)^2\right] = 0,$$

where the square brackets stand for "dimension of." We have stipulated that the coefficient α is dimensionless, so it follows that

$$\left[\phi^2 L^{d-2}\right] = 0,$$

and so

$$[\phi] = 1 - \frac{d}{2}.$$

However, as we have seen, the scale dimension is the inverse of this, and denoting the normal scale dimension by d_φ^0, we have:

$$d_\varphi^0 = \frac{d}{2} - 1. \tag{9.28}$$

9.5 Restatement of our objectives: numerical calculation of the critical exponents

Let us now remind ourselves of our basic RG programme. Our overall objective is to obtain numerical values for the six critical exponents. It is possible to derive four scaling laws for critical exponents. These may be written in various ways and as an example we state them as follows:

$$\alpha = 2 - \nu d;$$

$$\beta = \frac{\nu}{2}(d - 2 + \eta);$$

$$\gamma = \nu(2 - \eta); \tag{9.29}$$

$$\delta = \frac{d + 2 - \eta}{d - 2 + \eta}.$$

The first of these equations has been derived by us as eqn (8.47), and the remaining three can easily be derived by using the definitions (3.32)–(3.35) and (7.12)–(7.13). Detailed derivations can be found in the literature and among the most succinct are those in [4] or [13].

Thus of our six exponents, two are independent. We shall take these to be ν and η. As we know from Section 8.3, they are determined by the critical indices y_1 and y_2, respectively. We may identify these in the Ginsberg–Landau Hamiltonian as:

y_1(the thermal exponent) is related to the anomalous dimension of ϕ^2 (which is conjugate to $\theta_c = (T - T_c)/T_c$),
y_2(the magnetic exponent) is related to the anomalous dimension of ϕ (which is conjugate to B).

9.6 The Gaussian zero-order model

In order to carry out step A of the programme outlined in the previous section, we first write an explicit equation for the zero-order functional. From eqn (9.20) for the zero-order Hamiltonian we may write

$$Z_0[J] = \int \mathcal{D}\phi \exp\{-H_0[J]\}$$

$$= \int \mathcal{D}\phi \exp\left\{-\int d^d x \left[\frac{\alpha}{2}(\nabla\phi)^2 + \frac{u}{2}\phi^2 - J\phi\right]\right\}. \tag{9.30}$$

Note that we indicate the functional dependence of Z_0 on J explicitly as this lets us work out correlations using linear response theory. Now we need an "aside" on functional integration, followed by a discussion of how to evaluate the Gaussian functional integral.

9.6.1 Functional integration

Consider a particular functional $A[f(x)]$, as defined by

$$A[f(x)] = \exp\left[-k \int_{-L}^{L} f^2(x)\mathrm{d}x\right], \tag{9.31}$$

where $f(x)$ is a continuous function defined on the interval $-L \leq x \leq L$, and k is a constant. Then the *functional integral* I is defined as

$$I = \int \mathcal{D}f \exp\left[-k \int_{-L}^{L} f^2(x)\mathrm{d}x\right], \tag{9.32}$$

where $\int \mathcal{D}f$ stands for the integral over all possible functions $f(x)$. In order to carry out the integral, we:

- divide up the interval $[-L, L]$ into n steps.
- discretise $f(x)$ on the interval:

$$f(x) \rightarrow f_i \equiv f(x_i).$$

- replace $f(x)$ by the set of numbers f_i (but, by taking the limit $n \rightarrow \infty$, we can restore the continuous form).

Then an approximation to the functional integral can be written as

$$I_n = \int \cdots \int \mathrm{d}f_1 \cdots \mathrm{d}f_n \exp\left[-\frac{2kL}{n} \sum_{i=1}^{n} f_i^2\right], \tag{9.33}$$

where we require

$$\lim_{n \rightarrow \infty} I_n \rightarrow I.$$

Now we use the fact that the exponential form can be factored in order to interchange the operations of summation and integration. This allows us to work out I_n exactly, thus:

$$I_n = \prod_{i=1}^{n} \int \mathrm{d}f_i \exp\left[-\frac{2kL}{n} f_i^2\right] = \left(\frac{\pi n}{kL}\right)^{n/2}, \tag{9.34}$$

where the second equality follows from the standard result for a single-variable Gaussian integral. In principle we then get I by taking

$$\lim_{n \rightarrow \infty} I_n \rightarrow I.$$

Unfortunately this limit does not exist in general and in the present case is obviously wildly divergent! We note the following relevant points:

- In order to evaluate a functional integral we need some kind of normalization or restriction on admissible functions: this depends on the actual functional we are working with.

- However, the nature of the continuum limit depends on the particular physical system we are considering, so the above is not as bad as it might seem. We can make "physically appropriate" choices.

9.6.2   The Gaussian functional integral

Consider a standard form:

$$G[J] \equiv \int \mathcal{D}\phi \, \exp\left[-\int d^d x d^d y \phi(\mathbf{x}) M(\mathbf{x}, \mathbf{y}) \phi(\mathbf{y}) + \int d^d z \phi(\mathbf{z}) J(\mathbf{z})\right] \qquad (9.35)$$

where the kernel $M(\mathbf{x}, \mathbf{y})$ is symmetric, thus:

$$M(\mathbf{x}, \mathbf{y}) = M(\mathbf{y}, \mathbf{x}),$$

with real elements and positive eigenvalues. Introduce eigenfunctions ψ_n and eigenvalues λ_n of M, such that:

$$\int d^d y \, M(\mathbf{x}, \mathbf{y}) \psi_n(\mathbf{y}) = \lambda_n \psi_n(\mathbf{x}), \qquad (9.36)$$

or (in a standard operator notation)

$$M\psi_n = \lambda \psi_n. \qquad (9.37)$$

Now expand out ϕ in terms of the eigenfunctions ψ_n of M as:

$$\phi(\mathbf{x}) = \sum_n \phi_n \psi_n(\mathbf{x}), \qquad (9.38)$$

where the coefficients ϕ_n are given by

$$\phi_n = \int d^d x \, \psi_n^*(\mathbf{x}) \phi(\mathbf{x}) \equiv (\psi_n, \phi), \qquad (9.39)$$

where the latter equivalence introduces a standard notation for the inner product of two functions. Similarily, the external current $J(\mathbf{x})$ may be expanded out as

$$J(\mathbf{x}) = \sum_n J_n \psi_n(\mathbf{x}), \qquad (9.40)$$

where the coefficients J_n are given by

$$J_n = (\psi_n, J). \qquad (9.41)$$

With the substitution of (9.38) and (9.40), the exponent of (9.35) now becomes:

$$\text{Exponent of (9.35)} = \sum_{n,m} (\phi_n \psi_n, M \phi_m \psi_m) - \sum_{n,m} (J_n \psi_n, \phi_m \psi_m)$$

$$= \sum_{n,m} (\phi_n \psi_n, \phi_m \lambda_m \psi_m) - \sum_{n,m} (J_n \psi_n, \phi_m \psi_m) \qquad (9.42)$$

$$= \sum_n (\lambda_n \phi_n^2 - J_n \phi_n),$$

where we assume that the eigenfunctions ψ_n form a complete orthonormal set, such that

$$(\psi_n, \psi_m) = \delta_{nm},$$

and δ_{nm} is the Kronecker delta. Thus, eqn (9.35) becomes:

$$G[J] = \int \mathcal{D}\phi \exp[-\sum_n (\lambda_n \phi_n^2 - J_n \phi_n)]. \qquad (9.43)$$

Now let us *define* the set of functions over which we integrate. We say that $G_N[J]$ is the functional integral over the set $\{\phi(x)\}$ such that

$$\phi(\mathbf{x}) = \sum_{n=1}^{N} \phi_n \psi_n(\mathbf{x}) \quad \text{where } \phi_n = 0 \quad \text{when } n > N. \qquad (9.44)$$

Then the expression

$$G_N[J] = \int d^d\phi_1, \ldots, d^d\phi_N \exp\left[-\sum_{n=1}^{N} (\lambda_n \phi_n^2 - J_n \phi_n)\right], \qquad (9.45)$$

defines the functional integral.

How does G_N depend on N? To establish this, we factorize the expression by completing the square in the exponent:[39]

$$G_N[J] = \frac{\pi^{N/2}}{(\prod_{n=1}^{N} \lambda_n)^{1/2}} \exp\left[\sum_{n=1}^{N} \frac{J_n^2}{4\lambda_n}\right]$$

$$= \frac{\pi^{N/2}}{\sqrt{\det_N M}} \exp\left[\sum_{n=1}^{N} \frac{J_n^2}{4\lambda_n}\right], \qquad (9.46)$$

$$= G_N[0] \exp\left[\sum_{n=1}^{N} \frac{J_n^2}{4\lambda_n}\right],$$

where $\det_N M \equiv$ the product of first N eigenvalues of M.

[39] Remember: we used this trick in Section 1.3.4 and in Section 7.7.

The important feature is that the dependence on J has been factored out. If we restrict our attention to a "physical" choice of J, that is:

$$J(x) \text{ such that } J_n \to 0 \quad \text{as} \quad n \to \infty,$$

then the quotient $G_N[J]/G_N[0]$ will be well-defined in the limit as $N \to \infty$.

9.7   Partition function for the Gaussian model

We have the zero-order partition functional from (9.30) as

$$Z_0[J] = \int \mathcal{D}\phi \exp\{-H_0[J]\}$$

$$= \int \mathcal{D}\phi \exp\left\{-\int d^d x \left[\frac{\alpha}{2}(\nabla\phi)^2 + \frac{u}{2}\phi^2 - J\phi\right]\right\},$$

and our first step is to rewrite the exponent to make $Z_0[J]$ look like the standard Gaussian functional integral $G[J]$ as given by (9.35). To do this we write the zero-order Hamiltonian as:

$$H_0 = -\int\int d^d x d^d y \phi(y) \left[\frac{\delta(\mathbf{x} - \mathbf{y})}{2}(\alpha \nabla_x^2 - u)\right]\phi(\mathbf{x}) - \int d^d x J(\mathbf{x})\phi(\mathbf{x}). \qquad (9.47)$$

Note that the sifting property of the delta function may be used to perform the integration with respect to y and that this just turns the factor $\phi(y)$ back into $\phi(x)$. However, in order to get the term with coefficient α into this form it is necessary to do a partial integral with respect to x and thus accounts for the change of sign. We may show this for an arbitrary function $f(x)$ in one dimension, as follows:

$$\int_0^\infty dx f \frac{\partial^2 f}{\partial x^2} = f \frac{\partial f}{\partial x}\bigg|_0^\infty - \int_0^\infty dx \left(\frac{\partial f}{\partial x}\right)^2.$$

For the case where f and $\partial f/\partial x$ vanish at the limits $x = 0$ and $x = \infty$, we may write

$$\int_0^\infty dx \left(\frac{\partial f}{\partial x}\right)^2 = -\int_0^\infty dx f \frac{\partial^2 f}{\partial x^2}.$$

The extension to d dimensions is trivial.

Comparison of $Z_0[J]$, where H_0 is defined by eqn (9.47), with the standard form $G[J]$, as given by eqn (9.35), allows us to make the identification:

$$M(\mathbf{x}, \mathbf{y}) = \frac{1}{2}\delta(\mathbf{x} - \mathbf{y})(\alpha \nabla_x^2 - u). \qquad (9.48)$$

The eigenfunctions ψ_k and eigenvalues λ_k of $M(\mathbf{x}, \mathbf{y})$ are:

$$\psi_k = \exp\{i\mathbf{k}.\mathbf{x}\}$$

and

$$\lambda_k = -(\alpha k^2 + u)/2.$$

To ensure that the eigenfunctions form a denumerable set, we take the system to occupy a cube of volume $V = L^d$, with periodic boundary conditions. Then:

$$|\mathbf{k}| = \frac{2\pi p}{L},$$

for integer p.

We restrict the ψ_k to those for which $k < \Lambda$. We wish to work out $Z_0[J, \Lambda]$, and to do this we write

$$\phi(\mathbf{x}) = \frac{1}{L^d} \sum_{k<\Lambda} \phi(\mathbf{k}) \exp\{i\mathbf{k}.\mathbf{x}\}; \tag{9.49}$$

$$J(\mathbf{x}) = \frac{1}{L^d} \sum_{k<\Lambda} J(\mathbf{k}) \exp\{i\mathbf{k}.\mathbf{x}\}, \tag{9.50}$$

and substitute these into the exponent for $\phi(\mathbf{x})$ and $J(\mathbf{x})$:

$$H_0 = \frac{1}{L^d} \sum_{k<\Lambda} \left\{ \frac{1}{2} \left(\alpha k^2 + u \right) \phi(\mathbf{k})\phi(-\mathbf{k}) - \phi(\mathbf{k})J(-\mathbf{k}) \right\}. \tag{9.51}$$

Note the constraint that $\phi(\mathbf{x})$ and $J(\mathbf{x})$ are real. Hence they must satisfy the conditions:

$$\phi(-\mathbf{k}) = \phi^*(\mathbf{k}) \quad \text{and} \quad J(-\mathbf{k}) = J^*(\mathbf{k}).$$

where the asterisk denotes "complex conjugate." Therefore not all $\phi(\mathbf{k})$ and $J(\mathbf{k})$ are independent and we only have one half of the number of independent variables. Let the real part of $\phi(\mathbf{k})$ be

$$\mathrm{Re}\,\phi(\mathbf{k}) \equiv \phi_R(\mathbf{k}),$$

and the imaginary part

$$\mathrm{Im}\,\phi(\mathbf{k}) \equiv \phi_I(\mathbf{k}),$$

then we may write (9.51) as

$$H_0 = \frac{1}{L^d} \sideset{}{'}\sum_{k<\Lambda} \left\{ (\alpha k^2 + u)(\phi_R^2(\mathbf{k}) + \phi_I^2(\mathbf{k})) + 2[\phi_R(\mathbf{k})J_R(\mathbf{k}) + \phi_I(\mathbf{k})J_I(\mathbf{k})] \right\}, \tag{9.52}$$

where:

$$\sideset{}{'}\sum \equiv \sum \text{ over one half of the possible values of } k.$$

As a result, the $\phi(\mathbf{k})$ for different $\mathbf{k}$ are now decoupled from each other.

Next we work out

$$Z_0[J, \Lambda] = \prod_{k<\Lambda}' \int d\phi_R(\mathbf{k}) d\phi_I(\mathbf{k}) e^{-H_0},$$ (9.53)

where:

$$\prod' \equiv \text{only one half of the possible values of } k \text{ are being multiplied together.}$$

As before, we do the one-dimensional Gaussian integrals and multiply the results together:

$$Z_0[J, \Lambda] = \prod_{k<\Lambda}' \left(\frac{\pi L^d}{\alpha k^2 + u}\right) \exp\left[\frac{1}{L^d} \frac{J_R^2(\mathbf{k}) + J_I^2(\mathbf{k})}{\alpha k^2 + u}\right].$$ (9.54)

We take the factor πL^d outside the first exponent as it is part of the normalization.[40] Then we can re-write Z_0 as:

$$Z_0[J, \Lambda] = \exp\left[-\sum_{k<\Lambda}' \log(\alpha k^2 + u)\right] \times \exp\left[\frac{1}{L^d} \sum_{k<\Lambda}' \frac{J_R^2(\mathbf{k}) + J_I^2(\mathbf{k})}{\alpha k^2 + u}\right],$$ (9.55)

and reverse the earlier procedure for the second exponent:

$$Z_0[J, \Lambda] = \exp\left[-\frac{1}{2} \sum_{k<\Lambda} \log(\alpha k^2 + u)\right] \times \exp\left[\frac{1}{2L^d} \sum_{k<\Lambda} \frac{J(\mathbf{k})J(-\mathbf{k})}{\alpha k^2 + u}\right]$$ (9.56)

where the primes on the summations have been dropped and now *all* wave number variables are summed over for $|\mathbf{k}| < \Lambda$.

Finally, we take the limit $L \to \infty$,

$$Z_0[J, \Lambda] = \exp\left[-\frac{V}{2} \int_{0\le k\le\Lambda} d^d k \log(\alpha k^2 + u)\right] \times \exp\left[\frac{1}{2} \int_{0\le k\le\Lambda} d^d k \frac{J(\mathbf{k})J(-\mathbf{k})}{\alpha k^2 + u}\right],$$
(9.57)

where $V = L^d$ is the volume of the system.

Note that:

1. The J dependence has been factored out.

2. In general the limit $\Lambda \to \infty$ does not exist.

The first of these is what we were trying to achieve and will be used in conjunction with linear response theory in the next section. The second relates to the problem of ultraviolet divergences. This was discussed in Section 1.4.4 and we shall postpone further discussion until Chapter 10.

---

[40] Remember that our ultimate use for Z_0 is to form the bridge equation and obtain the free energy, which we then differentiate to obtain macroscopic thermodynamic quantities. It follows that any multiplicative constant in Z_0 can be ignored as it cannot contribute to the end result.

9.8 Correlation functions

The next stage is to use linear response theory to obtain correlation functions. To do this we write Z_0 as

$$Z_0[J] = Z_0[0] \times \exp\left[\frac{1}{2}\int_{0\leq k\leq\Lambda} d^d k \frac{J(\mathbf{k})J(-\mathbf{k})}{\alpha k^2 + u}\right], \quad (9.58)$$

where the prefactor $Z_0[0]$ is the first exponential factor in (9.57). Note that $Z_0[0]$ is independent of J and does not contribute to the correlations. Now let us rewrite the integral in the second factor as

$$\int_{0\leq k\leq\Lambda} d^d k \frac{J(\mathbf{k})J(-\mathbf{k})}{\alpha k^2 + u} = \int d^d x d^d y \, J(\mathbf{x})\Delta(\mathbf{x}-\mathbf{y})J(\mathbf{y}), \quad (9.59)$$

where

$$\Delta(\mathbf{x}-\mathbf{y}) \equiv \int_{0\leq k\leq\Lambda} d^d k \frac{e^{ik\cdot(\mathbf{x}-\mathbf{y})}}{\alpha k^2 + u} \equiv \int_{0\leq k\leq\Lambda} d^d k\, G^{(2)}(\mathbf{k}), \quad \text{for } u > 0, \quad (9.60)$$

and in passing we introduce $G^{(2)}(\mathbf{k})$ as the pair-correlation in wave number space. Note that the restriction on u is to avoid the denominator becoming zero at $k = 0$ which would cause the integral to diverge.

Now, as before, we obtain correlations by functional differentiation. We may define the general n-point correlation $G^{(n)}(\mathbf{x}_1, \ldots, \mathbf{x}_n)$ by an extension of eqn (7.6), thus:

$$G^{(n)}(\mathbf{x}_1, \ldots, \mathbf{x}_n) \equiv \langle \phi(\mathbf{x}_1), \ldots, \phi(\mathbf{x}_n)\rangle.$$

Then, by a generalization of previous formulae, as given in Section 7.4.1, we have:

$$G^{(n)}(\mathbf{x}_1, \ldots, \mathbf{x}_n) = \frac{1}{Z_0[J]}\frac{\delta}{\delta J(\mathbf{x}_1)} \cdots \frac{\delta}{\delta J(\mathbf{x}_n)}Z_0[J], \quad (9.61)$$

and for connected correlations:

$$G_c^{(n)}(\mathbf{x}_1, \ldots, \mathbf{x}_n) = \frac{\delta}{\delta J(\mathbf{x}_1)} \cdots \frac{\delta}{\delta J(\mathbf{x}_n)} \ln Z_0[J]. \quad (9.62)$$

9.8.1 Example: two-point connected Gaussian correlation

As an example, we shall work out the connected two-point or pair correlation for the Gaussian case, using eqn (9.62). Substituting (9.59) and (9.60) into (9.58) for $Z_0[J]$, and then taking logs, we have first

$$\ln Z_0[J] = \frac{1}{2} \int \int d^d x d^d y \, J(\mathbf{x}) \Delta(\mathbf{x} - \mathbf{y}) J(\mathbf{y})$$

and then, substituting this into (9.62),

$$G_c^{(2)}(\mathbf{x}_1, \mathbf{x}_2) = \frac{\delta}{\delta J(\mathbf{x}_1)} \frac{\delta}{\delta J(\mathbf{x}_2)} \cdot \frac{1}{2} \int \int d^d x d^d y \, J(\mathbf{x}) \Delta(\mathbf{x} - \mathbf{y}) J(\mathbf{y})$$

$$= \frac{1}{2} \int \int d^d x d^d y \frac{\delta J(\mathbf{x})}{\delta J(\mathbf{x}_1)} \Delta(\mathbf{x} - \mathbf{y}) \frac{\delta J(\mathbf{y})}{\delta J(\mathbf{x}_2)}$$

$$+ \frac{1}{2} \int d^d x d^d y \frac{\delta J(\mathbf{x})}{\delta J(\mathbf{x}_2)} \Delta(\mathbf{x} - \mathbf{y}) \frac{\delta J(\mathbf{y})}{\delta J(\mathbf{x}_1)},$$

and, interchanging dummy variables $\mathbf{x}$ and $\mathbf{y}$,

$$G_c^{(2)}(\mathbf{x}_1, \mathbf{x}_2) = 2 \times \frac{1}{2} \int \int d^d x d^d y \, \delta(\mathbf{x} - \mathbf{x}_1) \Delta(\mathbf{x} - \mathbf{y}) \delta(\mathbf{y} - \mathbf{x}_2) = \Delta(\mathbf{x}_1 - \mathbf{x}_2), \quad (9.63)$$

where the functional derivative gives in general:

$$\delta J(\mathbf{x}) / \delta J(\mathbf{y}) = \delta(\mathbf{x} - \mathbf{y}).$$

Higher-order connected correlations vanish for the Gaussian case. The connected pair-correlation $G_c^{(2)}(\mathbf{x} - \mathbf{y})$ falls off exponentially with $|(\mathbf{x} - \mathbf{y})|$ and has a correlation length ξ given by

$$\xi = \sqrt{\alpha/u}. \qquad (9.64)$$

Note that:

- $G_c^{(2)}$ for the Gaussian model $\equiv$ *bare* propagator;

- $G_c^{(2)}$ for the G–L model $\equiv$ *exact* propagator;

- Our objective is to express the exact propagator in terms of an expansion in bare propagators.

9.9   Fixed points for the Gaussian model

Now that we know how to work out the Gaussian functional integral we can apply the RGT, as defined by eqns (9.18) and (9.25), in wave number space.

9.9.1   The RG equations

Consider the Gaussian Hamiltonian in k-space. This is given by eqn (9.51), but now we take the limit of infinite system size (i.e. $L \to \infty$) and the Fourier sum is replaced by a Fourier integrals, as defined by eqn (9.5). Also, in order to simplify matters, we drop the term in J and restrict our attention to the case of zero external field. We then have:

$$H_0 = \int d^d k \frac{1}{2} \left(u + \alpha k^2 \right) \phi(\mathbf{k}) \phi(-\mathbf{k}) = \int d^d k \frac{1}{2} \left(u + \alpha k^2 \right) |\phi(\mathbf{k})|^2, \qquad (9.65)$$

where we recall that $\phi(-\mathbf{k}) \equiv \phi^*(\mathbf{k})$. The RGT is:

$$\exp\{-H'[\phi']\} = \left[\int \Pi_{\Lambda/b \leq k \leq \Lambda} d\phi(\mathbf{k}) \exp\{-H[\phi]\} \right]_{k=k'/b, \; \phi(k)=c\phi'(k')}, \qquad (9.66)$$

where

$$c = b^{d-d_\varphi}.$$

Integration over the $d\phi(\mathbf{k})$ just yields a constant and we obtain the new Hamiltonian as:

$$\begin{aligned}
H' &= \int_{k < \Lambda/b} d^d k \frac{1}{2} \left(u + \alpha k^2 \right) |\phi(\mathbf{k})|^2 \\
&= \int_{k' < \Lambda} d^d k' b^{-d} \frac{1}{2} \left(u + \alpha \frac{k'^2}{b^2} \right) b^{2d-2d_\varphi} |\phi'(\mathbf{k}')|^2,
\end{aligned} \qquad (9.67)$$

where the second equality is due to our having carried out the rescaling according to the second step of the RGT, along with eqn (9.25). In order to ensure form-invariance of the Hamiltonian, we absorb the scaling factors into the coupling constants. Thus, defining *renormalized coupling constants* as

$$u' = b^{d-2d_\varphi} u, \qquad (9.68)$$

and

$$\alpha' = b^{d-2d_\varphi-2} \alpha, \qquad (9.69)$$

we obtain for the renormalized Hamiltonian[41]

$$H' = \int_{k' < \Lambda} d^d k' \frac{1}{2} \left(u' + \alpha'(k')^2) \right) |\phi'(\mathbf{k}')|^2. \qquad (9.70)$$

Note that (9.68) and (9.69) relate coupling coefficients before and after the RGT and hence are the RG equations. Also note that the Hamiltonian is the same as that of eqn (9.65), with all the variables changed to primed form.

[41] At this point it may be helpful to recall eqn (8.6).

9.9.2   The fixed points

As usual, the fixed points are where $u = u'$ and $\alpha = \alpha'$. The RG eqns (9.68) and (9.69) allow for two fixed points:

1. $d - 2d_\varphi = 0$, with u arbitrary and $\alpha = 0$.
2. $d - 2d_\varphi - 2 = 0$, with α arbitrary and $u = 0$.

Now let us consider the two cases in turn:

Case (1): $d - 2d_\varphi = 0$. Equation (9.68) becomes

$$u' = u, \tag{9.71}$$

and so any (arbitrary) value of u corresponds to a fixed point. Equation (9.69) then becomes

$$\alpha' = b^{-2}\alpha. \tag{9.72}$$

Thus $\alpha = 0$ corresponds to the fixed point, and from the linearized analysis of Section 8.2.3, this corresponds to $y = -2$ and so α is an irrelevant field.
We can sum this up as follows:

$$\alpha \quad \text{is an irrelevant field under RGT.}$$

Hence the nearest-neighbor interaction plays no part and this is equivalent to a lattice of uncoupled sites. This fixed point therefore corresponds to the high-temperature ($T = \infty$) case, with the anomalous dimension

$$d - 2d_\varphi = 0 \quad \text{or} \quad d_\varphi = \frac{d}{2}.$$

Case (2): Equation (9.69) becomes

$$\alpha' = \alpha \tag{9.73}$$

and is satisfied by any α. Equation (9.68) becomes

$$u' = b^2 u. \tag{9.74}$$

Hence α is arbitrary and the anomalous dimension is given by:

$$d - 2d_\varphi - 2 = 0 \quad \text{or} \quad d_\varphi = \frac{d}{2} - 1.$$

The previous condition for the correlation at the critical point is given by eqn (9.27) as

$$d_\varphi = \frac{d}{2} - 1 + \frac{\eta}{2}$$

and the two taken together imply $\eta = 0$.

Also, from the linearized analysis of Section 8.2.3, and eqn (9.74),

$$u' = b^y u = b^2 u. \tag{9.75}$$

The scaling exponent $y = 2 > 0$, thus u is relevant. As $y = 2$, it follows from (8.48) that $v = 1/2$.

We should note the following points:

1. The parameter α plays no part in this at all, so we shall set $\alpha = 1$ hereafter.

2. The coupling constant takes the value $u = 0$ at $T = T_c$. Thus we do not have a low-temperature fixed point in the Gaussian model, as integrals fail to converge for $u < 0$. Or, on physical grounds, $\exp[-u\phi^2]$ would become $\exp[|u|\phi^2]$ for $u < 0$, and fluctuations would be unbounded.

3. The *anomalous dimension* of $\phi(x)$ is

$$d_\varphi = \frac{d}{2} - 1 = d_\varphi^0,$$

and is the same as the *normal* or *canonical (scaling) dimension* in this case: see Section 9.3.2.

9.9.3   Normal dimension of coupling constants

Again let us denote the scale dimension of any variable X by the notation $[X]$. As we saw before in Section 9.3.2, dimensional analysis gives us

$$[H] = 0, \quad [\phi] = d_\varphi^0 = \frac{d}{2} - 1,$$

where $d\varphi^0$ is the normal dimension.

Then, under RGT, we have: first,

$$[d^d k u \phi^2] = 0 = [b^{-d} u b^{d+2} d^d k' \phi'^2].$$

$$[u b^2] = 0,$$

thus

$$[u] = 2. \tag{9.76}$$

And second,

$$[d^d k \lambda \phi^4] = 0 = [b^{-d} \lambda b^{2d+4} d^d k' \phi'^4],$$

and so

$$[b^{d+4} \lambda] = 0,$$

hence

$$[\lambda] = 4 - d. \tag{9.77}$$

This is another indication, in addition to the work of Section 7.7.1, that the particular case $d = 4$ may be of some importance.

9.10 Ginsburg–Landau (GL) theory

Now let us turn our attention to the full model Hamiltonian as defined by eqns (9.19)–(9.21):

$$H = \int d^d x \left[\frac{1}{2}(\nabla\phi)^2 + \frac{u\phi^2}{2} + \frac{\lambda}{4!}\phi^4 \right] = H_0 + H_I, \tag{9.78}$$

where we have now put $\alpha = 1$. An immediate difficulty arises with the interaction term where Fourier transformation gives H_I in $\mathbf{k}$-space as,

$$H_I = \frac{\lambda}{4!}\int d^d x \phi^4(\mathbf{x}) = \frac{\lambda}{4!}\int d^d k_1 d^d k_2 d^d k_3 \phi(\mathbf{k}_1)\phi(\mathbf{k}_2)\phi(\mathbf{k}_3)\phi(-\mathbf{k}_1-\mathbf{k}_2-\mathbf{k}_3). \tag{9.79}$$

Thus we see that the ϕ^4 term couples normal modes together and so we cannot factorize into Gaussian forms.

In order to carry out the RGT, as defined by eqns (9.18) and (9.25), we make a decomposition of the scalar field $\phi(\mathbf{x})$, thus:

$$\phi(\mathbf{x}) = \phi^-(\mathbf{x}) + \phi^+(\mathbf{x}), \tag{9.80}$$

where $\phi^-(\mathbf{x})$ is the Fourier transform of $\phi(\mathbf{k})$ such that $0 \leq k \leq \Lambda/b$ and $\phi^+(\mathbf{x})$ is the Fourier transform of $\phi(\mathbf{k})$ such that $\Lambda/b \leq k \leq \Lambda$.

The explicit forms of $\phi^-(\mathbf{x})$ and $\phi^+(\mathbf{x})$ may be obtained from eqn (9.5) as:

$$\phi^-(\mathbf{x}) = \int_{0 \leq k \leq \Lambda/b} \frac{d^d k}{(2\pi)^d} \exp\{i\mathbf{k}\cdot\mathbf{x}\}\phi(\mathbf{k}), \tag{9.81}$$

and

$$\phi^+(\mathbf{x}) = \int_{\Lambda/b \leq k \leq \Lambda} \frac{d^d k}{(2\pi)^d} \exp\{i\mathbf{k}\cdot\mathbf{x}\}\phi(\mathbf{k}). \tag{9.82}$$

We shall find that the properties of the RGT depend on the lattice dimension d. After we have discussed the perturbative implementation of the RGT in the next sub-section, in the following three sections we consider the cases $d > 4$, $d < 4$, and $d = 4$ in turn.

9.10.1 Perturbative implementation of the RGT

In this section we carry out stage B of the programme outlined in Section 9.2. That is, we apply the RG transformation to the Ginsburg–Landau Hamiltonian in an approximate way by using perturbation theory. Denoting, as before, the RG transformation with spatial rescaling factor b by R_b, and taking the form given by eqn (9.19) for H_{GL}, we write the procedure symbolically as:

$$H'_{\text{GL}} = R_b H_{\text{GL}} = R_b H_0 + \langle R_b H_I \rangle_0 + \langle R_b H_I^2 \rangle_0 + \cdots$$

We begin by integrating over $\mathcal{D}\phi^+$, being guided by our previous experience with the Gaussian model. However, instead of (9.66), we now have:

$$e^{-H'[\phi^-]} = \int \mathcal{D}\phi^+ e^{-H[\phi^- + \phi^+]}$$

$$= \int \mathcal{D}\phi^+ e^{-H_0[\phi^- + \phi^+]}(1 - H_I[\phi^- + \phi^+] + \mathcal{O}(\lambda^2)),$$

where we have substituted $H = H_0 + H_I$, and expanded out e^{-H_I} to order λ. Now we use the fact that e^{-H_0} is Gaussian (or multivariate normal) and hence separable. So we are free to take out a factor $e^{-H_0[\phi^-]}$, which is unaffected by the integration $\int \mathcal{D}\phi^+$, thus:

$$e^{-H'[\phi^-]} = e^{-H_0[\phi^-]} \int \mathcal{D}\phi^+ e^{-H_0[\phi^+]}(1 - H_I[\phi^- + \phi^+] + \mathcal{O}(\lambda^2)),$$

$$= e^{-H_0[\phi^-]} \left[1 - \frac{\int \mathcal{D}\phi^+ e^{-H_0[\phi^+]} H_I[\phi^- + \phi^+] + \mathcal{O}(\lambda^2)}{\int \mathcal{D}\phi^+ e^{-H_0[\phi^+]}} \right] \int \mathcal{D}\phi^+ e^{-H_0[\phi^+]},$$

$$(9.83)$$

where we have also taken out the common factor $\int \mathcal{D}\phi^+ \exp\{-H_0[\phi^+]\}$. As a constant prefactor can only contribute an additive constant to the free energy, we are free to ignore it.[42] Accordingly dropping the constant prefactor, the new Hamiltonian, to first order in λ, is:

$$H'[\phi^-] = H_0[\phi^-] + \frac{\int \mathcal{D}\phi^+ \exp\{-H_0[\phi^+]\} H_I[\phi^- + \phi^+]}{\int \mathcal{D}\phi^+ \exp\{-H_0[\phi^+]\}},$$

$$= H_0[\phi^-] + \langle H_I[\phi^- + \phi^+] \rangle_0^+,$$

$$(9.84)$$

where the "+" superscript on the angle bracket indicate that the Gaussian average is only over the high-wave number modes.

Now substituting the decomposition $\phi = \phi^- + \phi^+$ into H_I and expanding out, we have:

$$\langle (\phi^- + \phi^+)^4 \rangle_0^+ = (\phi^-)^4 + 6(\phi^-)^2 \langle (\phi^+)^2 \rangle_0^+ + \langle (\phi^+)^4 \rangle_0^+, \qquad (9.85)$$

where it should be noted that we have used the fact that

$$\langle \phi^- \rangle_0^+ = \phi^- \quad \text{and} \quad \langle \phi^+ \rangle_0 = \langle (\phi^+)^3 \rangle_0 = 0,$$

as the average over the high-k modes leaves the low-k modes unaffected and as all odd-order moments of a Gaussian distribution vanish by symmetry.

---

[42] Remember! As we always differentiate the free energy to get thermodynamic quantities constant terms do not contribute.

The last term is constant (i.e. it does not depend on ϕ^-) and so we shall neglect it. Set

$$\langle(\phi^+)^2\rangle_0^+ \equiv G_0^+(0),$$

which we already know from the Gaussian model, hence:

$$H'[\phi^-] = \int d^d x \left[\frac{1}{2}(\nabla\phi^-)^2 + \frac{u}{2}(\phi^-)^2 + \frac{\lambda}{4!}(\phi^-)^4 + \frac{\lambda}{4}G_0^+(0)(\phi^-)^2\right]$$

$$= \int d^d x \left[\frac{1}{2}(\nabla\phi^-)^2 + \frac{1}{2}\left(u + \frac{\lambda}{2}G_0^+(0)\right)(\phi^-)^2 + \frac{\lambda}{4!}(\phi^-)^4\right].$$

In order to complete the second part of the RGT, we rescale lengths according to (9.18) and renormalize the field according to (9.25):

$$H'[\phi'] = \int d^d x' \left[b^{d-2-2d\varphi}\frac{1}{2}(\nabla'\phi')^2 + b^{d-2d\varphi}\frac{1}{2}\left(u + \frac{\lambda}{2}G_0^+(0)\right)\phi'^2 + b^{d-4d\varphi}\frac{\lambda}{4!}\phi'^4\right].$$

$$(9.86)$$

To keep the gradient term unaffected (i.e. to just have a coefficient $= 1/2$), we need:

$$b^{d-2-2d\varphi} = 1,$$

and so it follows that

$$d - 2 - 2d_\varphi = 0,$$

or, rearranging,

$$d_\varphi = \frac{d}{2} - 1. \qquad (9.87)$$

9.10.2   The Gaussian fixed point for $d > 4$

From (9.86), and the requirement of form invariance of the Hamiltonian, the RG equations are:

$$u' = b^2\left(u + \frac{\lambda}{2}G_0^+(0)\right);$$

$$\lambda' = b^{4-d}\lambda \equiv b^\epsilon\lambda \quad \text{where} \quad (\epsilon = 4 - d). \qquad (9.88)$$

We obtain $G_0^+(0)$ from the Gaussian result earlier. That is, eqn (9.60), but now we change our notation and make the replacement $\Delta(\mathbf{x} - \mathbf{y}) = G_0(\mathbf{x} - \mathbf{y})$, thus:

$$G_0(\mathbf{x} - \mathbf{y}) = \int_{0 \leq k \leq \Lambda} d^d k \frac{e^{ik(\mathbf{x}-\mathbf{y})}}{\alpha k^2 + u}. \qquad (9.89)$$

Setting $\alpha = 1$ (our new notation) and $x = y$:

$$G_0(0) = \int_{0 \leq k \leq \Lambda} \frac{d^d k}{k^2 + u},$$

and so

$$G_0^+(0) = \int_{\Lambda/b \le k \le \Lambda} \frac{d^d k}{k^2 + u} = \frac{S_d \Lambda^{d-2}}{(2\pi)^d (d-2)} (1 - b^{2-d}) + \mathcal{O}(u), \qquad (9.90)$$

where S_d is the surface of the d-dimensional sphere and the expression

$$G_0^+(0) \equiv C(1 - b^{2-d}) + \mathcal{O}(u),$$

defines C, which is

$$C = \frac{S_d \Lambda^{d-2}}{(2\pi)^d (d-2)}. \qquad (9.91)$$

9.10.2.1 The fixed point. Now, rewriting (9.88), and substituting from (9.90) for $G_0^+(0)$, we have the RG equations in the form:

$$u' = b^2 u + C(b^2 - b^\epsilon)\lambda/2 + \mathcal{O}(u\lambda),$$
$$\lambda' = b^\epsilon \lambda. \qquad (9.92)$$

The fixed point satisfies:

$$u^* = b^2 u^* + C(b^2 - b^\epsilon)\lambda^* + \mathcal{O}(u^*\lambda^*),$$
$$\lambda^* = b^\epsilon \lambda^*. \qquad (9.93)$$

and so is given by:

$$\lambda^* = 0,$$
$$u^* = 0. \qquad (9.94)$$

Linearizing about the fixed point, the RG transformation matrix as defined by eqn (8.22) becomes:

$$A_b = \begin{pmatrix} b^2 & C(b^2 - b^\epsilon) \\ 0 & b^\epsilon \end{pmatrix}, \qquad (9.95)$$

where, from (9.88) $\epsilon = 4 - d$, and with eigenvalues λ_i and eigenvectors $X^{(i)}$:

$$\lambda_1 = b^2 (y_1 = 2) \quad \text{with} \quad X^{(1)} = (1, 0)$$
$$\lambda_2 = b^\epsilon (y_2 = \epsilon) \quad \text{with} \quad X^{(2)} = (-C, 0) \qquad (9.96)$$

For a mixed (i.e. critical) fixed point, we must have y_1 and y_2 of different signs with one relevant and one irrelevant field. For this one must have $\epsilon < 0$ and hence $d > 4$. Then the critical exponents are identically the same as in the mean-field theory (or Gaussian result):

$$\nu = \frac{1}{y_1} = \frac{1}{2},$$

and from (9.27) and (9.87)

$$\eta = 2d_\varphi - d + 2 = 0.$$

9.10.3   Non-Gaussian fixed points for $d < 4$

If $d < 4$ then $\epsilon > 0$ and $y_2 = \epsilon > 0$. Hence the fixed point $(u^*, \lambda^*) = (0, 0)$ is unstable in both directions. As λ increases, we need to include higher orders in the perturbation expansion. It is a simple matter to extend the analysis in going from (9.84) to (9.85) to include terms of order λ^2 and we shall not give details here. Although we do discuss this order of approximation in Chapter 10, in the context of field-theoretic methods.

9.10.3.1   RG equations.   If we include terms of order λ^2 in our analysis leading to (9.92), it can be shown that the RGEs take the form:

$$u' = b^2 \left[u + \frac{\lambda}{16\pi^2} \left(\frac{1}{2} \Lambda^2 (1 - b^{-2}) - u \ln b \right) \right], \tag{9.97}$$

$$\lambda' = b^\epsilon \left[\lambda - \frac{3\lambda^2}{16\pi^2} \ln b \right]. \tag{9.98}$$

Note that these equations are only valid for small ϵ, and that the presence of the term in λ^2 means that the RGT is nonlinear.

A finite RGT can be built up from many infinitesimal transformations using the group composition property as discussed in Chapter 8. Near the fixed point, $u' \to u$, and in this way, the recursion relations can be turned into differential equations as follows.

Let $b = 1 + \rho$ where ρ is small and we shall neglect terms $\mathcal{O}(\rho^2)$. Then we have:

$$\ln b = \ln(1 + \rho) \simeq \rho,$$

and (9.97) can be written as:

$$u' = (1 + 2\rho) \left[u + \frac{\lambda}{16\pi^2} \frac{\Lambda^2}{2} \left(1 - \frac{1}{(1 + \rho)^2} \right) - \frac{\lambda u}{16\pi^2} \rho \right].$$

Rearranging, expanding to order ρ^2, and dividing across by ρ, gives:

$$\frac{u' - u}{\rho} = 2u + \frac{\Lambda^2 \lambda}{16\pi^2} - \frac{\lambda u}{16\pi^2}.$$

Then taking the limit $\rho \to 0$ as $u' - u \to 0$, we find that (9.97), and a similar procedure for (9.98), yield the differential RGEs as:

$$\frac{du(b)}{d \ln b} = 2u(b) + \frac{\Lambda^2 \lambda(b)}{16\pi^2} - \frac{u(b)\lambda(b)}{16\pi^2}, \tag{9.99}$$

$$\frac{d\lambda(b)}{d \ln b} = \epsilon \lambda(b) - \frac{3\lambda^2(b)}{16\pi^2}. \tag{9.100}$$

These equations can be solved for the position of the fixed point, based on the condition that

$$\frac{du^*}{d \ln b} = \frac{d\lambda^*}{d \ln b} = 0, \tag{9.101}$$

at the fixed point. Note that the coordinates found for the fixed point must be $\mathcal{O}(\epsilon)$ for the approximations to hold.

9.10.4   The beta-function

We can rewrite the above conditions by introducing new functions:

$$\beta_u = \frac{du(b)}{d \ln b};$$
(9.102)

$$\beta_\lambda = \frac{d\lambda(b)}{d \ln b},$$
(9.103)

and in general, for coupling constants K_α, we may write:

$$\beta_\alpha(K_\alpha(b)) = \frac{dK_\alpha(b)}{d \ln b}.$$
(9.104)

Then, the condition for a fixed point may be expressed as:

$$\beta_\alpha(K_\alpha^*) = 0.$$
(9.105)

These are examples of the Callan–Symanzik beta-function which occurs in quantum field theory, and which will be discussed in the next chapter.

9.10.4.1   The fixed point.   The differential RGEs are easily solved for (u^*, λ^*). Take λ^* first for simplicity. We have:

$$\beta_\lambda(\lambda^*) = 0 = \epsilon\lambda^* - \frac{3(\lambda^*)^2}{16\pi^2},$$
(9.106)

and so

$$\lambda^* = 0,$$
(9.107)

which is trivial, or:

$$\lambda^* = \frac{16\pi^2}{3}\epsilon$$
(9.108)

which is nontrivial.
    Assuming that u^* is also $\mathcal{O}(\epsilon)$, we can write

$$\beta_u(u^*, \lambda^*) = 0 = 2u^* + \frac{\Lambda^2\lambda^*}{16\pi^2} + \mathcal{O}(\epsilon^2),$$
(9.109)

and so find that

$$u^* = -\frac{\Lambda^2\epsilon}{6}.$$
(9.110)

That is, the nontrivial fixed point takes the form:

$$u^* = -\frac{\Lambda^2}{6}\epsilon, \qquad \lambda^* = \frac{16\pi^2}{3}\epsilon.$$
(9.111)

Now linearize the RGEs about the fixed point, with the result:

$$\frac{du}{d\ln b} = \left(2 - \frac{\epsilon}{3}\right)(u - u^*) + \frac{\Lambda^2}{16\pi^2}(\lambda - \lambda^*)(1 + \epsilon/6), \tag{9.112}$$

$$\frac{d\lambda}{d\ln b} = -\epsilon(\lambda - \lambda^*). \tag{9.113}$$

From which we conclude that:

$$y_1 = 2 - \frac{\epsilon}{3} > 0; \tag{9.114}$$

$$y_2 = -\epsilon < 0. \tag{9.115}$$

Hence this is a *mixed* fixed point and determines critical behavior for $\epsilon > 0$ and $d < 4$. Thus, from eqn (8.48),

$$\nu = \frac{1}{y_1} = \frac{1}{2} + \frac{\epsilon}{12} + \mathcal{O}(\epsilon^2) \tag{9.116}$$

and in general all critical exponents can be calculated as power series in ϵ. It should be noted that the analysis of Section 9.8.1 still applies and hence we still have $\eta = 0$.

9.10.5   The marginal case: $d = 4$

For $d = 4$, we have $\epsilon = 0$ and, by the analysis of Section 8.2.3, we have a marginal field which we identify as λ. Let us denote the marginal field λ by g and call the relevant field t, and define it as

$$t = u + \frac{\Lambda^2 g}{32\pi^2}. \tag{9.117}$$

The RGEs may then be written in terms of the variables g and t as:

$$\frac{dt(b)}{d\ln b} = 2t(b) - \frac{g(b)t(b)}{16\pi^2} + O(g^2), \tag{9.118}$$

$$\frac{dg(b)}{d\ln b} = -\frac{3}{16\pi^2}g^2(b). \tag{9.119}$$

Solving these, one finds logarithmic corrections to critical behavior; for example, the susceptibility is found to take the form

$$\chi(t, g, \Lambda) \sim |t|^{-1}|\ln(|t|)|^{-1/3}. \tag{9.120}$$

Any further discussion of the marginal case would take us beyond the scope of this book and reference should be made to the books listed at the end of the chapter.

9.10.6   Critical exponents to order ϵ

We already have a value for ν from (9.116) and to order ϵ, the exponent η remains the same as for the Gaussian fixed point in Section 9.5.2 or the Ginsburg–Landau fixed point for $d > 4$. That is, we take $\eta = 0$.

We have the choice of calculating α from either (9.29) or (8.46). The latter is perhaps more informative and we may write:

$$\alpha = 2 - \frac{d}{y_1} = 2 - \frac{4 - \epsilon}{2 - \epsilon/3} = \frac{\epsilon}{6} + \mathcal{O}(\epsilon^2), \tag{9.121}$$

where we substitute from (9.114) for y_1 and also put $d = 4 - \epsilon$.

Making use of the remaining relationships of eqn (9.29), in the same way we find

$$\beta = \frac{1}{2} - \epsilon/6; \tag{9.122}$$

$$\gamma = 2\nu = 1 + \epsilon/6; \tag{9.123}$$

and

$$\delta = 3 + \epsilon. \tag{9.124}$$

It should be noted that all these results neglect terms of order ϵ^2 and higher. This means that we are treating ϵ as a small parameter, when in reality we must have $\epsilon = 1$ to give results in $d = 4 - \epsilon = 3$ dimensions. Nevertheless, we set $\epsilon = 1$ in equations (9.116) and (9.121)–(9.124), and give the resulting values of the critical exponents in comparison with those obtained by mean-field theory, along with a numerical calculation of the Ising model in $d = 3$, using the high-temperature expansion [26], in Table 9.1. The improvement over mean-field theory is clear.

This is as far as we shall take these expansions: even to go to order ϵ^2 would require more elaborate techniques than those discussed in this chapter. In the next chapter we present an introduction to such techniques, mainly to give some idea of the "field theoretic" approach. Those who would like to see the results to ϵ^5 will find them summarized in [13] on pages 603–4. The results there are for a n-dimensional order parameter field and setting $n = 1$ produces values which may be compared with those in the present work. At the end of Chapter 10, we present a modified version of Table 9.1, in which we give numerical values of the critical exponents based on Borel summation of the ϵ-expansion.

TABLE 9.1 Comparison of values of critical exponents obtained to order ϵ with those obtained by mean-field theory and exact numerical calculation.

Exponents	Mean-field theory	RG to order ϵ	Numerical simulation of Ising model in $d = 3$
α	0	0.17	0.12
β	0.5	0.33	0.31
γ	1	1.17	1.25
δ	3	4	5.20
ν	0.5	0.58	0.64
η	0	0	0.06

Further reading

Comprehensive introductory treatments may be found in [15] and [5], which are both written from the "statistical physics" point of view. However, the latter reference does not shrink from the full field-theoretic treatment. There are numerous books in the field-theory camp and the trouble here is that the topics associated with critical phenomena tend to occupy only a very small part of such books. Having said that, the early chapters of [4] and [2] are very accessible while [13] and [33] are very clear and succinct.

9.11 Exercises

Solutions to these exercises, along with further exercises and their solutions, will be found at: *www.oup.com*

1. A particular model for the critical behavior of spins on a d-dimensional lattice leads to renormalization group equations of the form

$$p' = b^2 p + C(b^2 - b^\epsilon)q + O(pq);$$
$$q' = b^\epsilon q,$$

   where p and q are the coupling constants, C is a system constant which is positive and real, $\epsilon = 4-d$, and b is the usual (length) scaling factor. By linearizing about the fixed point $(p^*, q^*) = (0, 0)$, obtain the matrix of the RGT and show that the associated critical indices are $y_1 = 2$ and $y_2 = 4-d$. Briefly discuss the nature of this fixed point.

   Given that the critical indices for this model are the same as those for the mean-field theory of the Ising model, comment briefly on the validity or otherwise of mean-field theory.

2. A particular model for the critical behavior of spins on a d-dimensional lattice leads to renormalization group equations of the form

$$\frac{d\mu(b)}{d\ln b} = 2\mu(b) + \frac{\Lambda^2 \lambda(b)}{16\pi^2} - \frac{\mu(b)\lambda(b)}{16\pi^2};$$
$$\frac{d\lambda(b)}{d\ln b} = \epsilon\lambda(b) - \frac{3\lambda^2(b)}{16\pi^2},$$

   where μ and λ are the coupling constants, Λ is a system constant which is positive and real, $\epsilon = 4-d$, and b is the usual spatial rescaling factor. These equations are valid for small values of ϵ. Given that the criterion for a fixed point (μ^*, λ^*) is

$$\frac{d\mu^*(b)}{d\ln b} = \frac{d\lambda^*(b)}{d\ln b} = 0,$$

   verify that a fixed point (to order ϵ) is given by

$$\lambda^* = \frac{16\pi^2 \epsilon}{3}; \quad \text{and} \quad u^* = -\frac{\Lambda^2 \epsilon}{6}.$$

By linearizing about the fixed point, obtain the matrix of the RGT and show that the associated critical indices are $y_1 = 2 - \epsilon/3$ and $y_2 = -\epsilon$.

Briefly discuss the nature of the fixed point with particular reference to the dimensionality of the lattice.

Show that an expression for the critical exponent ν can be written as:

$$\nu = \frac{1}{2} + \frac{\epsilon}{12} + O(\epsilon^2).$$

10

FIELD-THEORETIC RENORMALIZATION GROUP

In Chapter 9 we worked out corrections to the mean-field values of the critical exponents by, in effect, cheating. We were able to work out the critical exponents to order ϵ for two reasons.

1. We made use of the four scaling relationships so that we could obtain values of all six exponents from (in principle) just two.

2. The exponent η is given by its mean-field value (i.e. $\eta = 0$) to order ϵ. The first correction to mean-field behavior comes at $\mathcal{O}(\epsilon^2)$. So in practice we only had to calculate one critical exponent.

A full theory would require us to work out every critical exponent as a power-series in ϵ and, although we shall not give every detail, in this chapter we shall explain how this is done. In the process we shall show how the field theory can be renormalized by the introduction of renormalized versions of the mass, the coupling constant and the scalar field. We then illustrate the use of the ϵ-expansion by using it to calculate one of the critical exponents (η) to order ϵ^2.

In this chapter we conclude our study of the application of renormalization group to microscopic systems by examining it in its most abstract and general form, as first formulated in quantum field theory. We shall find it convenient to work in a field-theoretic notation in order to explain the background. We begin by discussing the role of the Ginsburg–Landau model, which we have studied in Chapter 9, as a quantum field theory. Then, after considering the technical aspects of divergences and regularization, we present the basic idea of renormalization invariance and re-derive the Callan–Symanzik equations. In the process, we draw on the results of Chapter 9 to give an example of the Callan–Symanzik beta-function. After that we show how the techniques of quantum field theory may be used to recover the results of Chapter 9, but in a more concise, general and elegant way.

The remaining sections of the chapter are then devoted to perturbation theory in wave number (k) space, using renormalization, the ϵ-expansion and dimensional regularization to produce tractable calculations which can (in principle) be carried out to any order in ϵ. However, before beginning on this programme, we shall make some general remarks about the issues involved.

10.1 Preliminary remarks

10.1.1 Changes of notation

As we are now entering the territory of quantum field theory, there are definite advantages in changing to a more usual notation. However, there are two points which should be highlighted before we begin.

First, as we shall see in Section 10.2, changing to field theoretic notation means replacing the parameter u of Ginsburg–Landau theory by the square of the bare mass m_0^2. It may seem rather strange to work with a mass when we are not discussing particles but instead some phase transition. However, it is only a convention, and it simultaneously allows us to understand some quantum field theory as a bonus! Nevertheless, we should bear in mind that the parameter u is zero at the critical temperature and this means that our equivalent quantum field theory is one dealing with a massless particle. As we shall see in Section 10.3, this should alert us to the fact that we shall have to deal with an infrared (i.e. $k = 0$) divergence at the critical point.

Second, as we shall see in Section 10.4, a full progamme of renormalization requires the replacement of bare quantities m_0, λ_0, and ϕ_0 by renormalized (i.e. "exact" to some order in perturbation theory) quantities m, λ, and ϕ. Initially this is just a matter of relabeling these quantities where they occur in the Hamiltonian density and thereafter using the new notation. However, in the main part of this chapter, we do not actually work with the variable ϕ but rather with its statistical correlations. This means that the notational change can be accommodated by taking, for example $G_0^{(2)}(k)$ as the bare correlation (or propagator) and $G^{(?)}(k)$ as its renormalized counterpart. This allows us to leave the notation for fields unchanged in Section 10.5, where we continue to use only the notation $\langle \ldots \rangle_0$ to distinguish the bare case from $\langle \ldots \rangle$ as the exact case. This has the twin virtues of allowing us to compare the work of Section 10.5 with that of Chapter 9 and also avoiding a plethora of zero subscripts.

10.1.2 "Regularization" versus "renormalization"

Regularization is essentially a mathematical trick for handling divergent integrals in a theory. The basic idea is to introduce some temporary method of suppressing the divergence and then restore the *status quo* at the end of the calculation, when (one hopes) it does not matter. Usually one can introduce a fictitious mass, or a cut-off scale of some kind, and then set the fictitious quantity equal to zero after one has performed the integrals and (typically) summed the series. We shall discuss both those methods in Section 10.3 as being representative: other methods such as Schwinger regularization or Pauli–Villars regularization are merely more elegant versions of the basic idea.

However, we give most attention to dimensional regularization which is really quite a different technique and is crucial to the evaluation of integrals in the ϵ-expansion. This relies on the fact that certain divergent integrals depend on the dimension of the space and are in fact well-behaved in a nonintegral dimension. This is explained in detail in Section 10.3.2 and an example of its application is given in 10.3.3.

In contrast, renormalization may have a physical interpretation and may reduce the degree of divergence or even remove it altogether. For instance, in Chapter 1, we discussed the screening effect of other electrons on any one in an electron gas. This has the effect of modifying the relatively slowly varying Coulomb potential into a much more integrable exponential or screened potential. Later on in this chapter we shall see that it is possible to renormalize the mass and coupling constant in order to eliminate the ultraviolet divergence.

10.2 The Ginsburg–Landau model as a quantum field theory

In Section 1.4 we gave a brief introduction to quantum field theory in the form of the Klein–Gordon equation; both with, and without, interactions. From eqn (1.114) we deduce that the Green's function of the Klein–Gordon equation without interactions takes the form

$$G(k) = \frac{1}{k^2 + m_0^2},$$

where k is Fourier wave number and m_0 is the mass of the associated particle.[43]

In the theory of the Klein–Gordon equation *with* interactions, the above quantity is the bare propagator and eqn (1.111) (or, more strictly, its Fourier transform) is the perturbation series for the *exact propagator* in terms of the *bare propagator*.

In the previous chapter, we worked with a partition functional which was intended to provide a model system which had characteristics of both the microscopic Ising model and the macroscopic Landau model. From eqn (9.60), we can see that the "bare" propagator (which in this context is known as the Gaussian propagator) is given by

$$G^{(2)}(\mathbf{k}) = \frac{1}{\alpha k^2 + u}.$$

Hence the Klein–Gordon bare propagator is the same as the Ginsburg–Landau Gaussian propagator provided we make the identifications

$$\alpha = 1 \quad \text{and} \quad u = m_0^2.$$

We can also note that the Hamiltonian density for the Ginsburg–Landau model is formally the same as the Lagrangian density of the Klein–Gordon equation with interactions, once we have made the above identification of the constants.

10.3 Infrared and ultraviolet divergences

We have already discussed this topic in Section 1.4.4, but we repeat the main points here for convenience.

In our expansions, the single-point propagator $\Delta(0) \equiv \Delta(x - y)|_{x=y}$ crops up and from our Gaussian model[44] we have:

$$\Delta(0) = \int_{0 \le k \le \Lambda} G^{(2)}(\mathbf{k}) d^d k = \int_{0 \le k \le \Lambda} \frac{d^d k}{k^2 + u} \sim \int_0^\Lambda \frac{k^{d-1} dk}{k^2 + u}. \tag{10.1}$$

---

[43] We now identify this as the "bare" mass and re-name it m_0.

[44] In Section 9.8 we concluded that α was an irrelevant field and set it equal to unity.

1. For large Λ/u, the integral varies as Λ^{d-2}; where $\Lambda = 2\pi/a$ and a is the lattice constant. In quantum field theory one takes the continuum limit $a \to 0$ and so $\Lambda \to \infty$, and the integral diverges for $d > 2$. We refer to this as an *ultraviolet divergence*.

2. On the other hand, if $d \leq 2$ and $u = 0$ (zero mass), the integral diverges as $k \to 0$. This is known as an *infrared divergence*.

10.3.1 Example: the photon propagator

This is of course an example of a particle with zero mass. The infrared divergence can be cured by adding a fictitious mass to the photon. As the symbol u stands for the square of the mass, we denote the square of the fictitious photonic mass by u_{ph} and so we have:

$$\Delta(0) = \int_0^\Lambda \frac{k^{d-1}dk}{k^2 + u_{ph}}.$$

Then, at the end of the calculation, we take the limit $u_{ph} \to 0$. This is an example of *regularization*. We can also multiply by a function which depends on the cut-offs: this is known as a *convergence factor*. We have already met these techniques in our discussion of the use of perturbation theory to derive the Debye–Hückel theory in Section 4.6. In quantum field theory we may also resort to a technique known as *dimensional regularization*, and we now introduce this as follows.

10.3.2 Dimensional regularization

We take as an example a standard form of integral which occurs in quantum electrodynamics. We work in a four-dimensional space–time, with coordinates (x^0, x^1, x^2, x^3), and the inner product is defined as

$$x^2 = x_\alpha x^\alpha = (x^0)^2 - (x^1)^2 - (x^2)^2 - (x^3)^3 \equiv (x^0)^2 - x^i x^i, \tag{10.2}$$

where Greek indices take values $\alpha = 0, 1, 2, 3$, whereas Latin indices take values $i = 1, 2, 3$. The choice of signs is just a convention and the signs may be reversed.

Originally, the spacelike time coordinate was interpreted as $x^0 = ict$, where $i = \sqrt{-1}$, c is the speed of light *in vacuo*, and t is the time, but this is now seen as rather old-fashioned.

Integrals occur in forms like:

$$\int \frac{d^4k}{(k^2 - s + i\delta)^n} = i\pi^2(-1)^n \frac{\Gamma(n-2)}{\Gamma(n)} \frac{1}{s^{n-2}}, \quad n \geq 3, \tag{10.3}$$

where $\Gamma(n)$ is the Gamma function, and eventually one takes the limit $\delta \to 0$.

Now generalize from 4-dimensional to d-dimensional space, where d is a positive integer. The metric tensor of the old space is:

$$g^{00} = -1, \ g^{ii} = 1, \ i = 1, 2, 3, \ g^{\alpha\beta} = 0, \ \alpha \neq \beta. \tag{10.4}$$

while the new space has:

$$g^{00} = -1, \ g^{ii} = 1, \ i = 1, 2, 3, \ldots, d - 1, \ g^{\alpha\beta} = 0, \ \alpha \neq \beta. \tag{10.5}$$

The four-vector k^α becomes a d-vector, thus:

$$k^\alpha \equiv (k^0, k^1, k^2, \ldots, k^{d-1}), \tag{10.6}$$

with inner product

$$k^2 = k_\alpha k^\alpha = (k^0)^2 - \sum_{i=1}^{d-1} (k^i)^2. \tag{10.7}$$

Integrals in d dimensions are taken with respect to the volume element given by:

$$d^d k = dk^0 dk^1 dk^2, \ldots, dk^{d-1}. \tag{10.8}$$

The equation for the integral becomes:

$$\int \frac{d^d k}{(k^2 - s + i\delta)^n} = i\pi^{d/2}(-1)^n \frac{\Gamma(n - d/2)}{\Gamma(n)} \frac{1}{s^{n-d/2}}, \tag{10.9}$$

for *integer* values of $n > d/2$.

For the special case where $n = d/2$ (e.g. $d = 4$, $n = 2$), we have the following situation:

- The left-hand-side is log divergent.

- The right-hand-side is singular due to the pole of $\Gamma(z)$ at $z = 0$.

However, for noninteger d, the right hand side is well-behaved and finite. Accordingly, we use the right-hand-side to define a *generalization* of the integral on the left-hand-side to noninteger d. In particular, we take $d = 4 - \epsilon$, where $0 \leq \epsilon \leq 1$, and for a 4-dimensional result, we simply let $\epsilon \to 0$.[45]

10.3.3 Examples of regularization

Suppose we want to evaluate the integral

$$\Pi(s) = \int \frac{d^4 k}{(k^2 - s + i\delta)^2}, \tag{10.10}$$

which is just a specific case of the form in (10.3), with the choice $n = 2$. We do it by the cut-off method first. If we impose a large but finite cut-off Λ, then we see that the integral behaves like

$$\Pi \sim \int^\Lambda \frac{k^3 dk}{k^4} \sim \int^\Lambda \frac{dk}{k} \sim \ln \Lambda. \tag{10.11}$$

For $\Lambda \to \infty$, this is log divergent.

---

[45] Contrast this with the work of the previous chapter where we worked with $d = 4 - \epsilon$ and then let $\epsilon \to 1$ for a 3-dimensional result.

Alternatively, multiply through by a convergence factor,

$$f = \left(\frac{-\Lambda^2}{k^2 - \Lambda^2} \right),\qquad(10.12)$$

where $\lim_{\Lambda \to \infty} f(k^2, \Lambda^2) \to 1$, and evaluate:

$$\Pi_\Lambda(s) = \int \frac{d^4 k}{(k^2 - s + i\delta)^2} \cdot \frac{-\Lambda^2}{k^2 - \Lambda^2}.\qquad(10.13)$$

Now $\Pi_\Lambda(s)$ is still log divergent as $\Lambda \to \infty$. *But* after renormalization[46] $s_0 \to s$, thus we have to deal with the difference:

$$\Pi_\Lambda(s) - \Pi_\Lambda(s_0),$$

and it can be shown that

$$\lim_{\Lambda \to \infty} \{ \Pi_\Lambda(s) - \Pi_\Lambda(s_0) \} = -i\pi^2 \ln(s/s_0)\qquad(10.14)$$

with a well-defined limit, independent of Λ.

Now use the dimensional method: generalize the integral formula eqn (10.9) to $d = 4 - \epsilon$,

$$\Pi_\epsilon(s) = K^\epsilon \int \frac{d^{4-\epsilon} k}{(k^2 - s + i\delta)^2} = i\pi^{2-\epsilon/2} \frac{\Gamma(\epsilon/2)}{\Gamma(2)} \left(\frac{s}{K^2} \right)^{-\epsilon/2},\qquad(10.15)$$

where K is a mass scale which keeps the natural dimension of $\Pi_\epsilon(s)$ independent of ϵ.

At this stage we call upon two standard mathematical identities which, for $\epsilon \to 0$, may be written as

$$\Gamma(\epsilon/2) = \frac{2}{\epsilon} - \gamma + \cdots,$$

where $\gamma \equiv$ Euler's constant, and

$$x^{-\epsilon/2} = 1 - \frac{\epsilon}{2} \ln x + \cdots,$$

and put $x = (s/K^2)$ in order to work out the last term on the right-hand-side of (10.15). Hence we find

$$\Pi_\epsilon(s) = i\frac{2\pi^2}{\epsilon} - i\pi^2 \left[\gamma + \ln\left(\frac{\pi s}{K^2} \right) \right],\qquad(10.16)$$

and so

$$\Pi_\epsilon(s) - \Pi_\epsilon(s_0) = i\frac{2\pi^2}{\epsilon} - i\pi^2 \left[\gamma + \ln\left(\frac{\pi s}{K^2} \right) \right]$$

$$- i\frac{2\pi^2}{\epsilon} + i\pi^2 \left[\gamma + \ln\left(\frac{\pi s_0}{K^2} \right) \right]$$

$$= -i\pi^2 \ln(s/s_0),\qquad(10.17)$$

which is the same as for the cut-off method.

---

[46] We shall discuss this in detail in Section 10.8.

The fact that predictions are independent of the mass scale can be seen as the basis of renormalization group (RG) in quantum field theory. It is also worth noting that the technique of dimensional regularization is used to evaluate the integrals necessary to carry out a full ϵ-expansion. We will look at this further in Section 10.9.

10.4 Renormalization invariance

As we have just seen the problem of diverging integrals can be solved by introducing a cutoff, but this introduces a new scale to the problem. For a universally valid theory this procedure must not depend in any significant way on an *ad hoc* choice of this new scale. This requirement motivated the introduction of the hypothesis of *renormalization invariance*, which essentially states that there is no dependence on the choice of the cutoff scale. This invariance can be tested by scaling transformations, and Stuckelberg and Petermann first recognized that a set of such transformations makes up a continuous group: hence the name *renormalization group*. Later Gell-Mann and Low used the idea of renormalization invariance in QED to find asymptotic expressions for photon Green's functions.

In the 1970s, these ideas were revived by Callan and Symanzik who derived what are now called the Callan–Symanzik equations.

10.4.1 The Callan–Symanzik equations

Let us consider the Hamiltonian density $h(x)$ in quantum field-theoretic form:

$$h(x) = \frac{1}{2}(\nabla \phi_0)^2 + \frac{1}{2}m_0^2 \phi_0^2 + \frac{\lambda_0}{4!}\phi_0^4. \qquad (10.18)$$

This is just eqn (9.13) with ϕ replaced by ϕ_0. We expect that we are going to renormalize "bare" to "exact" quantities, for instance:

$$m_0 \to m \quad \text{and} \quad \lambda_0 \to \lambda.$$

Then $h(x)$ has the required form invariance under the transformations:

$$m_0^2 \to m^2 = \mathcal{Z}_\phi m_0^2;$$

$$\lambda_0 \to \lambda = \mathcal{Z}_\phi^2 \lambda_0;$$

$$\phi_0 \to \phi \equiv \mathcal{Z}_\phi^{-1/2} \phi_0,$$

where ϕ is the *renormalized field*, $\mathcal{Z}_\phi$ is the *renormalization constant*, and we impose the condition

$$(\nabla \phi) \equiv \text{invariant}.$$

The n-particle "bare" propagator is

$$G_0^{(n)}(\mathbf{k}_1, \mathbf{k}_2, \ldots, \mathbf{k}_n) = \langle \phi_0(\mathbf{k}_1)\phi_0(\mathbf{k}_2) \ldots \phi_0(\mathbf{k}_n) \rangle, \qquad (10.19)$$

and the n-particle "exact" propagator is

$$G^{(n)}(\mathbf{k}_1, \mathbf{k}_2, \ldots, \mathbf{k}_n) = \langle \phi(\mathbf{k}_1)\phi(\mathbf{k}_2)\ldots\phi(\mathbf{k}_n)\rangle. \tag{10.20}$$

Hence, if we use the same functional measure to sum over the fields $\phi(\mathbf{k})$ as the fields $\phi_0(\mathbf{k})$, it follows that $G^{(n)}$ transforms like the product $|\phi|^n$ and so:

$$G_0^{(n)}(\mathbf{k}_1 \ldots \mathbf{k}_n; \lambda_0, m_0, \epsilon) = \mathcal{Z}_\phi^{n/2} G^{(n)}(\mathbf{k}_1 \ldots \mathbf{k}_n; \lambda, m, b, \epsilon). \tag{10.21}$$

Note that the bare propagator is parameterized by an unrenormalized or bare coupling constant λ_0, the bare mass m_0 of the free particle and $\epsilon = d - 4$, which occurs in eqn (9.88). The scaling transformation leads to the exact n-point propagator, in this case parameterized on the renormalized coupling constant λ, the renormalized mass m, the spatial rescaling factor b and (again) $\epsilon = d - 4$.

Now introduce the proper vertex function $\Gamma^{(n)}(\mathbf{k}_1 \ldots \mathbf{k}_n; \lambda_0, m_0, \epsilon)$ which has an inverse relationship to $G^{(n)}$ and which therefore scales inversely. Hence it follows that the above equation can be written as:

$$\Gamma_0^{(n)}(\mathbf{k}_1 \ldots \mathbf{k}_n; \lambda_0, m_0, \epsilon) = \mathcal{Z}_\phi^{-n/2}\Gamma^{(n)}(\mathbf{k}_1 \ldots \mathbf{k}_n; \lambda, m, b, \epsilon) \tag{10.22}$$

where b is referred to as the "scale" of the renormalization point and $\mathcal{Z}_\phi$ is the "renormalization constant." In our own terminology for RG in critical phenomena, b is the scale enlargement ratio or spatial rescaling factor.

The left hand side is independent of b, as λ_0 and m_0 are "free-field" or bare values. So, differentiating both sides of (10.22), for fixed λ_0, m_0, and ϵ, we may write:

$$b\frac{d}{db}\Gamma_0^{(n)} = 0 = b\frac{d}{db}(\mathcal{Z}_\phi^{-n/2}\Gamma^{(n)}). \tag{10.23}$$

By (10.22), and using the chain rule of differentiation (10.23) implies that

$$\left[b\frac{\partial}{\partial b} + b\frac{\partial\lambda}{\partial b}\cdot\frac{\partial}{\partial\lambda} + b\frac{\partial m}{\partial b}\cdot\frac{\partial}{\partial m} - n\gamma_\phi(\lambda, \epsilon)\right]\Gamma^{(n)} = 0,$$

or

$$\left[b\frac{\partial}{\partial b} + \beta(\lambda, \epsilon)\frac{\partial}{\partial\lambda} + \gamma_m(\lambda, \epsilon)\frac{\partial}{\partial m} - n\gamma_\phi(\lambda, \epsilon)\right]\Gamma^{(n)} = 0, \tag{10.24}$$

with:

$$\beta(\lambda, \epsilon) = b\frac{\partial\lambda}{\partial b} = \frac{\partial\lambda}{\partial\ln b}; \tag{10.25}$$

$$\gamma_m(\lambda, \epsilon) = b\frac{\partial m}{\partial b} = \frac{\partial m}{\partial\ln b}; \tag{10.26}$$

$$\gamma_\phi(\lambda, \epsilon) = \mathcal{Z}_\phi^{-1/2}\left[b\frac{\partial}{\partial b}\mathcal{Z}_\phi^{1/2}(\lambda, \epsilon)\right]. \tag{10.27}$$

This is the RGE or Callan–Symanzik equation, where:

- $\beta(\lambda, \epsilon)$ is the Callan–Symanzik beta-function;
- γ_m is the anomalous dimension of m;
- γ_ϕ is the anomalous dimension of ϕ.

We have previously met the beta-function in Section 9.6.4, but attention is drawn to the slight notational change in the form employed here.

The overall objective, as in critical phenomena, is to calculate β as a power series in λ and ϵ for a given Hamiltonian.

10.4.2   Example: the beta-function for ϕ^4 theory in dimension $d < 4$

The Callen–Symanzik beta-function is often studied in quantum field theory, with fixed points being classed as *infrared stable* or *ultraviolet stable*. While our objective here is to obtain values of critical exponents, it may be of general interest to give a brief discussion of a particular example in field theory. In Chapter 9 we obtained beta-functions for the Ginsberg–Landau model for $d < 4$.

As a particular example, we consider eqn (9.100), which we write as

$$\beta(\lambda, \epsilon) = \epsilon\lambda - \frac{3\lambda^2}{16\pi^2},\tag{10.28}$$

and, from (9.107) and (9.108), we have identified the fixed points as:

$$\lambda^* = 0 \quad \text{and} \quad \lambda^* = \frac{16\pi^2\epsilon}{3}.\tag{10.29}$$

The general behavior of the beta-function is easily established. Obviously as λ becomes larger the squared term will dominate, with

$$\beta(\lambda, \epsilon) \propto \lambda^2.$$

We can also obtain the position of the turning point between the two fixed points, as:

$$\frac{\partial\beta}{\partial\lambda} = -\epsilon + \frac{3\lambda}{8\pi^2} = 0$$

for

$$\lambda = \frac{8\pi^2\epsilon}{3}.\tag{10.30}$$

The general behavior of the beta-function is sketched in Fig 10.1.

The beta-function has positive slope at the nontrivial fixed point. This means that a value of λ which starts anywhere between $\lambda = 0$ and $\lambda = 16\pi^2\epsilon/3$ will flow toward the point $\lambda = 16\pi^2\epsilon/3$, as $b \to \infty$. This behavior controls the small-wave number, large-distance behavior of the system and such a fixed point is termed *infrared stable*.

Similarly, as $b \to 0$, a point will flow to $\lambda = 0$ and this fixed point is termed *ultraviolet stable*.

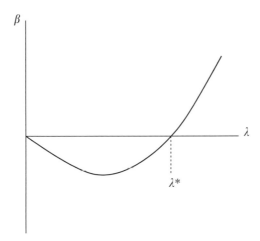

FIG. 10.1 The beta-function for the ϕ^4 theory when $d < 4$.

10.5 Perturbation theory in x-space

We can extend our earlier work on the Gaussian model to set up a more formal approach to the ϕ^4 scalar field theory. From now on we shall use λ_0 rather than λ to denote the bare coupling constant.

10.5.1 The generating functional for correlations

For the Gaussian model, we have the partition functional:

$$\mathcal{Z}_0[J] = \int \mathcal{D}\phi \, e^{-H_0[J]}, \tag{10.31}$$

and for the ϕ^4 or Ginsburg–Landau model

$$\mathcal{Z}[J] = \int \mathcal{D}\phi \, e^{-H[J]} = \int \mathcal{D}\phi \, e^{-H_0[J]} \exp[- \int d^d x \lambda_0 \phi^4 / 4!]. \tag{10.32}$$

Now we expand the second exponential in powers of λ_0, thus:

$$\mathcal{Z}[J] = \int \mathcal{D}\phi e^{-H_0[J]} \left[\sum_{n=0}^{\infty} \frac{1}{n!} \left(-\frac{\lambda_0}{4!} \int d^d x \phi^4 \right)^n \right]. \tag{10.33}$$

Then we exchange the order of summation and integration,

$$\mathcal{Z}[J] = \sum_{n=0}^{\infty} \frac{1}{n!} \int \mathcal{D}\phi e^{-H_0[J]} \left[\left(-\frac{\lambda_0}{4!} \int d^d x \phi^4 \right)^n \right]; \tag{10.34}$$

and multiply above and below by $\mathcal{Z}_0[J]$,

$$\mathcal{Z}[J] = \sum_{n=0}^{\infty} \frac{1}{n!} \mathcal{Z}_0[J] \int \mathcal{D}\phi\, e^{-H_0[J]} \left(\frac{-\lambda_0}{4!} \int \mathrm{d}^d x \phi^4 \right)^n \Big/ \mathcal{Z}_0[J]$$

$$= \mathcal{Z}_0[J] \sum_{n=0}^{\infty} \frac{1}{n!} \left\langle \left(-\frac{\lambda_0}{4!} \int \mathrm{d}^d x \phi^4 \right)^n \right\rangle_0, \tag{10.35}$$

where $\langle \ldots \rangle_0$ denotes an average against the Gaussian distribution.

Two points should be noted:

1. This series is not convergent but it is believed to be asymptotic.

2. We have expressed the full partition function in terms of an expansion in Gaussian correlations: $G_0^{(4)} = \langle \phi^4 \rangle_0$.

10.5.2   Gaussian n-point correlations

Now, as discussed in Section 9.4.1, from linear response theory, in particular eqn (9.61), we have the *bare* n-point propagators:

$$G_0^{(n)}(\mathbf{x}_1 \ldots \mathbf{x}_n) = \frac{1}{\mathcal{Z}_0[J]} \frac{\delta}{\delta J(\mathbf{x}_1)} \cdots \frac{\delta}{\delta J(\mathbf{x}_n)} \mathcal{Z}_0[J] \Big|_{J=0}. \tag{10.36}$$

This allows us to replace ϕ factors in the exponent by differentiating $\delta/\delta J(\mathbf{x})$ and so rewrite eqn (10.33) as:

$$\mathcal{Z}[J] = \sum_{n=0}^{\infty} \frac{1}{n!} \left(-\frac{\lambda_0}{4!} \int \mathrm{d}^d x \frac{\delta^4}{\delta J^4(\mathbf{x})} \right)^n \mathcal{Z}_0[J] = \exp\left[-\frac{\lambda_0}{4!} \int \mathrm{d}^d x \frac{\delta^4}{\delta J^4(\mathbf{x})} \right] \mathcal{Z}_0[J].$$
$$\tag{10.37}$$

Also, from our earlier treatment of the Gaussian model, we have from eqns (9.58) and (9.60),

$$\mathcal{Z}_0[J] = \mathcal{Z}_0[0] \exp\left[\frac{1}{2} \int_{0 \leq k \leq \lambda_0} \mathrm{d}^d k \frac{J(\mathbf{k}) J(-\mathbf{k})}{k^2 + u} \right], \tag{10.38}$$

where

$$\int_{0 \leq k \leq \Lambda} \mathrm{d}^d k \frac{J(\mathbf{k}) J(-\mathbf{k})}{\alpha^2 k^2 + u} = \int \mathrm{d}^d x \int \mathrm{d}^d y\, J(\mathbf{x}) \Delta(\mathbf{x} - \mathbf{y}) J(\mathbf{y}). \tag{10.39}$$

and so (10.38) may be written as

$$\mathcal{Z}_0[J] = \mathcal{Z}_0[0] \exp\left[\frac{1}{2} \int \int \mathrm{d}^d x \mathrm{d}^d y\, J(\mathbf{x}) \Delta(\mathbf{x} - \mathbf{y}) J(\mathbf{y}) \right] = \mathcal{Z}_0[0] \exp\left[\frac{1}{2} J \Delta J \right].$$
$$\tag{10.40}$$

Then, from eqn (10.35), the exact partition functional becomes

$$\mathcal{Z}[J] = \mathcal{Z}_0[0] \exp\left[\frac{-\lambda_0}{4!} \int d^d x \frac{\delta^4}{\delta J^4(\mathbf{x})}\right] \exp\left[\frac{1}{2} \int \int d^d x d^d y J(\mathbf{x}) \Delta(\mathbf{x} - \mathbf{y}) J(\mathbf{y})\right],$$
(10.41)

and, by expanding out the exponentials, we can obtain a power series for $\mathcal{Z}$.
   We should note the following points:

1. Each time a $\delta/\delta J$ acts on the exponential it brings down a factor J. Hence a second $\delta/\delta J$ must act to eliminate this so that we get a nonzero contribution when we set $J = 0$.

2. Thus we only get correlations of *even* numbers of ϕs. Correlations like $\langle \phi \rangle_0$, $\langle \phi^3 \rangle_0$, $\langle \phi^5 \rangle_0$, etc. are zero.

3. In this way the differentiation reproduces the correct behavior for the evaluation of Gaussian correlations.

10.5.3 Wick's theorem: evaluation of the $2p$-point Gaussian correlation

Consider the $2p$-point correlation evaluated against the Gaussian (zero-order) distribution:

$$G_0^{(2p)}(\mathbf{x}_1 \ldots \mathbf{x}_{2p}) = \langle \phi(\mathbf{x}_1) \ldots \phi(\mathbf{x}_{2p}) \rangle_0.$$
(10.42)

Then the preceding formalism gives us

$$\left\langle \prod_{i=1}^{2p} \phi(\mathbf{x}_i) \right\rangle_0 = \prod_{i=1}^{2p} \frac{\delta}{\delta J(\mathbf{x}_i)} \exp\left[\frac{1}{2} J \Delta J\right]$$

$$= \sum_{\{1,2,\ldots,2p\}} \Delta(\mathbf{x}_1 - \mathbf{x}_2) \Delta(\mathbf{x}_3 - \mathbf{x}_4) \ldots \Delta(\mathbf{x}_{2p-1} - \mathbf{x}_{2p}),$$
(10.43)

where the notation $\{1, 2, \ldots, 2p\}$ indicates that the summation is over all possible pairings of that set. In field theory, this is known as *Wick's theorem*.
   For a specific example, we consider $p = 2$, $2p = 4$, thus:

$$\langle \phi(\mathbf{x}_1)\phi(\mathbf{x}_2)\phi(\mathbf{x}_3)\phi(\mathbf{x}_4) \rangle_0 = \Delta(\mathbf{x}_1 - \mathbf{x}_2) \Delta(\mathbf{x}_3 - \mathbf{x}_4)$$

$$+ \Delta(\mathbf{x}_1 - \mathbf{x}_3) \Delta(\mathbf{x}_2 - \mathbf{x}_4)$$

$$+ \Delta(\mathbf{x}_1 - \mathbf{x}_4) \Delta(\mathbf{x}_2 - \mathbf{x}_3).$$
(10.44)

Or, alternatively, one can think of a set of pairwise contractions:

$$\langle \phi(\mathbf{x}_1)\phi(\mathbf{x}_2)\phi(\mathbf{x}_3)\phi(\mathbf{x}_4) \rangle_0 = \langle \phi(\mathbf{x}_1)\phi(\mathbf{x}_2) \rangle_0 \langle \phi(\mathbf{x}_3)\phi(\mathbf{x}_4) \rangle_0$$

$$+ \langle \phi(\mathbf{x}_1)\phi(\mathbf{x}_3) \rangle_0 \langle \phi(\mathbf{x}_2)\phi(\mathbf{x}_4) \rangle_0$$

$$+ \langle \phi(\mathbf{x}_1)\phi(\mathbf{x}_4) \rangle_0 \langle \phi(\mathbf{x}_2)\phi(\mathbf{x}_3) \rangle_0,$$
(10.45)

which is just the same result.

Then $2p = 6$ can be expressed as three sets of pairwise contractions, and so on for $2p = 8$ and higher orders. In statistics this property is sometimes summed up in the phrase: "a Gaussian distribution can be expressed in terms of its second moment." In fact this is just the same property as we used for the classical problem in Chapter 5: see eqn (5.32).

10.6   Perturbation expansion in x-space

As we have seen in earlier chapters, in many-body perturbation expansions the number of terms increases rapidly with the order. Thus, even when one has a small expansion parameter, a series may not be convergent because of the rapid build-up of terms: what one might call a *combinational divergence*. In this chapter, we no longer truncate the expansion at first-order (as we did in Chapter 9) but try to establish general properties of the series, some of which will be valid to all orders. In this section we find out that one class of terms is exactly cancelled when we also expand the normalization of the ensemble average. In Section 10.7 we go into wave number space and establish some other general results.

10.6.1   Evaluation of the exact n-point correlation

To obtain the *exact* n-point correlation, we simply differentiate $\mathcal{Z}[J]$, as given by (10.41), n times with respect to J and then set $J = 0$. Hence we have:

$$G^{(n)}(\mathbf{x}_1 \ldots \mathbf{x}_n) = \text{sum of an expansion in bare two-point Gaussian correlations.}$$

More formally, we make use of eqn (9.61), with $\mathcal{Z}_0[J]$ replaced by $\mathcal{Z}[J]$, thus:

$$G^n(\mathbf{x}_1, \ldots, \mathbf{x}_n) = \frac{1}{\mathcal{Z}[J]} \frac{\delta}{\delta J(\mathbf{x}_1)} \cdots \frac{\delta}{\delta J(\mathbf{x}_n)} \mathcal{Z}[J]. \tag{10.46}$$

10.6.2   Example: the two-point correlation

Consider the exact two-point correlation

$$G^{(2)}(\mathbf{x}_1, \mathbf{x}_2) = \langle \phi(\mathbf{x}_1)\phi(\mathbf{x}_2) \rangle. \tag{10.47}$$

We first do this symbolically[47] to zero order in λ_0:

$$\langle \phi(\mathbf{x}_1)\phi(\mathbf{x}_2) \rangle = \frac{1}{\mathcal{Z}[J]} \frac{\delta}{\delta J(\mathbf{x}_1)} \frac{\delta}{\delta J(\mathbf{x}_2)} \mathcal{Z}[J] \Big|_{J=0}$$

$$= \frac{\mathcal{Z}_0[0]}{\mathcal{Z}[J]} \frac{\delta}{\delta J(\mathbf{x}_1)} \frac{\delta}{\delta J(\mathbf{x}_2)} \exp\left[\frac{-\lambda_0}{4!} \int \frac{\delta^4}{\delta J^4(\mathbf{x})} d^d\mathbf{x} \right] \exp\left[\frac{1}{2} J\Delta J \right] \Big|_{J=0}$$

$$= \frac{\mathcal{Z}_0[0]}{\mathcal{Z}[J]} \frac{\delta}{\delta J(\mathbf{x}_1)} \frac{\delta}{\delta J(\mathbf{x}_2)} [1 + \mathcal{O}(\lambda_0)] \exp\left[\frac{1}{2} J\Delta J \right] \Big|_{J=0}. \tag{10.48}$$

---

[47] That is, we drop various vector labels in order to keep the algebra as simple as possible.

Now, in the denominator:

$$\mathcal{Z}[J] = \mathcal{Z}_0[J][1 + \mathcal{O}(\lambda_0)]. \tag{10.49}$$

Doing the differentiation, and setting $J = 0$, we have:

$$\langle \phi(\mathbf{x}_1)\phi(\mathbf{x}_2) \rangle = \frac{\mathcal{Z}_0[0]}{\mathcal{Z}_0[0]} \cdot \frac{[1 + \mathcal{O}(\lambda_0)]}{[1 + \mathcal{O}(\lambda_0)]} \times 2 \times \frac{1}{2} \times \Delta(\mathbf{x}_1 - \mathbf{x}_2)$$

$$= \Delta(\mathbf{x}_1 - \mathbf{x}_2). \tag{10.50}$$

Hence the zero-order contribution to the *exact* propagator is the *bare* propagator, or:

$$\langle \phi(\mathbf{x}_1)\phi(\mathbf{x}_2) \rangle = \Delta(\mathbf{x}_1 - \mathbf{x}_2) + \mathcal{O}(\lambda_0). \tag{10.51}$$

Note that $\mathbf{x}_1$ and $\mathbf{x}_2$ must appear as free variables on both sides of the equation. Thus one must have integrals over the other variables.

Next, to order λ_0 we have:

$$G^{(2)}(\mathbf{x}_1, \mathbf{x}_2) = \frac{\mathcal{Z}_0[0]}{\mathcal{Z}[J]} \left[\Delta(\mathbf{x}_1 - \mathbf{x}_2) - \frac{\lambda_0}{4!} \frac{\delta^2}{\delta J(\mathbf{x}_1)\delta J(\mathbf{x}_2)} \int d^d \mathbf{x} \frac{\delta^4}{\delta J^4(\mathbf{x})} e^{\left[\frac{1}{2} \int J\Delta J\right]} + \cdots \right]_{J=0}$$

$$= \frac{\mathcal{Z}_0[0]}{\mathcal{Z}[J]} \left[\Delta(\mathbf{x}_1 - \mathbf{x}_2) - \frac{\lambda_0}{8} \left(\int d^d \mathbf{x} \Delta^2(\mathbf{x} - \mathbf{x}) \right) \Delta(\mathbf{x}_1 - \mathbf{x}_2) \right.$$

$$\left. - \frac{\lambda_0}{2} \int d^d \mathbf{x} \Delta(\mathbf{x}_1 - \mathbf{x})\Delta(\mathbf{x} - \mathbf{x})\Delta(\mathbf{x} - \mathbf{x}_2) + \cdots \right]_{J=0}. \tag{10.52}$$

Take out $\Delta(\mathbf{x}_1 - \mathbf{x}_2)$ as a common factor:

$$G^{(2)}(\mathbf{x}_1, \mathbf{x}_2) = \frac{\mathcal{Z}_0[0]\Delta(\mathbf{x}_1 - \mathbf{x}_2)}{\mathcal{Z}[0]} \left[1 - \frac{\lambda_0}{8} \left[\int d^d \mathbf{x} \Delta^2(\mathbf{x} - \mathbf{x}) \right] \right.$$

$$\left. - \frac{\lambda_0}{2\Delta(\mathbf{x}_1 - \mathbf{x}_2)} \int d^d \mathbf{x} \Delta(\mathbf{x}_1 - \mathbf{x})\Delta(\mathbf{x} - \mathbf{x})\Delta(\mathbf{x} - \mathbf{x}_2) + \cdots \right]. \tag{10.53}$$

Now consider the expansion of $\mathcal{Z}[0]$ in powers of λ_0:

$$\mathcal{Z}[0] = \mathcal{Z}_0[0] \sum_{n=0}^{\infty} \frac{1}{n!} \left\langle \left(-\frac{\lambda_0}{4!} \int d^d \mathbf{x} \phi^4(\mathbf{x}) \right)^n \right\rangle_0$$

$$= \mathcal{Z}_0[0] \left[1 - \frac{\lambda_0}{4!} \int d^d \mathbf{x} \langle \phi^4(\mathbf{x}) \rangle_0 + \frac{\lambda_0^2}{2!4!} \left\langle \left(-\int d^d \mathbf{x} \phi^4(\mathbf{x}) \right)^2 \right\rangle_0 \cdots \right]. \tag{10.54}$$

To order λ_0, from Wick's theorem we have:

$$\langle \phi^4(\mathbf{x}) \rangle_0 = 3\langle \phi^2(\mathbf{x}) \rangle_0^2 = 3\Delta^2(\mathbf{x} - \mathbf{x}), \tag{10.55}$$

and so

$$\mathcal{Z}[0] = \mathcal{Z}_0[0]\left[1 - \frac{\lambda_0}{8}\int d^d x \Delta^2(x-x) + \cdots\right].\tag{10.56}$$

Substituting this back into the expansion for $G^{(2)}$ we obtain an expression of the form:

$$G \sim \frac{1 + \lambda_0 A + \lambda_0 B}{1 + \lambda_0 A} \sim (1 + \lambda_0 A + \lambda_0 B)(1 - \lambda_0 A)$$

$$= (1 + \lambda_0 A - \lambda_0 A + \lambda_0 B + \mathcal{O}(\lambda_0^2)) = 1 + \lambda_0 B.\tag{10.57}$$

Hence:

$$G^{(2)}(x_1, x_2) = \Delta(x_1 - x_2) + \frac{\lambda_0}{2}\int d^d x \Delta(x_1 - x)\Delta(x - x)\Delta(x - x_2) + \mathcal{O}(\lambda_0^2).\tag{10.58}$$

This is starting to take the form:

Exact propagator = bare propagator + expansion in convolutions of bare propagators.

At each order, the "convolution" must "propagate" from x_1 to x_2. At first order, we had a term which was not of this form but this was cancelled by expanding $\mathcal{Z}[0]$ in the denominator.

 This leads on to a general result: terms which do not "propagate" are cancelled to all orders by expansion of the normalization:

$$N = \mathcal{Z}^{-1}[0].\tag{10.59}$$

In diagrams, such terms are known as *vacuum bubbles* and we shall consider these further in Section 10.6.3.

 This procedure becomes more complicated at higher orders, but now that we know the rules, we can drop the differentials and use ϕs with contractions instead. For example, at $\mathcal{O}(\lambda_0^2)$, we have

$$\langle \phi(x_1)\phi(x_2)\frac{\lambda_0^2}{2!}\cdot\frac{1}{4!}\int d^d v \phi^4(v)\cdot\frac{1}{4!}\int d^d w \phi^4(w)\rangle_0.$$

 Apply Wick's theorem:

1. Contract $\phi(x_1)$ with any ϕ in either of the two interaction terms: this gives eight choices.

2. Contract $\phi(x_2)$ with any ϕ in the unchosen interaction term: this gives four choices.

3. The three remaining fields in either interaction term can be paired off with any permutation of the other three fields: this gives 3! choices.

Hence the numerical factor associated with this term is:

$$\frac{1}{2!}\frac{1}{(4!)^2} \times 8 \times 4 \times 3! = \frac{1}{6},$$ (10.60)

and the term is:

$$\frac{1}{6}\lambda_0^2 \int d^d v \int d^d w \, \Delta(\mathbf{x}_1 - \mathbf{v})\Delta^3(\mathbf{v} - \mathbf{w})\Delta(\mathbf{w} - \mathbf{x}_2).$$

There are two other terms at this order but, instead of treating them algebraically, at this stage it is helpful to introduce Feynman diagrams.

10.6.3   Feynman diagrams

It should now be clear that terms of any order can be worked out by following a simple set of rules, with the numerical coefficients being determined by purely combinatoric operations. For this reason it is only the "topology" of a term which matters, as we have seen in Chapters 4 and 5, and this fact can be exploited if we replace the algebraic terms by diagrams (sometimes called graphs).

We may construct these diagrams by following a set of rules:

1. Represent a coordinate (e.g. $\mathbf{x}$ or $\mathbf{x}_1$) by a dot.

2. Represent a propagator Δ by a line segment joining two dots.

3. Any dot which is common to two or more segments is a variable which is integrated over.

Using these rules one can write down the perturbation series for any propagator to any order as a series of diagrams. As an example diagrams in the graphical expansion for $G^{(2)}(\mathbf{x}, \mathbf{y})$ are shown to order λ_0^2 in Fig 10.2. The first two terms on the right hand side correspond to the two terms on the right hand side of eqn (10.58). The first second-order diagram corresponds

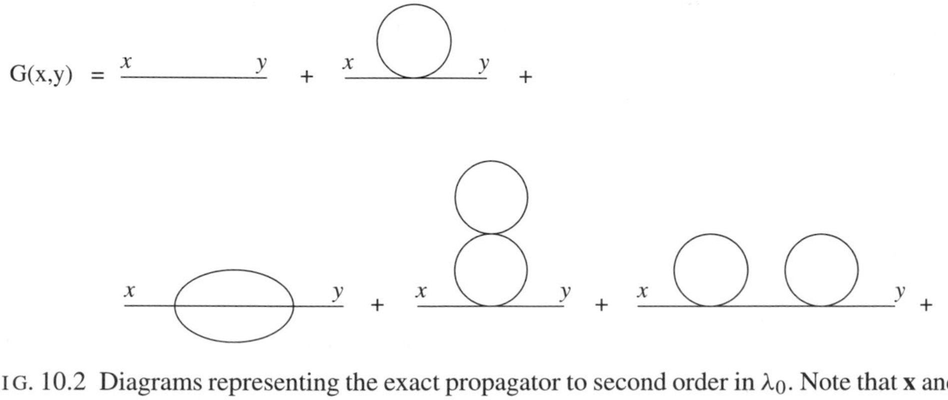

FIG. 10.2 Diagrams representing the exact propagator to second order in λ_0. Note that $\mathbf{x}$ and $\mathbf{y}$ in the figure correspond to $\mathbf{x}_1$ and $\mathbf{x}_2$ in the text.

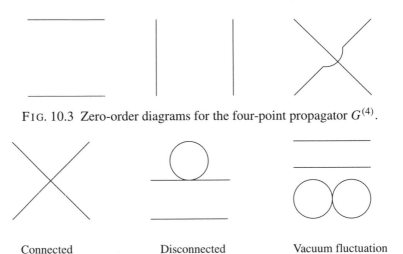

FIG. 10.3 Zero-order diagrams for the four-point propagator $G^{(4)}$.

Connected   Disconnected   Vacuum fluctuation

FIG. 10.4 The first-order diagrams in the expansion of $G^{(4)}$.

to the second-order term given after eqn (10.60). This is an example of the famous Feynmann diagrams.[48]

As another example, we shall work out the zero-order part of $G^4(\mathbf{x}_1, \mathbf{x}_2, \mathbf{x}_3, \mathbf{x}_4)$, purely by diagrams. This is done as follows

1. Put down 4 points.

2. Join them up pairwise in all possible ways.

The resulting zero-order diagrams in the expansion for $G^{(4)}(\mathbf{x}_1, \mathbf{x}_2, \mathbf{x}_3, \mathbf{x}_4)$ are given in Fig. 10.3, while the first-order diagrams are given in Fig. 10.4. An interesting feature of the first-order diagrams is the presence of three different types of diagrams, thus: *connected diagram, disconnected diagram*, and a *vacuum fluctuation*. We shall deal with the first two types of diagram in Section 10.7.

10.6.4 Vacuum fluctuations or bubbles

If we look at our expression for $G^{(2)}$ at order λ_0, as given by eqn (10.53), we have two first-order terms, one of which was cancelled by expanding the normalization and hence does not appear in eqn (10.58). This latter term is known as a *vacuum fluctuation* or, in diagrams as a *vacuum bubble*. Diagramatically we can represent both of the $\mathcal{O}(\lambda_0)$ terms as shown in Fig. 10.5.

We can then write down the vacuum fluctuations to second order as shown in Fig. 10.6. This is the perturbation expansion of the normalization $\mathcal{Z}[0]$.

---

[48] Actually diagrammatic methods of handling terms in the many-body perturbation series were introduced some years before the Feynmann diagrams. In Chapter 4 we discussed Mayer diagrams and in the literature the reference [18] is normally cited, but in fact it is the second part of that work where they first appeared. It is only in Mayer and Mayer that something like the familiar modern form appears.

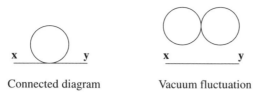

Connected diagram   Vacuum fluctuation

FIG. 10.5 Both first-order diagrams for $G^{(2)}$ showing the vacuum bubble.

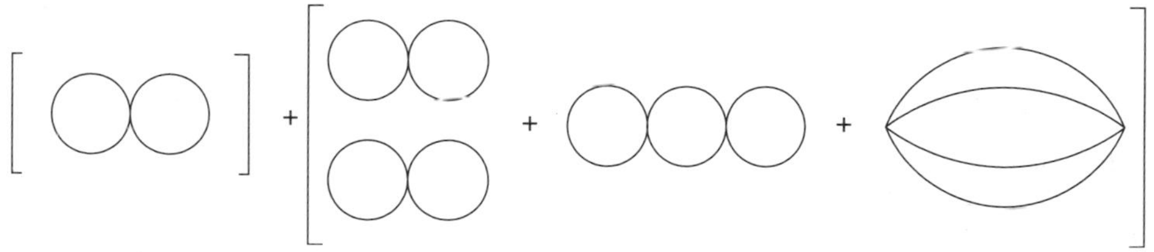

FIG. 10.6 Equivalence of the sum of the vacuum bubbles and the normalization to second order.

10.7 Perturbation expansion in k-space

We will continue to use diagrams as necessary. A helpful feature is that the diagrams remain the same in both real and wave number space. Only the labeling is different. For instance, now every line corresponds to a wave vector, instead of joining two positions or coordinates.

Our first-order expression for $G^{(2)}(\mathbf{x}_1, \mathbf{x}_2)$ as given by eqn (10.58), can be re-written in wave number space as

$$G^{(2)}(k) = G_0(k) + \frac{\lambda_0}{2} G_0(k) \left[\int d^d q\, G_0(q)\right] G_0(k), \qquad (10.61)$$

where q is a dummy wave number. This result follows readily from Fourier transformation and the use of the convolution theorem.

Rules for labeling diagrams in k-space:

1. A vertex $\leftrightarrow$ a factor λ_0.

2. A line $\leftrightarrow$ a factor $G_0(k)$.

3. A "loop" $\leftrightarrow$ an integral $\int d^d q$.

4. Every graph is multiplied by a symmetry factor, for example, as worked out in eqn (10.60).

As another example we can write down the k-space form of our term at $\mathcal{O}(\lambda_0^2)$, as given in the Section 10.5.1 after eqn (10.60),

$$G_2^{(2)}(k) = \frac{\lambda_0^2}{6} G_0(k) \left[\int d^d q \int d^d p\, G_0(q) G_0(p) G_0(k - q - p)\right] G_0(k) \qquad (10.62)$$

$+ 2$ others as in x-space. The diagram for this term is shown in Fig. 10.7.

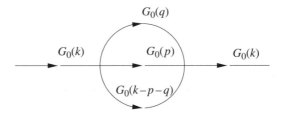

FIG. 10.7 The "Saturn" diagram in wave number space.

10.7.1 Connected and disconnected diagrams

We may make further progress by noting (e.g. from, Fig. 10.4) that in addition to the vacuum fluctuations, some diagrams are *disconnected*: that is, made up of a number of sub-diagrams which are not connected together.

The correspondence (in real space) between terms and diagrams is:

- some terms multiplied together ↔ *disconnected diagram*

- all terms convolved together ↔ *connected diagram*.

For example, in Fig. 10.4, we distinguish "connected" diagrams form those that are "disconnected" and also from vacuum fluctuations.

It can be shown that the expansion of $\ln \mathcal{Z}[J]$ contains only *connected diagrams*. Hence there is an equivalence between *connected correlations*,[49] as defined earlier, and *connected diagrams*. We highlight this point as follows:

- $\ln \mathcal{Z}[J]$ is the generating functional for connected correlations $G_c^{(n)}(\mathbf{k}_1 \ldots \mathbf{k}_n)$.

To sum up, we can reduce the size of the problem by concentrating only on connected correlations and then disconnected diagrams can be ignored.

10.7.2 Reducible and irreducible diagrams

In principle, we now have $\mathcal{Z}[J]$ as a power series in λ_0. This also applies to $\ln \mathcal{Z}$ and the free energy F. Hence all critical exponents can be calculated as power series in λ_0. We can improve the properties of these expansions by various forms of partial summation.

To do this, we need to identify the basic "building blocks," and we begin with the fact that diagrams are reducible or irreducible, thus:

- *One-particle reducible* $(1 - \mathrm{PR})$ diagram: if cutting a single internal line produces two disconnected diagrams.

- *One-particle irreducible* $(1 - \mathrm{PI})$ diagram: if it cannot be separated by cutting a single internal line.

---

[49] Note that in statistics, connected correlations are known as cumulants.

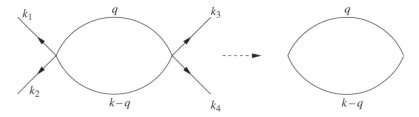

FIG. 10.8 Creation of an "amputated" diagram.

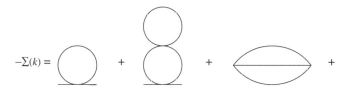

FIG. 10.9 Expansion for the self energy.

This leads on to the important building block known as the proper vertex, thus:

- *Proper vertex*: 1 − PI diagram with its *external lines* cut off.

A proper vertex is also referred to as an "amputated" diagram and we show an example in Fig. 10.8.

10.7.3 The self-energy $\sum(k)$

In order to partially sum the series we introduce a new quantity which is the sum of all the proper vertex or amputated diagrams. This is the self-energy and is denoted by $\sum(k)$. We define it as:

$$-\sum(k) = \text{sum of all two-point } 1 - \text{PI diagrams without external links.}$$

We now concentrate on connected correlations and, referring to Fig. 10.2, we see that at first order, $G_c^{(2)}$ has a $1 - \text{PI}$ diagram, at second order two $1 - \text{PI}$ diagrams (third diagram is $1 - \text{PR}$), so to third order the self-energy is as shown in Fig. 10.9.

The correlation function $G_c^{(2)}$ can then be written in terms of $\sum(k)$ as:

$$G_c^{(2)}(k) = G_0(k) - G_0(k)\Sigma(k)G_0(k) + G_0(k)\Sigma(k)G_0(k)\Sigma(k)G_0(k) + \cdots, \quad (10.63)$$

and re-arranging,

$$G^{(2)}(k) = G_0(k)[1 - (\Sigma G_0) + (\Sigma G_0)^2 - \cdots]$$

$$= \frac{G_0(k)}{1 + \sum(k)G_0(k)} = \frac{1}{G_0^{-1}(k) + \sum(k)}, \quad (10.64)$$

where we summed the series, and now substituting for $G_0^{-1}(k)$,

$$G_c^{(2)}(k) = \frac{1}{k^2 + m_0^2 + \Sigma(k)}. \tag{10.65}$$

10.7.4　Vertex functions

Define two-point vertex function $\Gamma^{(2)}(k)$ by:

$$\Gamma^{(2)}(k) = G_0^{-1}(k) + \Sigma(k). \tag{10.66}$$

Then eqn (10.64) implies

$$G_c^{(2)}(k) = \frac{1}{\Gamma^{(2)}(k)}, \tag{10.67}$$

and this can be shown as a general case by using the generating functional in the following subsection.

10.7.5　Generating functional for vertex functions $\Gamma[\phi]$

Write the dimensionless Gibbs function as:

$$\Gamma[\phi] \equiv \beta G[\phi] = \int d^d \mathbf{x} J(\mathbf{x}) \bar{\phi}(\mathbf{x}) - \ln \mathcal{Z}[J]. \tag{10.68}$$

This corresponds to the usual Legendre transformation in thermodynamics, thus:

$$\bar{\phi}(\mathbf{x}) \equiv \langle \phi(\mathbf{x}) \rangle = \frac{\delta \ln \mathcal{Z}[J]}{\delta J(\mathbf{x})}. \tag{10.69}$$

Above the critical temperature $T = T_c$, the n-point vertex functions are defined as coefficients in the Taylor series for $\Gamma(\phi)$ in powers of ϕ.

$$\Gamma^{(n)}(\mathbf{k}_1 \ldots \mathbf{k}_n) = \frac{\delta}{\delta \bar{\phi}(-\mathbf{k}_1)} \cdots \frac{\delta}{\delta \bar{\phi}(-\mathbf{k}_n)} \Gamma[\bar{\phi}] \Big|_{\bar{\phi}=0}. \tag{10.70}$$

10.8　The UV divergence and renormalization

We now concentrate on dealing with the UV divergences which arise when we let the upper cutoff $\Lambda \to \infty$ in the loop integrals. In this work we will find it helpful to keep in mind the topic of reversion of power series, as discussed in Section 5.8.

10.8.1 Mass renormalization: $m_0 \to m_1$ at one-loop order

Consider the second-order vertex function as defined by eqn (10.66), thus:

$$\Gamma^{(2)}(k) = m_0^2 + k^2 + \frac{\lambda_0}{2} \int_{0 \leq q \leq \Lambda} \frac{d^d q}{q^2 + m_0^2} + \mathcal{O}(\lambda_0^2). \tag{10.71}$$

We note that m_0^2 is not observable, whereas:

$$\Gamma^{(2)}(k)\bigg|_{k=0} = \beta/\chi_T, \tag{10.72}$$

where χ_T is the susceptibility and is observable.

Now define m_1^2 by:

$$m_1^2 \equiv \Gamma^{(2)}(0) = m_0^2 + \frac{\lambda_0}{2} \int_{0 \leq q \leq \Lambda} \frac{d^d q}{q^2 + m_0^2} + \mathcal{O}(\lambda_0^2). \tag{10.73}$$

Then from (10.71) and (10.73), the equation for $\Gamma^{(2)}(k)$ in terms of m_1^2 is, by definition:

$$\Gamma^{(2)}(k) = \Gamma^{(2)}(k) - \Gamma^{(2)}(0) + m_1^2,$$

as we are adding and subtracting the same thing, and so

$$\Gamma^{(2)}(k) = \left(m_0^2 + k^2 + \frac{\lambda_0}{2} \int \frac{d^d q}{q^2 + m_0^2} + \cdots \right) - \left(m_0^2 + \frac{\lambda_0}{2} \int \frac{d^d q}{q^2 + m_0^2} + \cdots \right) + m_1^2,$$

and hence

$$\Gamma^{(2)}(k) = m_1^2 + k^2 + \mathcal{O}(\lambda_0^2), \tag{10.74}$$

while

$$G^{(2)}(k) = \frac{1}{k^2 + m_1^2} + \mathcal{O}(\lambda_0^2).$$

Cancellation is not perfect at two-loops, but the dependence on Λ can be reduced by inserting $m_0 \to m_1$ in the non-cancelling terms.

Note: the *renormalization condition* is:

$$\Gamma^{(2)}(0) = m_1^2. \tag{10.75}$$

10.8.2   Coupling constant renormalization: $\lambda_0 \to \lambda$ at one-loop order

Using (10.70) we can calculate the 4-point vertex to two-loop order as:

$$\Gamma^{(4)}(k_1, k_2, k_3) = \lambda_0 - \frac{\lambda_0^2}{2} \int_\Lambda \frac{d^d q}{(q^2 + m_0^2)} \left[\frac{1}{(k_1 + k_2 - q)^2 + m_0^2} \right.$$

$$\left. + \frac{1}{(k_2 + k_3 - q)^2 + m_0^2} + \frac{1}{(k_3 + k_1 - q)^2 + m_0^2} \right] + \mathcal{O}(\lambda_0^3). \quad (10.76)$$

By analogy with the previous mass renormalization, we require $\Gamma^{(4)}$ to be finite at some chosen point, for example, $k_i = 0$.

10.8.2.1   Renormalization condition:

$$\Gamma^{(4)}(0, 0, 0) = \lambda. \quad (10.77)$$

Re-write eqn (10.76) as:

$$\Gamma^{(4)}(k_1, k_2, k_3) = \lambda_0 - \frac{\lambda_0^2}{2} M_\Lambda(k_1, k_2, k_3, m_0^2), \quad (10.78)$$

which defines the function M_Λ. Then the renormalization condition becomes:

$$\lambda = \Gamma^{(4)}(0, 0, 0) = \lambda_0 - \frac{\lambda_0^2}{2} \int_\Lambda \frac{d^d q}{q^2 + m_0^2}$$

$$= \lambda_0 - \frac{\lambda_0^2}{2} M_\Lambda(0, 0, 0, m_0^2) + \mathcal{O}(\lambda_0^3). \quad (10.79)$$

Conversely, we can invert this as:

$$\lambda_0 = \lambda + \frac{\lambda^2}{2} M_\Lambda(0, m_1^2) + \mathcal{O}(\lambda^3). \quad (10.80)$$

Then substitute it for λ_0 into eqn (10.78):

$$\Gamma^4(k_i) = \lambda - \frac{\lambda^2}{2} [M_\Lambda(k_i, m_1^2) - M_\Lambda(0, m_1^2)] \quad (10.81)$$

and this combination remains finite as $\Lambda \to \infty$.
  Note

- Cancellation of leading order term as $1/q^4$ as $\Lambda \to \infty$
- Self-consistent replacement of $m_0 \to m_1$ in $M_\Lambda(k_i, m_0^2)$.

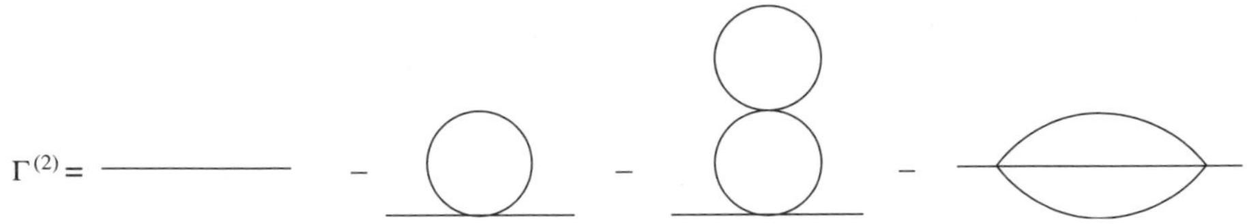

FIG. 10.10 Vertex function to second order.

10.8.3   Field renormalization: $\Gamma^{(2)}$ to two-loop order

Let us now extend (10.71) for the two-point vertex function to second order:

$$\Gamma^{(2)}(k) = k^2 + m_0^2 + \frac{\lambda_0}{2} \int_\Lambda \frac{d^d q}{(q^2 + m_0^2)}$$

$$- \frac{1}{4}\lambda_0^2 \int \frac{d^d q}{(q^2 + m_0^2)} M_\Lambda(0, m_0^2) - \frac{1}{6}\lambda_0^2 K_\Lambda(k^2, m_0^2), \qquad (10.82)$$

where Fig. 10.10 shows the same expression in diagrams and

$$K_\Lambda = \int \int d^d p\, d^d q\, \frac{1}{(p^2 + m_0^2)} \cdot \frac{1}{(q^2 + m_0^2)} \frac{1}{(k - p - q)^2 + m_0^2}. \qquad (10.83)$$

The previous mass renormalization does not work at this level: we have introduced it as an intermediate step:

$$\Gamma^{(2)}(0) = m_1^2. \qquad (10.84)$$

Applying this condition at second order now gives us:

$$m_1^2 = m_0^2 + \frac{\lambda_0}{2} D(\Lambda, m_0^2) - \frac{\lambda_0^2}{4} D(\Lambda, m_1^2) M_\Lambda(0, m_1^2) - \frac{\lambda_0^2}{6} K_\Lambda(0, m_1^2) + \mathcal{O}(\lambda_0^3).$$

$$(10.85)$$

Note:

1. We can replace $m_0^2 \to m_1^2$ in the second and third correction terms as the resulting corrections are $\mathcal{O}(\lambda_0^3)$ or higher.

2. The first correction term is only to first order so we need to extend to second order to be consistent. To do this, we use the earlier mass renormalization:

$$D(\Lambda, m_0^2) = \int \frac{d^d q}{q^2 + m_0^2} = \int \frac{d^d q}{q^2 + m_1^2 - \lambda_0 D(\Lambda, m_1^2)/2}$$

$$= \int \frac{d^d q}{q^2 + m_1^2} \left[1 + \frac{\lambda_0}{2} D(\Lambda, m_1^2) \frac{1}{q^2 + m_1^2} + \cdots \right]$$

$$= D(\Lambda, m_1^2) + \frac{\lambda_0}{2} D(\Lambda, m_1^2) \int \frac{d^d q}{(q^2 + m_1^2)^2} + \cdots . \tag{10.86}$$

Hence

$$D(\Lambda, m_0^2) = D(\Lambda, m_1^2) + \frac{\lambda_0}{2} D(\Lambda, m_1^2) M_\Lambda(0, m_1^2) + \cdots \tag{10.87}$$

Substitute back into eqn (10.85) for m_1^2:

$$m_1^2 = m_0^2 + \frac{\lambda_0}{2} D(\Lambda, m_1^2) + \frac{\lambda_0^2}{4} D(\Lambda, m_1^2) M_\Lambda(0, m_1^2),$$

$$- \frac{\lambda_0^2}{4} D(\Lambda, m_1^2) M_\Lambda(0, m_1^2) - \frac{\lambda_0^2}{6} K_\Lambda(0, m_1^2). \tag{10.88}$$

Make the cancellation and re-arrange:

$$m_0^2 = m_1^2 - \frac{\lambda_0}{2} D(\Lambda, m_1^2) + \frac{\lambda_0^2}{6} K_\Lambda(0, m_1^2). \tag{10.89}$$

Then the two-point vertex function becomes:

$$\Gamma^{(2)}(k) = k^2 + m_1^2 - \frac{\lambda_0}{2} D(\Lambda, m_1^2) + \frac{\lambda_0^2}{6} K_\Lambda(0, m_1^2) + \frac{\lambda_0}{2} D(\Lambda, m_1^2)$$

$$+ \frac{\lambda_0^2}{4} D(\Lambda, m_1^2) M_\Lambda(0, m_1^2) - \frac{\lambda_0^2}{4} D(\Lambda, m_1^2) M_\Lambda(0, m_1^2)$$

$$- \frac{\lambda_0^2}{6} K_\Lambda(k_1^2, m_1^2). \tag{10.90}$$

Make the cancellations (in two pairs):

$$\Gamma^{(2)}(k) = k^2 + m_1^2 + \frac{\lambda^2}{6} [K_\Lambda(0, m_1^2) - K_\Lambda(k^2, m_1^2)]. \tag{10.91}$$

Note:

1. The term in square brackets is still UV divergent.

2. It is consistent to replace $\lambda_0^2 \rightarrow \lambda^2$ to $\mathcal{O}(\lambda^3)$.

This is where we bring in the field of renormalization. We cure the remaining UV divergence by renormalizing the vertex functions (which is the same as renormalizing the field), thus:

$$\Gamma_R^{(2)}(k^2, m^2, \lambda) = \mathcal{Z}_\phi \Gamma^{(2)}(k^2, m_0^2, \lambda_0, \Lambda). \tag{10.92}$$

and $\mathcal{Z}_\phi$ is the renormalization constant, as in Section 10.4.

For $\Gamma_R^{(2)}$ to be finite, we impose the condition:

$$\frac{d}{dk^2} \Gamma_R^{(2)}(k^2) \Big|_{k^2=0} = 1 \equiv \Gamma_R'^{(2)}(0). \tag{10.93}$$

The choice of renormalization point is arbitrary but $k^2 = 0$ is convenient.

Then, to apply this condition, we differentiate eqn (10.92) and use (10.91) to obtain:

$$\frac{d}{dk^2} \Gamma_R^{(2)}(k^2) \Big|_{k^2=0} = \mathcal{Z}_\phi \left[1 - \frac{\lambda^2}{6} K_\Lambda'(k^2, m_1^2) \Big|_{k^2=0} \right]$$

$$= \mathcal{Z}_\phi \left[1 - \frac{\lambda^2}{6} K_\Lambda'(0, m_1^2) \right]. \tag{10.94}$$

Applying the renormalization condition:

$$\mathcal{Z}_\phi^{-1} = \left[1 - \frac{\lambda^2}{6} K_\Lambda'(0, m_1^2) \right], \tag{10.95}$$

where we note that $\mathcal{Z}_\phi$ is unity to one-loop order.

Multiplying (10.91) across by $\mathcal{Z}_\phi$ and setting $\mathcal{Z}_\phi = 1$ in the third term on the right hand side, we have

$$\mathcal{Z}_\phi \Gamma^{(2)}(k) = \mathcal{Z}_\phi k^2 + \mathcal{Z}_\phi m_1^2 + \frac{\lambda^2}{6} [K_\Lambda(0, m_1^2) - K_\Lambda(k^2, m_1^2)] + \mathcal{O}(\lambda^3). \tag{10.96}$$

Then, dividing across by $\mathcal{Z}_\phi$ and expanding $\mathcal{Z}_\phi^{-1}$ to second order in λ, we obtain

$$\Gamma_R^{(2)}(k) = k^2 + \mathcal{Z}_\phi m_1^2 + \frac{\lambda^2}{6} [K_\Lambda(0, m_1^2) - K_\Lambda(k^2, m_1^2) + k^2 K_\Lambda'(0, m_1^2)]. \tag{10.97}$$

We fix m_1^2 by imposing the condition:

$$m^2 = \mathcal{Z}_\phi m_1^2. \tag{10.98}$$

Note that $m_1^2 = m^2$ to $\mathcal{O}(\lambda^3)$ so we can make the replacement $m_1^2 \to m^2$ in K_Λ and hence:

$$\Gamma_R^{(2)}(k) = k^2 + m^2 - \tfrac{1}{6}\lambda^2[K_\Lambda(0, m^2) - K_\Lambda(k^2, m^2) + k^2 K_\Lambda'(0, m^2)], \qquad (10.99)$$

and this cures the UV divergence by cancellation. This leaves the theory in good shape for tackling the IR divergence in the next section.

10.9   The IR divergence and the ϵ-expansion

We wish to show how the epsilon expansion can be used to calculate critical exponents and as an example we will calculate η to $\mathcal{O}(\epsilon^2)$. From the definition of η, we have

$$G^{(2)}(k) \sim k^{-2+\eta} \quad \text{as } k \to 0, \quad \text{and} \quad T = T_c.$$

or, in terms of the vertex function,

$$\Gamma^{(2)}(k) \sim k^{2-\eta} \quad \text{as } k \to 0, \quad \text{and} \quad T = T_c.$$

But at the critical temperature $T = T_c$, m^2 (or u) is zero therefore we are faced with a "massless" theory at $T = T_c$ and one can have IR divergences as $k \to 0$. To deal with the IR divergence we re-expand in $\epsilon = 4 - d$.

10.9.1   Modified coupling constant

To begin with let us consider a typical integral:

$$I = \int \frac{d^d q}{q^2(q^2 + m^2)}; \quad 0 \le q \le \Lambda. \qquad (10.100)$$

Set $\mathbf{q} = m\mathbf{x}$ and rewrite as:

$$I = \int \frac{m^d d^d x}{m^2 x^2(m^2 x^2 + m^2)}; \quad 0 \le x \le \Lambda/m,$$

$$= m^{-\epsilon} \int \frac{d^d x}{x^2(x^2 + 1)}. \qquad (10.101)$$

The integral is now well-behaved, as $m \to 0$, even for $\Lambda/m \to \infty$, as we have fixed the UV divergences. The IR divergence now shows up as

$$\frac{1}{m^\epsilon} \to \infty \quad \text{as } m \to 0.$$

This behavior produces a factor of $m^{-\epsilon}$ at each order and we can redefine the coupling to absorb it, thus:

$$\lambda_0 = \lambda + \frac{1}{2}\lambda^2 M_\Lambda(0, m^2) = \lambda - \frac{\lambda^2}{2} \int_\Lambda \frac{\mathrm{d}^d q}{(q^2 + m^2)^2}. \tag{10.102}$$

Now set $mx = q$:

$$\lambda_0 = \lambda - \frac{\lambda^2}{2} m^{-\epsilon} \int_{\Lambda/m} \frac{\mathrm{d}^d x}{(x^2 + 1)^2}. \tag{10.103}$$

Divide across by m^ϵ:

$$\lambda_0 m^{-\epsilon} = \lambda m^{-\epsilon} - \frac{\lambda^2}{2} m^{-2\epsilon} \int_{\Lambda/m} \frac{\mathrm{d}^d x}{(x^2 + 1)^2} \tag{10.104}$$

and further setting

$$\tilde\lambda = m^{-\epsilon}\lambda, \tag{10.105}$$

so that (10.104) becomes

$$\tilde\lambda_0 = \tilde\lambda - \frac{\tilde\lambda^2}{2} \int_{\Lambda/m} \frac{\mathrm{d}^d x}{(x^2 + 1)^2} \tag{10.106}$$

gives the modified coupling constant $\tilde\lambda$.

10.9.2 Calculation of η

Now put $m = 0$, for $T = T_c$, in $\Gamma^{(2)}(k)$ as given by eqn (10.99):

$$\Gamma^{(2)}(k, 0) = k^2 - \frac{\lambda^2}{6}[K_\Lambda(0, 0) - K_\Lambda(k^2, 0) + k^2 K'_\Lambda(k^2, 0)] + \mathcal{O}(\lambda^3). \tag{10.107}$$

Take logs of both sides of the equation defining η:

$$(2 - \eta)\ln k = \ln \Gamma^{(2)}(k) \tag{10.108}$$

and differentiate:

$$2 - \eta = \lim_{k \to 0} \frac{\partial \ln \Gamma^{(2)}(k)}{\partial \ln k}. \tag{10.109}$$

Re-write $\Gamma^{(2)}(k, 0)$ as:

$$\Gamma^{(2)}(k, 0) = k^2 \left[1 - \frac{\lambda^2}{6k^2} X_\Lambda(k^2, 0) \right] + \mathcal{O}(\lambda^3), \tag{10.110}$$

which defines X_Λ.

Take logs of both sides:

$$\ln \Gamma^{(2)}(k, 0) = 2 \ln k + \ln \left[1 - \frac{\lambda^2}{6k^2} X_\Lambda(k^2, 0) \right] + \mathcal{O}(\lambda^3) \tag{10.111}$$

and expand $\ln[\ldots]$ to $\mathcal{O}(\lambda^3)$,

$$\ln \Gamma^{(2)}(k, 0) = 2 \ln k - \frac{\lambda^2}{6k^2} X_\Lambda(k^2, 0) + \mathcal{O}(\lambda^3). \tag{10.112}$$

Then, from the defining relation for η, as rewritten in the form:

$$2 - \eta = \lim_{k \to 0} \frac{\partial}{\partial \ln k} 2 \ln k - \lim_{k \to 0} k \frac{\partial}{\partial k} \left[\frac{\lambda^2}{6k^2} X_\Lambda(k^2, 0) \right] + \mathcal{O}(\lambda^3), \tag{10.113}$$

we have

$$\eta = \lim_{k \to 0} k \frac{\partial}{\partial k} \left[\frac{\lambda^2}{6k^2} X_\Lambda(k^2, 0) \right] + \mathcal{O}(\lambda^3). \tag{10.114}$$

Now X_Λ scales as K_Λ, so let us examine

$$\frac{1}{k^2} K_\Lambda(k^2, 0) = \frac{1}{k^2} \int \frac{d^d p \, d^d q}{p^2 q^2 (k - q)^2}. \tag{10.115}$$

We set

$$\mathbf{x} = \mathbf{p}/k; \quad \mathbf{y} = \mathbf{q}/k, \tag{10.116}$$

$$\tfrac{1}{2} K_\Lambda(k^2, 0) = k^{-2\epsilon+2} K_\Lambda(x, y, 0). \tag{10.117}$$

From this, write (10.114) for η as:

$$\eta = \lim_{k \to 0} k \frac{\partial}{\partial k} \left[\frac{1}{6} (\lambda k^{-\epsilon})^2 X_\Lambda(x, y, 0) \right] + \mathcal{O}(\lambda^3)$$

$$\tag{10.118}$$

$$= \lim_{k \to 0} \frac{k}{6} (-2\epsilon)(\lambda k^{-\epsilon})^2 k^{-1} X_\Lambda(x, y, 0) + \mathcal{O}(\lambda^3),$$

and so

$$\eta = - \lim_{k \to 0} \frac{\epsilon \bar{\lambda}^2}{3} X_\Lambda(x, y, 0) + \mathcal{O}(\lambda^3) \tag{10.119}$$

with

$$\bar{\lambda} = k^{-\epsilon} \lambda \tag{10.120}$$

and

$$\eta = -\epsilon \frac{\bar{\lambda}_c^2}{3} X_\Lambda(x, y, 0) + \mathcal{O}(\lambda_c^3) \tag{10.121}$$

where $k \to 0$ takes $\bar{\lambda}$ in to the critical point.

The next step is to solve for the zeros of

$$\beta(\bar{\lambda}) = \partial\bar{\lambda}/\partial \ln k. \tag{10.122}$$

We use

$$\bar{\lambda} = \bar{\lambda}_0 - \frac{3}{2}\bar{\lambda}_0^2 \int_0^\infty \frac{d^d x}{x^2(K_1 - K_2 - x)^2} + \mathcal{O}(\lambda_0^3), \tag{10.123}$$

where K_1, K_2 are associated with a specially chosen renormalization point: in $d = 3$ the values are usually taken as:

$$|\mathbf{K}_1| = |\mathbf{K}_2| = \sqrt{3}/4. \tag{10.124}$$

Then, differentiating both sides of eqn (10.123) with respect to $\ln k$,

$$\beta(\bar{\lambda}) = \partial\bar{\lambda}/\partial \ln k = -\epsilon\bar{\lambda}_0 + 3\epsilon\bar{\lambda}_0^2 \int_0^\infty \frac{d^d x}{x^2(K_1 - K_2 - x)^2} + \mathcal{O}(\lambda_0^3)$$

$$= -\epsilon\bar{\lambda} + \frac{3\epsilon}{2}\bar{\lambda}^2 \int_0^\infty \frac{d^d x}{x^2(K_1 - K_2 - x)^2} + \mathcal{O}(\lambda^3). \tag{10.125}$$

We need to evaluate the integral and then obtain the roots:

$$\bar{\lambda}_c = 0 \quad \text{and} \quad \bar{\lambda}_c = \frac{2}{3\pi^3}. \tag{10.126}$$

The integrals making up $X_\Lambda(x, y, 0)$ can be evaluated by dimensional regularization: this amounts to evaluation in $\epsilon = d - 4$.

If we evaluate $\bar{\lambda}_c$ the same way we find

$$\bar{\lambda}_c = \frac{\epsilon}{3\pi^2} + \mathcal{O}(\epsilon^2) \tag{10.127}$$

so that our perturbation expansion is now in powers of ϵ, which is the $\mathcal{O}(\epsilon)$ approximation to the previous value.

It can be shown that we can write the expression for η as

$$\eta = \frac{1}{3}\epsilon\bar{\lambda}_c^2[I(0) - I(1)] \tag{10.128}$$

where I is just X_Λ, with the field renormalization term $K'_\Lambda(0, m^2)$ omitted as it does not contribute, being independent of k and vanishing when we differentiate with respect to $\ln k$.

It can also be shown that

$$I(k) = -\frac{1}{4}\frac{\pi^4 k^2}{\epsilon} + \mathcal{O}(1), I(1) \qquad\qquad = -\frac{1}{2}\frac{\pi^4}{\epsilon}, \tag{10.129}$$

hence

$$\eta = \frac{1}{3}\epsilon\left(\frac{\epsilon^2}{3^2\pi^4}\right)\left[\frac{1}{2}\frac{\pi^4}{\epsilon}\right] = \frac{\epsilon^2}{54} + \mathcal{O}(\epsilon^3). \tag{10.130}$$

For $d = 3$, $\epsilon = 1$, and so $\eta \simeq 0.019$.

TABLE 10.1 Comparison of values of critical exponents obtained from the Borel-summed ϵ-expansion with those obtained to order ϵ, by mean-field theory and by numerical calculation of the Ising model in $d = 3$.

Exponents	Mean-field theory	RG to order ϵ	Borel-summed ϵ-expansion	Numerical simulation
α	0	0.17	0.10	0.12
β	0.5	0.33	0.33	0.31
γ	1	1.17	1.24	1.25
δ	3	4	4.81	5.20
ν	0.5	0.58	0.63	0.64
η	0	0	0.04	0.06

10.9.3 Values of the critical exponents

The above value of η is obviously an improvement on the previous result of $\eta = 0$ as given in Chapter 9. However, when compared with the value obtained from numerical simulation which is $\eta = 0.06$, as shown in Table 9.1, clearly one still has some way to go.

The main problem now lies in the fact that the ϵ-expansion is not simply convergent. Taking higher powers of ϵ does not guarantee an improvement and at any order the situation may actually be worse.

Despite this it is possible to obtain good results because the expansion is asymptotically convergent and hence "Borel summable," as discussed in Section 1.3.5. Values of critical exponents (rounded to only two decimal places for sake of simplicity) are shown in Table 10.1. This is just Table 9.1 with an extra column showing the critical exponents obtained from the Borel-summed ϵ-expansion (see [5], [33]).

10.10 The pictorial significance of Feynman diagrams

In this account we have discussed Feynman diagrams solely from the point of view of illustrating (or even calculating) the terms in a perturbation series. In this sense we have taken exactly the same view as we did when considering the diagrams in the high-temperature or virial cluster expansions in Chapter 4 or in the perturbation expansion of the Navier–Stokes equation in Chapter 5. However, the true importance of Feynman diagrams is that they can be used to represent events involving fundamental particles.

For instance, the exact two-point propagator could describe an event where a particle is created at a point $\mathbf{x}'$ at time t' and propagates to a point $\mathbf{x}$ where it is annihilated at time t. The perturbation expansion then expresses this event as a result of many interactions of bare particles, which are themselves not observable.

Such "physical" interpretations can also be made for the other perturbation expansions. In the case of the virial cluster expansion the physical interpretation is quite straightforward whereas for turbulence one probably has to settle for the "mode coupling" or "nonlinear mixing" interpretation. But the reason why the Feynman diagrams are so important is that they provide a language for the subject of high-energy particle physics.

Although one can learn more about these diagrams from the books listed in the next section, an interesting and informal account from the point of view of their pictorial significance can be found in [17].

Further reading

One may use the techniques discussed here to carry on the ϵ-expansion, to higher orders. We have already seen in Chapter 9 that, even at first order, the ϵ-expansion gives an improvement on mean-field theory. In practice, going to a higher order can lead to worse results, but ultimately the ϵ expansion can be used to obtain results that are very good indeed. As indicated in the previous chapter, [5] is probably the most accessible introduction for the reader who is not a field theorist, but [4] is also helpful and while [2, 13, 33] give fewer details of calculations, their treatment does let the overall structure stand out.

11

DYNAMICAL RENORMALIZATION GROUP APPLIED TO CLASSICAL NONLINEAR SYSTEMS

In Chapters 9 and 10, we have discussed the application of renormalization group (RG) to static problems involving continuous fields. Now we extend the concept to dynamical systems governed by nonlinear Langevin equations, as discussed in Chapter 5. The pioneering work in this area is usually taken to be the theory of ferromagnets in dimension $d = 6 - \epsilon$ which is due to Ma and Mazenko, and we begin with a brief discussion of this. Then we go on to consider how the same algorithm could be applied to the Navier–Stokes equation (NSE), and examine the technical problems involved. We conclude that a perturbative approach would only be possible at either low wave numbers or at high wave numbers.

In Section 11.3, we summarize the theory of Forster, Nelson, and Stephen, which is closely analogous to the methods presented in Chapter 9, and is a perturbative treatment of stirred fluid motion with coefficients evaluated as Gaussian averages. This work relies on asymptotic freedom as $k \rightarrow 0$, and therefore has no application to real fluid turbulence. However it may have some application to problems in soft condensed matter and we close the section with some remarks on that subject.

The remainder of the chapter is devoted to fluid turbulence. In Section 11.4, we discuss the practical technique of reducing the number of degrees of freedom requiring numerical simulation. This is known as large-eddy simulation and requires a theoretical model to represent the nonlinear coupling between those eddies which are explicitly resolved and those whose length scales lie below the resolution of the computational grid. This seems a natural application for RG, but nontrivial difficulties lie in the way, not least being the need for a conditional average. In Section 11.5 we discuss the formulation and evaluation of a conditional average as an approximation. Then in Section 11.6, we present a practical calculation of an effective viscosity based on the technique of iterative conditional averaging.

11.1 The dynamical RG algorithm

Now we extend the analysis presented in Chapters 8 and 9 to the case where the spin field is dependent on time and we can make the replacement $\phi(\mathbf{x}) \rightarrow \phi(\mathbf{x}, t)$. Previously, in static problems, we averaged out the effect of the high-wave number (short wavelength) modes in order to give a coarse-grained description of the low wave number modes (long wavelengths). Now, in dynamical problems, we consider slow variations in time of the order parameter and eliminate the effect of quicker variations ("fast variables") from the equation of motion.

The equation of motion is normally modeled as a nonlinear Langevin equation. A typical form is

$$\frac{\partial \phi}{\partial t} = M\phi \times \mathbf{B_E} - \Gamma \nabla^2 \mathbf{B_E} + \mathbf{g}, \tag{11.1}$$

where M and Γ are constants, $\mathbf{B_E}$ is the local magnetic field and $\mathbf{g}$ is a random noise which models the effect of thermal agitation of the spins. As in Section 3.1, we can write

$$B_E = B + B',$$

where B is the externally applied field and B' is the field due to the spins. From the bridge equation of statistical mechanics we have

$$\beta F = -\ln \mathcal{Z} = \beta H,$$

where F is the free energy and H is the Hamiltonian, and so the useful relationship:

$$F = H.$$

The Ginsburg–Landau model (see eqns (9.10) and (9.11)) gives us a suitable form for the relationship between the free energy and the spin field, thus:

$$F = \int d^d x \left(\frac{1}{2}\alpha|\nabla \phi|^2 + \frac{1}{2}u|\phi|^2 + \frac{\lambda}{4!}|\phi|^4 \right). \tag{11.2}$$

(Note that because of the vector product in the equation of motion we now have to retain the vector nature of the field.)

We can also write

$$\mathbf{B}' = \delta F / \delta \phi, \tag{11.3}$$

and rewrite the equation of motion so that it becomes

$$\frac{\partial \phi}{\partial t} = M\phi \times \frac{\delta F}{\delta \phi} - \Gamma \nabla^2 \frac{\delta F}{\delta \phi} + \mathbf{g}. \tag{11.4}$$

Or, in Fourier components,

$$\frac{\partial \phi(\mathbf{k}, t)}{\partial t} = M \int d^3 j\, \phi(\mathbf{k} - \mathbf{j}, t) \times \frac{\delta F}{\delta \phi(\mathbf{j}, t)} - \Gamma k^2 \frac{\delta F}{\delta \phi(\mathbf{k}, t)} + \mathbf{g}(\mathbf{k}, t), \tag{11.5}$$

where the random noise is taken to have a Gaussian distribution with covariance given by

$$\langle g_\alpha(\mathbf{k}, t)g_\beta(\mathbf{k}', t')\rangle = 2\Gamma k^2 \delta_{\alpha\beta}\delta(\mathbf{k} + \mathbf{k}')\delta(t - t'). \tag{11.6}$$

In extending the RG transformation from the static case, we note that the parameter set μ, as defined by eqn (8.11), now takes the analogous form

$$\mu = \{\bar{M}, \alpha, u, \lambda, B_E\}, \quad \bar{M} = M/\Gamma, \tag{11.7}$$

although from the analysis of Chapter 9 we can expect that it is possible to set $\alpha = 1$. The actual RGT may be adapted from that given in Section 9.2, as follows:

1. Eliminate the variables $\phi^>(\mathbf{k}, t)$, such that $\Lambda/b \leq k \leq \Lambda$, by solving the equations of motion, substituting back into the equations of motion for $\phi^<(\mathbf{k}, t)$ and then averaging over the statistics of the random noise $\mathbf{g}(\mathbf{k}, t)$.

2. Replace the $\phi^<(\mathbf{k}, t)$, on $k < \Lambda/b$, by $b^{1-n/2} \phi(b\mathbf{k}, tb^{-z})$ in the equations of motion, where the constants η and z remain to be fixed.

Steps 1 and 2 define the RGT, as denoted by R_b in Chapter 8, and under such transformations we can expect the parameter set $\boldsymbol{\mu}$ to change according to:

$$\boldsymbol{\mu}' = R_b \boldsymbol{\mu}. \tag{11.8}$$

The constants η and z in Step 2 are chosen such that (11.8) has a solution.

We shall not take this further. It becomes very complicated and to some extent this is signalled by the fact that simple solutions are only possible for space dimension d close to 6. Thus one ends up with a theory in $d = 6 - \epsilon$, which may be compared with the idea of a theory in $d = 4 - \epsilon$, as in Chapter 9. Instead, we shall turn to the application of the basic algorithm to macroscopic equations and, in particular, the Navier–Stokes equation or NSE.

11.2 Application to the Navier–Stokes equation

From eqn (5.2), we have the NSE[50] in the form:

$$\left(\frac{\partial}{\partial t} + \nu_0 k^2\right) u_\alpha(\mathbf{k}, t) = \lambda_0 M_{\alpha\beta\gamma}(\mathbf{k}) \int d^3 j u_\beta(\mathbf{j}, t) u_\gamma(\mathbf{k} - \mathbf{j}, t) + f_\alpha(\mathbf{k}, t), \tag{11.9}$$

where, from eqn (6.10),

$$M_{\alpha\beta\gamma}(\mathbf{k}) = (2i)^{-1}[k_\beta D_{\alpha\gamma}(\mathbf{k}) + k_\gamma D_{\alpha\beta}(\mathbf{k})]. \tag{11.10}$$

Note that we have changed the symbol for kinematic viscosity ν to ν_0 to indicate that it is unrenormalized. The reason for this notational change will soon become obvious. To be consistent, we have also changed the notation for the book-keeping parameter from λ to λ_0, although it should be borne in mind that each of them is equal to unity. The random force $\mathbf{f}$ is taken to have a multivariate normal distribution, with covariance chosen to be of the form:

$$\langle f_\alpha(\mathbf{k}, t) f_\beta(\mathbf{k}', t') \rangle = 2W(k)(2\pi)^d D_{\alpha\beta}(\mathbf{k}) \delta(\mathbf{k} + \mathbf{k}') \delta(t + t'). \tag{11.11}$$

The factor $W(k)$ is a measure of the rate at which the stirring force does work on the fluid and for stationarity this must equal the dissipation rate:

$$\int W(k) dk = \varepsilon. \tag{11.12}$$

---

[50] With random stirring forces $\mathbf{f}$ which are analogous to the noise $\mathbf{g}$.

The specification of the problem is completed by fixing the number of degrees of freedom in the system. We do this by introducing a maximum wave number k_{max} through a modification of the dissipation integral as

$$\varepsilon = \int_0^\infty dk 2v_0 k^2 E(k) \simeq \int_0^{k_{max}} dk 2v_0 k^2 E(k), \tag{11.13}$$

where $E(k)$ is the energy spectrum and v_0 is the kinematic viscosity of the fluid.

11.2.1 The RG transformation: the technical problems

We can adapt the dynamical RG algorithm to the NSE by dividing up the velocity field at $k = \Lambda/b$ as follows:

$$u_\alpha(\mathbf{k}, t) = \begin{cases} u_\alpha^<(\mathbf{k}, t) & \text{for } 0 \le k \le \Lambda/b \\ u_\alpha^>(\mathbf{k}, t) & \text{for } \Lambda/b \le k \le \Lambda, \end{cases} \tag{11.14}$$

where b is the usual spatial rescaling parameter such that $b > 1$.

In principle, the RG transformation now involves two stages:

1. Solve the NSE on $\Lambda/b \le k \le \Lambda$. Substitute that solution for the mean effect of the high-k modes into the NSE on $0 \le k \le \Lambda/b$. This results in an increment to the viscosity δv_0. That is, $v_0 \rightarrow v_1 = v_0 + \delta v_0$.

2. Rescale the basic variables, so that the NSE on $0 \le k \le \Lambda/b$ looks like the original NSE on $0 \le k \le \Lambda$.

The algorithm is then applied to successively lower wave numbers.

Substituting the decomposition given by (11.14) into (11.9) results in the two filtered equations of motion:

$$\left(\frac{\partial}{\partial t} + v_0 k^2 \right) u_\alpha^<(\mathbf{k}, t) = M_{\alpha\beta\gamma}^< \int d^3 j \left\{ u_\beta^<(\mathbf{j}, t) u_\gamma^<(\mathbf{k} - \mathbf{j}, t) \right.$$

$$+ 2 u_\beta^<(\mathbf{j}, t) u_\gamma^>(\mathbf{k} - \mathbf{j}, t)$$

$$\left. + u_\beta^>(\mathbf{j}, t) u_\gamma^>(\mathbf{k} - \mathbf{j}, t) \right\} + f_\alpha^<(\mathbf{k}, t); \tag{11.15}$$

$$\left(\frac{\partial}{\partial t} + v_0 k^2 \right) u_\alpha^>(\mathbf{k}, t) = M_{\alpha\beta\gamma}^> \int d^3 j \left\{ u_\beta^<(\mathbf{j}, t) u_\gamma^<(\mathbf{k} - \mathbf{j}, t) \right.$$

$$+ 2 u_\beta^<(\mathbf{j}, t) u_\gamma^>(\mathbf{k} - \mathbf{j}, t)$$

$$\left. + u_\beta^>(\mathbf{j}, t) u_\gamma^>(\mathbf{k} - \mathbf{j}, t) \right\} + f_\alpha^>(\mathbf{k}, t). \tag{11.16}$$

The relevant statistical quantity (see Section 5.3 for a fuller discussion) is the band-filtered spectral density, thus:

$$Q^>(k) = \langle u^>(\mathbf{k}, t)u^>(-\mathbf{k}, t) \rangle. \tag{11.17}$$

11.2.1.1   Technical problem 1: Need for a conditional average.   The normal ensemble average is illustrated by Fig. 11.1 where a lot of independent realizations of the (squared) turbulence field are averaged together to give the mean turbulence spectral density. Later, we shall

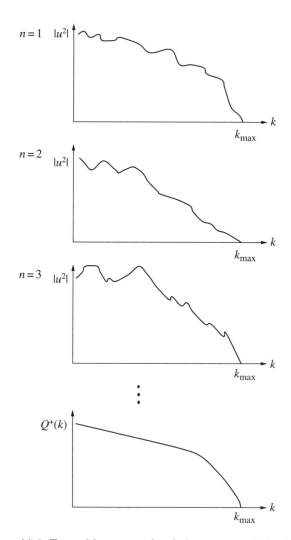

FIG. 11.1 Ensemble-averaged turbulence spectral density.

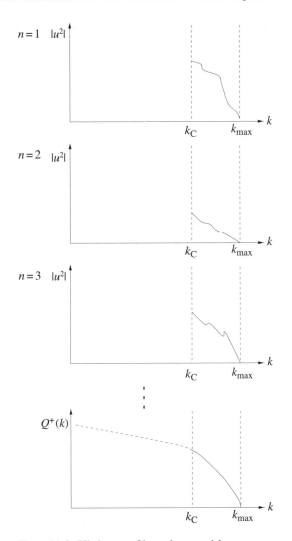

FIG. 11.2 High-pass filtered ensemble average.

modify this picture in order to define a conditional average. In practice, RG is often applied to turbulence using a filtered average, as in Fig. 11.2. Denote the filtered ensemble average by $\langle \ldots \rangle^>$. This operation has the properties

$$\langle u_\alpha^<(\mathbf{k}, t) \rangle^> = u_\alpha^<(\mathbf{k}, t);$$

$$\langle u_\alpha^>(\mathbf{k}, t) \rangle^> = 0.$$

Now average the low-k equation of motion using the filtered average, thus:

$$\left(\frac{\partial}{\partial t} + v_0 k^2\right) u_\alpha^<(\mathbf{k}, t) = M_{\alpha\beta\gamma}^<(k) \int d^3 j \left\{ u_\beta^<(j) u_\gamma^<(\mathbf{k} - \mathbf{j}) \right.$$

$$+ 2 u_\beta^<(j) \langle u_\gamma^>(\mathbf{k} - \mathbf{j}) \rangle\rangle^>$$

$$\left. + \langle u_\beta^>(\mathbf{j}) u_\gamma^>(\mathbf{k} - \mathbf{j}) \rangle\rangle^> \right\} + \mathbf{f}^<. \qquad (11.18)$$

Evidently the second term on the right hand side equals zero, as $\langle u^> \rangle^> = 0$, and the third term on the right hand side equals zero, from homogeneity, as $M(\mathbf{k}) Q^>(\mathbf{j}) \delta(\mathbf{j} + \mathbf{k} - \mathbf{j}) \sim M(0) = 0$. Hence a filtered ensemble average gives the strange result that there is no turbulence problem! Terms involving $u^>$ are just averaged away.

As we shall see in Section 11.3, in a restricted application of RG at low wave numbers, the filtered ensemble average can be carried out at zero order of perturbation theory (i.e. as a Gaussian average, as discussed in Chapter 9), but it is important to recognize the need for a nontrivial conditional average in general.

11.2.1.2 Technical problem 2: Higher-order nonlinearities. In order to implement the first step of the RGT we wish to eliminate the $u^>$ from eqn (11.15) for the $u^<$. If we "solve' the high-k eqn (11.16), by inverting the linear operator $(\partial/\partial t + v_0 k^2) \equiv L_0$, then we find:

$$u_{\mathbf{k}}^> = L_0^{-1} M_{\mathbf{k}}^> \int d^3 j [u_j^< u_{k-j}^< + 2 u_j^< u_{k-j}^> + u_j^> u_{k-j}^>].$$

Then, substituting back into the low-k equation as given by (11.15), we obtain:

$$L_0 u_{\mathbf{k}}^< = M_k^< [u_j^< u_{k-j}^< + 2 u_j^< L_0^{-1} M_{k-j} u_{k-j}^< u_{k-j-\gamma}^< + \cdots].$$

Note the new ocurrence of the triple nonlinearity $u^< u^< u^<$ which breaks the form-invariance of the NSE. That is, there is no way in which we can make the new NSE look like the old NSE.

11.2.2 Overview of perturbation theory

As we saw in Chapter 5, a perturbation expansion of the velocity field in the NSE is divergent. The number of terms rises with the "order," and some form of regularization or renormalization is required. The expansion parameter (or coupling constant) for the NSE is always a Reynolds number, and most definitions of Reynolds number have a value greater than unity when turbulence is present.

We can define a local, spectral Reynolds number at wave number k as

$$R(k) = \frac{k^2 V(k)}{v_0} = \frac{[E(k)]^{1/2}}{v_0 k^{1/2}}. \qquad (11.19)$$

Here v_0 is the kinematic viscosity and $V(k)$ is the root mean square Fourier coefficient. From an examination of the turbulence energy spectrum, as shown schematically in Fig. 11.3, we

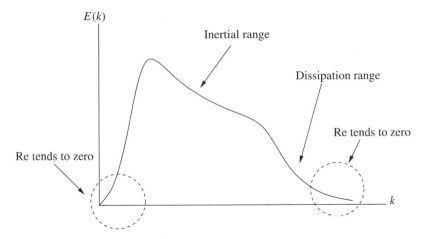

$$\text{F\scriptsize IG}. \; 11.3 \quad \text{Regions of wave number space where the local Reynolds number is small.}$$

FIG. 11.3 Regions of wave number space where the local Reynolds number is small.

see that there are two circumstances under which $R(k)$ is small:

1. $R(k) \to 0 \quad$ as $\; k \to 0.$

2. $R(k) \to 0 \quad$ as $\; k \to \infty.$

If we take the value of the wave number in eqn (11.19) to be the lower cut-off $k = \Lambda/b$ for the first step of the RG algorithm, then this provides a basis for a perturbative elimination of modes in the range $\Lambda/b \le k \le \Lambda$. Clearly $k = \Lambda/b$ must be chosen within one of the dotted circles in Fig. 11.3 for there to be any hope of a rational basis for low-order truncation of the expansion.

11.2.3   The application of RG at small wave numbers

In order to apply the RGT to turbulence we would have to choose the ultraviolet cut-off to equal the largest wave number associated with the turbulence modes, which we may take to be k_{max}, as defined by eqn (11.13). Although the Kolmogorov length scale does provide a small scale for turbulence, its inverse may differ by as much as an order of magnitude from k_{max}, as calculated from (11.13), and so it does not have the unambiguous significance of the lattice constant a in the problems discussed in Chapters 8 and 9.

We shall return to this point shortly, but here we shall consider the problem posed when we choose Λ to be small enough to exclude the inertial range and the dissipation region. By operating in the wave number range indicated schematically by the left-hand dotted circle in Fig. 11.3, we can formulate a problem that resembles the scalar field theory discussed in Chapter 9. The essential simplification is that we can solve eqn (11.16) perturbatively about a zero-order solution

$$u_{\alpha,0}^{>}(\mathbf{k}, t) = \int \mathrm{d}t' G^{(0)}(\mathbf{k}, t - t') f_{\alpha}^{>}(\mathbf{k}, t'), \qquad (11.20)$$

where, as in eqn (5.15), $G^{(0)}$ is the Green's function which allows one to invert the linear operator on the left hand side of eqn (11.16).

The perturbation solution for $u_\alpha^>$ follows essentially the same procedure as in Chapter 5. However, we can now restate the RG transformation given in Section 11.1 in a form more specific to this application as:

1. Solve eqn (11.16) perturbatively for the high-k modes and substitute the result into eqn (11.15) for the low-k modes. Average over the stirring forces $f_\alpha^>$.

2. Rescale $\mathbf{k}$, t, $\mathbf{u}^<$, and $\mathbf{f}^<$ so that the new equation looks like the original NSE. This step involves the introduction of renormalized transport coefficients.

We note that all the averages are against the distribution of the $f_\alpha^<$, and this has been chosen to be multivariate normal or Gaussian. Accordingly, as in Chapter 9, the corrections to Gaussian behavior are evaluated perturbatively with all the coefficients as Gaussian averages. The procedure here is different in that we now work directly with the moments of the field variables rather than with the partition functional. However, this is purely a formality. Even although fluid motion is neither Hamiltonian nor microscopic, one can still use a functional formalism analogous to that used in Chapters 9 and 10, although we shall not pursue that here. We shall continue our discussion of this application of RG at low wave numbers in Section 11.3.

11.2.4   The application of RG at large wave numbers

We discussed the problem of many length or time scales in Section 1.5 and now we again consider the relevance of self-similarity, but this time in the context of turbulence. If we represent the kinematic viscosity of the fluid by ν_0, then the corresponding Kolmogorov dissipation wave number, as given by eqn (6.6), may be denoted by $k_d^{(0)}$. Now, if we have a procedure in which modes in a small band of wave numbers are averaged out and their mean effect replaced by an increment to the viscosity, such that we make the replacement $\nu_0 \to \nu_1 = \nu_0 + \delta\nu_0$. The NSE is then on a reduced set of wave numbers, but with an increased viscosity and a correspondingly reduced Kolmogorov dissipation wave number $k_d^{(1)}$. Accordingly, solutions which formerly scaled as $f(k/k_d^{(0)})$ may now scale as $f(k/k_d^{(1)})$, so that rescaling appropriately can lead to scale-invariance. In fact, as we shall see in Section 11.5, it usually takes about 5 or 6 iteration cycles before scale invariance is demonstrated by the recursion reaching a fixed point.

In order to distinguish between this and the previous case, we make a change of notation and replace the $<$ and $>$ superscripts by $-$ and $+$. We begin by dividing up the velocity field at $k = k_1$ as follows:

$$u_\alpha(\mathbf{k}, t) = \begin{cases} u_\alpha^-(\mathbf{k}, t) & \text{for } 0 \leq k \leq k_1 \\ u_\alpha^+(\mathbf{k}, t) & \text{for } k_1 \leq k \leq k_0, \end{cases}$$

where k_1 is defined by

$$k_1 = (1 - \eta)k_0,$$

and the band width parameter η satisfies.

$$0 \leq \eta \leq 1.$$

Note that the factor $1 - \eta$ plays the same part as b^{-1} in the previous section.

In principle, the RG algorithm still involves two stages:

1. Solve the NSE on $k_1 \leq k \leq k_0$. Substitute that solution for the mean effect of the high-k modes into the NSE on $0 \leq k \leq k_1$. This results in an increment to the viscosity $\nu_0 \rightarrow \nu_1 = \nu_0 + \delta\nu_0$.

2. Rescale the basic variables, so that the NSE on $0 \leq k \leq k_1$ looks like the original NSE on $0 \leq k \leq k_0$.

The algorithm is then applied to successively lower wave numbers:

$$k_1 = (1 - \eta)k_0$$

$$k_2 = (1 - \eta)k_1$$

$$\vdots$$

where $0 < \eta < 1$, as illustrated in Fig. 11.4.

As we shall see, the difficulty here is that the conditional average cannot, unlike at low wave numbers, be evaluated perturbatively about a Gaussian zero-order average. We shall return to this topic in Section 11.4, when we consider the application of RG to large-eddy simulation.

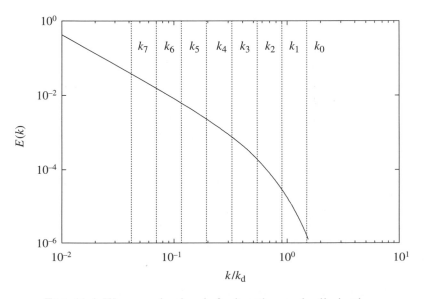

FIG. 11.4 Wave number bands for iterative mode elimination.

11.3   Application of RG to stirred fluid motion with asymptotic freedom as $k \to 0$

Forster, Nelson, and Stephen [10] adapted the RG theory of dynamical critical phenomena due to Ma and Mazenko to the case of stirred fluid motion.[51] In order to apply the theory, they chose the wave number cut-off to exclude cascade effects. We shall now give an outline discussion of this work.

At this stage we adopt a compact notation in which wave vector, time, and cartesian tensor index are all combined in a single symbol. Thus eqn (11.20) becomes

$$u_0^>(k) = G_0(k) f^>(k), \tag{11.21}$$

where k now stands for $\mathbf{k}, t$, and subscript α. We then follow the same procedure as in Chapter 5, by substituting the perturbation expansion

$$u^>(k) = u_0^>(k) + \lambda_0 u_1^>(k) + \lambda_0^2 u_2^>(k) + \cdots \tag{11.22}$$

into (11.16), equating coefficients at each order of λ_0, and expressing $u_1^>, u_2^>, \ldots$ in terms of $u_0^>$.

The next step is to replace each $u^>$ on the right hand side of eqn (11.15) by its perturbation expansion and formally write the solution for the low-wave number modes as

$$u^<(k) = G_0(k) f^<(k) + \lambda_0 G_0(k) M^<(k) \int d^3 j u^<(j) u^<(k-j)$$

$$+ 2\lambda_0 G_0(k) M^<(k) \int d^3 j u^<(j) \left\{ u_0^>(k-j) + \lambda_0 u_1^>(k-j) \right.$$

$$+ \lambda_0^2 u_2^>(k-j) + \cdots \left. \right\} + \lambda_0 G_0(k) M^<(k) \int d^3 j \left\{ u_0^>(j) u_0^>(k-j) \right.$$

$$+ 2\lambda_0 u_0^>(j) u_1^>(k-j) + \lambda_0^2 u_1^>(j) u_1^>(k-j)$$

$$+ 2\lambda_0^2 u_2^>(j) u_0^>(k-j) + \cdots \left. \right\}. \tag{11.23}$$

Then we substitute explicit forms for $u_1^>, u_2^>, \ldots$, and average over the $u_0^>$, which from (11.20) is just the prescribed distribution of the $f^>$. In doing so, we note that as the distribution of the random force f is multivariate normal, or Gaussian, one can average the $u^>$ independently of the $f^<$. However, it is not the same thing to average the $u^>$ independently of the $u^<$, as eqn (11.23) gives the $u^<$ a dependence on the $f^>$ in addition to their explicit dependence on the $f^<$. Nevertheless, we shall follow Forster et al. and ignore this nicely for

---

[51] They did not claim that this was a theory of turbulence. Despite this, numerous further investigations have been made along these lines and are claimed to be theories of turbulence!

the moment.[52] With some detailed algebra, one ends up with

$$\left(\frac{\partial}{\partial t} + \nu_1 k^2\right) u_\alpha^<(\mathbf{k}, t) = f_\alpha^<(\mathbf{k}, t) + \lambda_0 M_{\alpha\beta\gamma}^<(\mathbf{k}) \int_{j\leq\Lambda} d^3 j u_\beta^<(\mathbf{j}, t) u_\gamma^<(\mathbf{k}-\mathbf{j}, t)$$

$$+ 2\lambda_0^2 M_{\alpha\beta\gamma}^<(\mathbf{k}) \int_{j\leq\Lambda} d^3 j \int_{p\leq\Lambda} d^3 p M_{\rho\rho\sigma}^>(\mathbf{k}-\mathbf{j})$$

$$\times G_0(\mathbf{k}-\mathbf{j}) u_\gamma^<(\mathbf{k}-\mathbf{j}, t) u_\rho^<(\mathbf{p}, t) u_\sigma^<(\mathbf{k}-\mathbf{j}-\mathbf{p});$$

or, alternatively, Fourier transforming with respect to time as well as wave number, we can write the explicit-scales equation of motion as

$$(i\omega + \nu_1 k^2) u_\alpha^<(\mathbf{k}, \omega) = f_\alpha^<(\mathbf{k}, \omega)$$

$$+ \lambda_0 M_{\alpha\beta\gamma}^<(\mathbf{k}) \int_{j\leq\Lambda} d^3 j \int d\Omega u_\beta^<(\mathbf{j}, \Omega) u_\gamma^<(\mathbf{k}-\mathbf{j}, \omega-\Omega)$$

$$+ \text{a term involving} \quad u_\gamma^<(\mathbf{k}-\mathbf{j}) u_\rho^<(\mathbf{p}) u_\sigma^<(\mathbf{k}-\mathbf{j}-\mathbf{p}), \quad (11.24)$$

where ω is the angular temporal frequency and Ω is a dummy variable. The renormalized viscosity is given by

$$\nu_1 = \nu_0 + \Delta\nu_0, \quad (11.25)$$

and the increment to the viscosity is

$$\Delta\nu_0(k) = 8\lambda_0^2 k^{-2} M_{\rho\beta\gamma}^<(\mathbf{k}) \int d^3 j \int d\Omega$$

$$\times G_0(|\mathbf{k}-\mathbf{j}|, \omega-\Omega)|G_0(j, \Omega)|^2 M_{\gamma\sigma\rho}^<(\mathbf{k}-\mathbf{j}) D_{\beta\sigma}(\mathbf{j}) W(j), \quad (11.26)$$

where $D_{\beta\sigma}(\mathbf{j})$ is the transverse projector as defined by (5.13), and $W(j)$ is the "strength" of the force covariance, as given by (11.11).

At this stage the dominant role of the stirring forces (or noise) is apparent. The result for the renormalization of the viscosity depends on the arbitrarily chosen force covariance. Following Forster *et al.*, we can specify a class of models by choosing

$$W(k) = W_0 k^{-y}. \quad (11.27)$$

It can be shown that the increment to the viscosity is given by

$$\Delta\nu_0(0) = \nu_0 \left\{ 1 + K(d)\bar{\lambda}_0^2 \frac{\exp\{\epsilon l\} - 1}{\epsilon} \right\} \quad (11.28)$$

---

[52] See endnote number 25 in [10].

where the modified strength parameter $\bar{\lambda}_0$ is given by

$$\bar{\lambda}_0 = \frac{\lambda_0 W_0^{1/2}}{v_0^{3/2} \Lambda^{\epsilon/2}}, \tag{11.29}$$

and

$$K(d) = \frac{A(d) S_d}{(2\pi)^d}; \tag{11.30}$$

$$A(d) = \frac{d^2 - d - \epsilon}{2d(d+2)}; \tag{11.31}$$

$$S_d = \frac{2\pi^{d/2}}{\Gamma(d/2)}; \tag{11.32}$$

where Γ is the usual Gamma function, and

$$\epsilon = 4 + y - d. \tag{11.33}$$

The important thing to note is that the fixed point is found to exist in the limit $k \to 0$. Under these circumstances, the triple term involving $u^< u^< u^<$ on the right hand side of eqn (11.24) (its appearance anticipated in our earlier discussion of technical difficulties) vanishes and hence is an irrelevant variable. Of course, the limit $k \to 0$ has no relevance to real fluid turbulence. We shall consider the problems involved in tackling turbulence later in this chapter.

11.3.1 Differential RG equations

At this stage we note that the zero-order propagator

$$G_0(k, \omega) = \frac{1}{i\omega + v_0 k^2}, \tag{11.34}$$

(where we again work in terms of wave number and frequency ω) can be replaced by a modified or renormalized propagator

$$G_1(k, \omega) = \frac{1}{i\omega + v_1 k^2}. \tag{11.35}$$

Now we want to make (11.24) look like the original NSE and we do this by scaling the variables. As k is now defined on the interval $0 < k < \Lambda \exp(-l)$, we introduce the new variable $\tilde{k}$, where

$$\tilde{k} = k \exp(l), \tag{11.36}$$

such that $\tilde{k}$ is defined on the interval $0 < \tilde{k} < \Lambda$.

We shall require the coefficient of the time derivative (that is, $i\omega$) in (11.24) to remain unity, in order to have form invariance. So we introduce a general scaling of the temporal frequency

$$\tilde{\omega} = \omega \exp\{a(l)\}, \tag{11.37}$$

along with the velocity field

$$\tilde{u}_\alpha(\tilde{\mathbf{k}}, \tilde{\omega}) = u_\alpha^<(\mathbf{k}, \omega) \exp\{-c(l)\}, \tag{11.38}$$

where $a(l)$ and $c(l)$ are to be determined. Substituting those into (11.24), and insisting on form invariance gives us

$$\tilde{f}_\alpha(\tilde{k}, \tilde{\omega}) = f_\alpha^<(\mathbf{k}, \omega) \exp(a - c); \tag{11.39}$$

$$\nu(l) = \nu_1 \exp(a - 2l); \tag{11.40}$$

$$\lambda(l) = \lambda_0 \exp\{c - (d+1)l\}. \tag{11.41}$$

Recall the similar procedure in Chapter 9.

If we then apply these transformations to the covariance of the stirring forces, as given by (11.11) and (11.27), we have the homogeneity requirement:

$$2c = 3a + (y + d)l. \tag{11.42}$$

Working with infinitesimal wave number bands, l can be turned into a continuous variable and from eqns (11.25), (11.28), (11.29), (11.40), and (11.41), we have

$$\frac{d\nu}{dl} = \nu(l)\{z - 2 + K(d)\bar{\lambda}^2\}; \tag{11.43}$$

$$\frac{dW_0}{dl} = 0; \tag{11.44}$$

$$\frac{d\lambda}{dl} = \lambda(l)\left(\frac{3z}{2} - 1 - \frac{d-y}{2}\right); \tag{11.45}$$

where z is defined by

$$z = \frac{da}{dl}, \tag{11.46}$$

$\bar{\lambda}$ is defined by

$$\bar{\lambda}^2 = \frac{\lambda^2 W_0}{\nu^3 \Lambda^\epsilon}, \tag{11.47}$$

and the recursion relation for $\bar{\lambda}$ takes the form:

$$\frac{d\bar{\lambda}}{dl} = \frac{\bar{\lambda}}{2}\{\epsilon - 3K(d)\bar{\lambda}^2\}. \tag{11.48}$$

The nature of the solutions to eqn (11.48) depends on the value of $\epsilon = 4 + y - d$, where y is the exponent in the arbitrarily chosen covariance of the stirring forces, as given by eqn (11.27). As is usual in these situations, we can distinguish three cases as follows:

$\epsilon < 0$: $\bar{\lambda}(l)$ tends to zero as $l \to \infty$.

$\epsilon = 0$: the marginal case. As in the Ginsburg–Landau theory of Chapter 9, thus leads to logarithmic corrections.

$\epsilon > 0$: $\bar{\lambda}$ tends to the fixed point $\bar{\lambda}^*$.

The last case is the interesting one and the solution for $\bar{\lambda}$ at the fixed point is given by

$$\bar{\lambda}^* = \left\{ \frac{\epsilon}{3A(d)} \right\}^{1/3}, \tag{11.49}$$

as $l \to \infty$. If we substitute this result into (11.43) then it is easily seen that the renormalized viscosity becomes independent of l at the fixed point, provided only that

$$z = 2 - \epsilon/3. \tag{11.50}$$

This fixes the remaining free parameter $a(l)$ through eqn (11.46).

The exponent y in the power-law for the force covariance is an arbitrary choice. But for $d = 3$, the above analysis is only valid for $y > -1$. Also, in general, the argument that the triple term $u^< u^< u^<$ is an irrelevant variable depends on $y < d$. Thus, for $d = 3$, we have the restriction on our choice of exponent

$$-1 < y < 3. \tag{11.51}$$

11.3.2 Application to other systems

The analysis of Forster *et al.* [10] has been applied to other nonlinear Langevin equations such as the Burgers equation and the Kardar–Parisi–Zhang (KPZ) equation, as discussed in Chapter 5. This is an active field of research and it is impossible to give any clear conclusions at this stage. However, it seems to be a real possibility that non-trivial results can be obtained in areas other than macroscopic fluid motion, with current interest centered on the existence or otherwise of an upper critical dimension for the KPZ equation [8]. For background reading on this topic, we suggest the readable book [3] and the encylopaedic review article [11].

11.4   Relevance of RG to the large-eddy simulation of turbulence

In most areas of physics involving a range of length and time scales, the possibility of carrying out a direct numerical simulation on a computer is limited by the ability of the computer to resolve the entire range of scales. In simple terms, the "size" of the computer determines the "size" of the problem which can be tackled. Thus, in general there is an interest in reducing the number of degrees of freedom of such problems and, as we have seen in Chapters 8 and 9, this is precisely what the renormalization group does.

In the specific case of turbulence, the largest length scale L (say) can be taken to be the diameter of the pipe or the width of the jet, as limiting the size of the largest possible eddy.

The smallest scale is then set by the ability of the fluid viscosity ν_0 to damp out the smallest eddies (i.e. by dissipation of kinetic energy of fluid motion into kinetic energy of molecular motion or heat). The corresponding largest wave number, may be estimated by assuming it to be of the order of the Kolmogorov dissipation wave number k_d, as given by eqn (6.6). We redefine this here as

$$k_d^{(0)} = (\epsilon/\nu_0^3)^{1/4}, \tag{11.52}$$

as we are now using ν_0 for the (un-renormalized) molecular viscosity and accordingly we have put superscript "0" on our symbol for the Kolmogorov wave number.

To be more specific, in a spectral simulation of turbulence, we need to resolve wave numbers over the range:

$$2\pi/L < k < k_{\text{max}} \sim k_d^{(0)}. \tag{11.53}$$

Strictly speaking, the maximum resolved wave number mode can only be determined approximately and eqn (11.13) is one way of doing this. We shall return to this point shortly when we consider how a simulation can represent both the kinetic energy and the dissipation rate. However, it is clear from both (11.53) and (11.52) that the number of degrees of freedom to be simulated must increase as we either increase the rate at which we do work on the fliud (i.e. ε) or reduce the molecular viscosity. Either of these actions is equivalent to increasing the Reynolds number and so the number of degrees of freedom will increase with increasing Reynolds number.

In turbulence, *ad hoc* proposals for reducing the number of degrees of freedom antedated the modern form of RG by almost a decade and come under the umbrella title of *large eddy simulation*, or LES for short. The general idea is to work (in real space) with a grid or lattice on which the equations of motion are represented in finite difference form. Then, if we economize on computer size by choosing the grid such that the smallest eddies are not resolved, we may hope that the simulation of large eddies is not too sensitive to the way in which we model the nonlinear coupling between resolved eddies and *subgrid-scale* (or just *subgrid*) eddies. The need for such a model is often seen as a "reduced" form of the turbulence problem and clearly RG is a natural way to tackle it.

We confine our attention here to the subgrid modeling problem for homogeneous, isotropic, stationary turbulence, with zero mean velocity. Our main emphasis is on recognizing the importance of conditionally averaging the subgrid stresses. We also distinguish the relative importance of different categories of subgrid stress according to their effect on the following characteristics of the resolved scales: (a) the evolution of the velocity field; (b) the evolution of the energy spectrum; and (c) the inertial transfer of energy and the dissipation rate. The turbulence which we consider only occurs as an approximation in nature but can be represented experimentally by numerical simulation. This may make it seem rather an artificial problem but in fact it forces us to confront some aspects of LES which can be obscured in more realistic formulations. In particular, there is the problem of representing both the kinetic energy and the dissipation rate while maintaining the shape of the spectrum.

11.4.1   Statement of the problem

The statistical problem is formulated in terms of the Eulerian velocity field $u_\alpha(\mathbf{x}, t)$, and, as we are interested in spectral simulations, we work in wave number $(\mathbf{k})$ space, thus:

$$u_\alpha(\mathbf{k}, t) = \int d^3 x \, u_\alpha(\mathbf{x}, t) e^{-i \mathbf{k} \cdot \mathbf{x}}.$$

We wish to simulate the energy spectrum on the interval $0 \le k \le k_C$. Interactions with the *subgrid* modes in the interval $k_C \le k \le k_{max}$ are represented analytically by a subgrid model. Note that we are only concerned with high-wave number modes in the inertial and dissipation ranges. To have a well-posed problem, we specify ε, the rate of energy transfer through the inertial range of wave numbers.

The situation is illustrated schematically in Fig. 11.5. In general, the number of degrees of freedom may be reduced by filtering. We start by introducing a sharp cutoff filter at wave number $k = k_C$ where $k_C \ll k_{max}$. This filter is defined using the two unit step functions:

$$\theta^-(k - k_C) = \begin{cases} 1 & \text{for } 0 \le k \le k_C; \\ 0 & \text{for } k_C < k \le k_{max}, \end{cases} \tag{11.54}$$

and

$$\theta^+(k - k_C) = \begin{cases} 0 & \text{for } 0 \le k \le k_C; \\ 1 & \text{for } k_C < k \le k_{max}. \end{cases} \tag{11.55}$$

Using these functions we can decompose the velocity field into $u_\alpha^-(\mathbf{k}, t)$ and $u_\alpha^+(\mathbf{k}, t)$, where

$$u_\alpha^-(\mathbf{k}, t) = \theta^-(k - k_C) u_\alpha(\mathbf{k}, t); \tag{11.56}$$

$$u_\alpha^+(\mathbf{k}, t) = \theta^+(k - k_C) u_\alpha(\mathbf{k}, t). \tag{11.57}$$

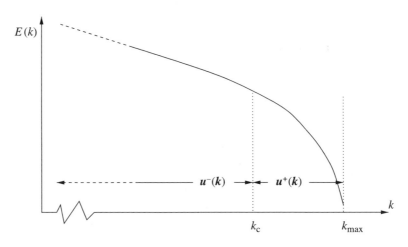

FIG. 11.5 Filtered energy spectrum.

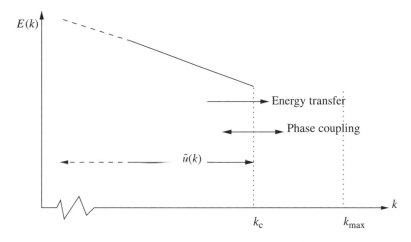

FIG. 11.6 Truncated or LES energy spectrum.

Evidently we can study the relationship between $\mathbf{u}^-(\mathbf{k}, t)$ and the $\mathbf{u}^+(\mathbf{k}, t)$ fields by imposing a notional filter on either a numerical simulation or an analytical model of spectral turbulence, as shown in Fig. 11.5. However, if we actually truncate the range of modes, as shown in Fig. 11.6, then we have to in some way represent the effects on energy transfer to, and phase coupling with, the missing modes. At this stage, only the first of these has received significant attention and it is only the "subgrid" energy transfer that we shall discuss here.

Also, the fact that wave number truncation in itself eliminates modes corresponding to small-scale motions means that an LES cannot have the correct values for extensive quantities like the total kinetic energy. Since both $E(k)$ and ν_0 are positive definite, we have:

$$\int_0^{k_C} dk\, E(k) < E_{\text{tot}} \simeq \int_0^{k_{\max}} dk\, E(k) \tag{11.58}$$

and

$$\int_0^{k_C} dk\, 2\nu_0 k^2 E(k) < \varepsilon \simeq \int_0^{k_{\max}} dk\, 2\nu_0 k^2 E(k). \tag{11.59}$$

In other words, truncation reduces both the amount of energy in the system and the rate at which energy is dissipated.[53]

The amount of such losses as a function of the cut-off wave number can be estimated from either fully resolved numerical simulations or semi-empirical models of the turbulent spectrum. Typical results are shown in Fig. 11.7. It can be seen that, for instance, if we take $k_C \sim 0.5k_d$, then the energy of the turbulence is virtually unaffected whereas the dissipation rate would be reduced to less than 40% of its correct value.

This discrepancy is what holds out hope for LES as a practical technique. At a pragmatic level, we can reduce the number of degrees of freedom by at least a factor of 2^3, without

---

[53] Compare this situation with the situation in microscopic physics studied earlier, where it is an essential feature of the method that applying RG should leave the free energy F unaffected, despite a reduction in the number of degrees of freedom.

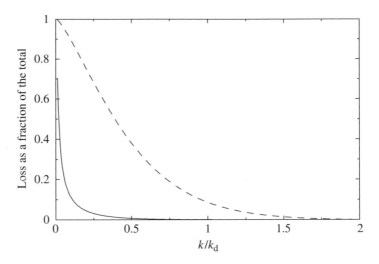

FIG. 11.7 Loss of energy and dissipation rate due to spectral truncation: The continuous line represents turbulent kinetic energy while the dashed line is the dissipation rate.

affecting either the total energy or the shape of the spectrum at resolved wave numbers. All the subgrid model is required to do is to ensure the correct dissipation rate.[54]

11.4.2 Conservation equations for the explicit scales $k \leq k_c$

We begin by decomposing the NSE, in terms of the filtered variables. We obtain an equation of motion for the low-k modes, thus:[55]

$$\left(\frac{\partial}{\partial t} + v_0 k^2\right) u_\alpha^-(\mathbf{k}, t) = M_{\alpha\beta\gamma}^-(\mathbf{k}) \int d^3 j \{u_\beta^-(\mathbf{j}, t) u_\gamma^-(\mathbf{k} - \mathbf{j}, t)$$

$$+ 2u_\beta^-(\mathbf{j}, t) u_\gamma^+(\mathbf{k} - \mathbf{j}, t) + u_\beta^+(\mathbf{j}, t) u_\gamma^+(\mathbf{k} - \mathbf{j}, t)\} \qquad (11.60)$$

Each term in this equation is defined on the wave number interval $0 \leq k \leq k_C$. For example,

$$M_{\alpha\beta\gamma}^-(\mathbf{k}) = \theta^-(k - k_C) M_{\alpha\beta\gamma}(\mathbf{k}). \qquad (11.61)$$

We are still left with a version of the turbulence closure problem in which the dynamical equation for the retained modes depends directly on modes in the range which we wish to filter out. This equation expresses the principle of conservation of momentum for the explicit scales. We may use it, as we used the full NSE, to derive conservation relations for kinetic energy and energy dissipation for the explicit scales and this is our next step.

[54] In fact, there are phase coupling effects to be taken into account as well, but we are ignoring these in the present account.

[55] This is, of course, just eqn (11.15), but it is worth restating it in the notation that we are using in this section.

11.4.2.1 The ad hoc *effective viscosity.* The effects of coupling between the retained and eliminated modes are often represented by an enhanced, wave number-dependent viscosity $\nu(k|k_C)$ acting upon the explicitly simulated modes. If we make the hypothesis that the effect of coupling between explicit and implicit scales can be represented by a turbulent effective viscosity $\nu_T(k)$, then the defining relation can be written as

$$\nu_T(k)k^2 u^-(\mathbf{k}, t) = -M_{\alpha\beta\gamma}^-(\mathbf{k}) \int d^3 j \{2u_\beta^-(\mathbf{j}, t)u_\gamma^+(\mathbf{k} - \mathbf{j}, t)$$

$$+ u_\beta^+(\mathbf{j}, t)u_\gamma^+(\mathbf{k} - \mathbf{j}, t)\} + S_\alpha^-(\mathbf{k}|k_C, t), \qquad (11.62)$$

where $S_\alpha^-(\mathbf{k}|k_C; t)$ has been added as a correction term. This term reflects the inadequacy of the effective viscosity concept, and incorporates additional effects of the kind which are often referred to as "eddy noise" or "backscatter". We can combine the molecular and (hypothetical) turbulent coefficients of viscosity into a single coefficient thus

$$\nu(k|k_C) = \nu_0 + \nu_T(k), \qquad (11.63)$$

where the notation on the left hand side indicates the dependence on the cut-off wave number k_C. Then, the general dynamical equation for use in a low-wave number simulation can be written in the form

$$\left(\frac{\partial}{\partial t} + \nu(k|k_C)k^2\right) u_\alpha^-(\mathbf{k}, t) = M_{\alpha\beta\gamma}^-(\mathbf{k}) \int d^3 j \, u_\beta^-(\mathbf{j}, t)u_\gamma^-(\mathbf{k} - \mathbf{j}, t) + S_\alpha^-(\mathbf{k}|k_C; t).$$

$$(11.64)$$

This equation is exact and as yet no approximation has been made. We may then obtain a modified energy balance equation, for the low-wave number simulation by multiplying (11.64) through by $u_\alpha^-(-\mathbf{k}, t)$ and averaging unconditionally over the general turbulence ensemble, thus:

$$2\nu(k|k_C)k^2 E^-(k) = W(k) + T^-(k) + 8\pi k^2 \langle \hat{S}_\alpha^-(\mathbf{k}|k_C; t)u_\alpha^-(-\mathbf{k}, t)\rangle, \qquad (11.65)$$

where $T^-(k)$ is the energy transfer spectrum, as defined by eqn (6.31), but here filtered to $0 \le k \le k_C$, and the energy input (forcing) spectrum $W(k)$ has been added.

We have assumed that $\kappa \ll k_C$, where κ is the upper endpoint of any forcing spectrum.[56] Integrating both sides of this equation with respect to k over the interval 0 to k_C, we then obtain a modified form of the energy dissipation relation as:

$$\int_0^{k_C} dk \, 2\nu(k|k_C)k^2 E(k) = \varepsilon + \int_0^{k_C} dk \, 8\pi k^2 \langle \tilde{S}_\alpha^-(\mathbf{k}|k_C; t)u_\alpha^-(-\mathbf{k}, t)\rangle, \qquad (11.66)$$

which may be compared to eqn (11.13) which is the usual form of this relationship. The reduction in the upper limit from $k_{max} \to k_c$ is supposed to be compensated for by the increase in viscosity $\nu_0 \to \nu(k|k_c)$.

---

[56] We choose a forcing spectrum which is bounded by wave number $\kappa \ll k_{max}$, so that we can study inertial and dissipation wave number ranges which do not depend on the stirring forces and hence are universal in character.

Note that in these three equations, we write the correction term as S^-, $\hat{S}^-$, and $\tilde{S}^-$, respectively. This is because we should allow for the possibility that S^- contains terms which vanish under the average or under the subsequent wave number integration. It can be shown that this is, in fact, the case. The effective viscosity, as it appears in the momentum equation is not (to employ the language of quantum physics) an observable. It may, however, be an observable in the equation for the dissipation rate, provided there exist circumstances under which the term in $\tilde{S}^-$ vanishes.

11.4.2.2 Local energy transfer? Another limitation on the concept of a "subgrid" eddy viscosity arises in the equation for the energy spectrum. Consider the case of local transfers near the cut-off wave number k_C. With wave number triads subject to the restrictions:

$$0 \leq k \leq k_C \qquad k_C \leq j, \ |\mathbf{k} - \mathbf{j}| \leq k_{max}.$$

It follows from an exact symmetry that energy transfer is zero for the case:

$$k \approx j \approx |\mathbf{k} - \mathbf{j}| \approx k_C.$$

Evidentally the term "local" must be interpreted with some caution in this context. It also follows, that a change from finite recursion relations to differential RGEs may not actually be valid for the NSE.

11.5   The conditional average at large wave numbers

The need for a conditional average when eliminating modes has been widely, if not universally, recognized. However, the formulation of a conditional average for the NSE is not as straightforward as one might think. We can see this as follows. From the Fourier transform we have

$$u_\alpha^-(\mathbf{k}, t) = \int d^3x \theta^-(k - k_C) u_\alpha(\mathbf{x}, t) e^{-i\mathbf{k}\cdot\mathbf{x}}; \tag{11.67}$$

and

$$u_\alpha^+(\mathbf{k}, t) = \int d^3x \theta^+(k - k_C) u_\alpha(\mathbf{x}, t) e^{-i\mathbf{k}\cdot\mathbf{x}}. \tag{11.68}$$

Evidently, if we average either of $u_\alpha^\pm(\mathbf{k}, t)$, the only thing which is actually averaged is the field $u_\alpha(\mathbf{x}, t)$. Therefore, if in any kind of averaging process $u_\alpha^-(\mathbf{k}, t)$ is left unaveraged, it follows immediately that $u_\alpha^+(\mathbf{k}, t)$ is also left unaveraged. This leaves us with the question: how does one average out the effect of high-k modes while leaving the low-k modes unaffected?

The answer may lie in the concept of unpredictability and, particularly, in formulating and evaluating the conditional average as an *approximation*. This involves specifying the condition for the average with some degree of imprecision in the low-k modes and then relying on the chaotic nature of turbulence to amplify this uncertainty in the high-k modes. We have referred to this property of turbulence as *local chaos*. This hypothesis has been examined in high-resolution numerical experiments with encouraging results.

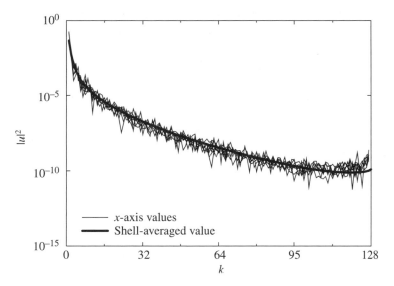

FIG. 11.8 Chaotic behavior of instantaneous energy spectra. The individual realizations of $|\mathbf{u}(\mathbf{k})|^2$ have been taken along the x-axis of the computational box.

In view of the determinism of the equations of motion, it can be helpful to see just how chaotic the resulting solutions appear to be, and we show here some results obtained from a numerical simulation of the NSE on a computer. In Fig. 11.8 we show several realizations of the squared velocity field along with the ensemble-averaged spectrum.

It is interesting to note how small the "scatter" is about the mean value. However, unpredictability criteria will be based on the instantaneous field $\mathbf{u}(\mathbf{k}, t)$ and not on its square, so we present the same set of realizations of the instantaneous velocity field in Fig. 11.9. Both these figures seem to be quite reassuring: evidently the turbulence velocity fields are quite chaotic.

Previously we have defined the conditional average in terms of a conditionally sampled ensemble, and the basic idea is illustrated in Fig. 11.10. From the full ensemble of turbulence realizations we select a subset of realizations. The criterion for selection is that the low-wave number part of the field $|\mathbf{u}^-(\mathbf{k}, t)|^2$ lies between specified limits which are drawn as dotted lines on the region with $k \leq k_C$ on each graph. Evidently if one takes an ensemble average on the conditionally sampled ensemble, the high-wave number part is a complete unknown. So, not only do we have to consider the question of how do we formulate a meaningful conditional average; but we also have to consider the question of how do we evaluate it? Our answer to the first question is that we can adapt the usual definition of conditional average but write it in terms of a limit. Let us consider a fluctuation in the low-k modes, thus:

$$u_\alpha^-(\mathbf{k}, t) \to u_\alpha^-(\mathbf{k}, t) + \delta u_\alpha^-(\mathbf{k}, t). \tag{11.69}$$

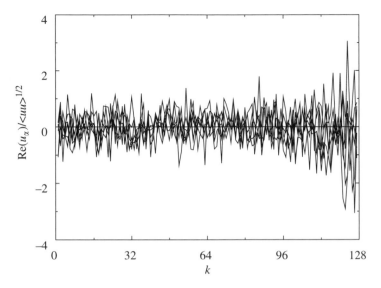

FIG. 11.9 Chaotic behavior of Fourier components of velocity.

We may define the conditional average of any functional $F[u_\alpha(\mathbf{k}, t)]$ in the limit of $\delta u_\alpha^-(\mathbf{k}, t) \to 0$, as

$$\langle F[u_\alpha(\mathbf{k}, t)]\rangle_c = \langle F[u_\alpha(\mathbf{k}, t)]|u_\alpha^-(\mathbf{k}, t)\rangle, \qquad (11.70)$$

such that

$$\langle u_\alpha^-(\mathbf{k}, t)\rangle_c = u_\alpha^-(\mathbf{k}, t), \qquad (11.71)$$

where the subscript "c" on the angle brackets stands for "conditional average." Note that we employ the standard notation for a conditional average on the right hand side of eqn (11.70), where the vertical bar separates the argument from the "condition."

This allows us to introduce the concept of a "fuzzy" conditional average, as illustrated schematically in Fig. 11.11. We assume that the small uncertainty for wave numbers $k \le k_C$, leads to a degree of unpredictability which grows with increasing k, for wave numbers $k > k_C$.

11.5.1 The asymptotic conditional average

It is reasonable to assume that the conditional average will become asymptotically free for wave numbers $k \gg k_C$ and hence can be evaluated as an ordinary ensemble average in some limit of large scale separation. This view is reinforced by the fact that the triangle condition on wave numbers means that there can be no direct coupling between wave numbers $k \le k_C$ and $k > 2k_C$. Hence, for suitably chosen values of k_C, we must have a band $2k_C \le k \le k_{max}$ for which the conditional average becomes unconditional.

If we now consider the band of wave numbers in which the conditional average is being performed, we may picture a situation as depicted in Fig. 11.12. Essentially we assume that unpredictability develops with increasing values of $k > k_C$. Near the cut-off $k = k_C$, one may suppose the u^+ modes to be strongly coupled to the u^- modes. However, for large wave

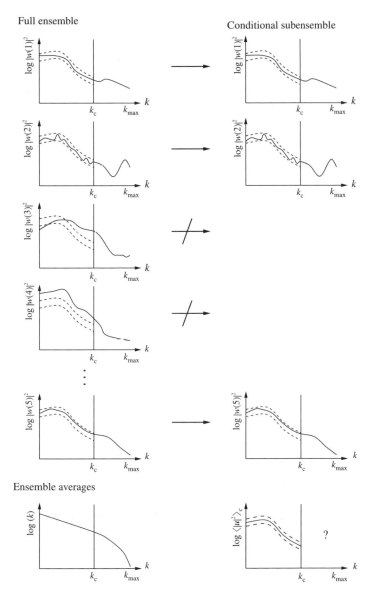

FIG. 11.10 A conditionally sampled ensemble.

numbers we can expect to find a region where the coupling may be seen in some sense, to be weak. In this region, we argue that a *quasi-stochastic estimate* of a conditional average could be a reasonable approximation.

By quasi-stochastic estimate we mean, that eqn (11.16) is used to derive an equation of motion for the required conditional average. Then an assumption of asymptotic freedom at

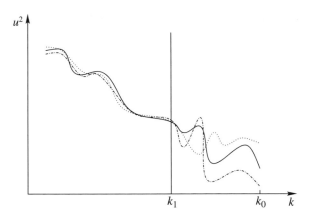

FIG. 11.11 Fuzzy conditional average.

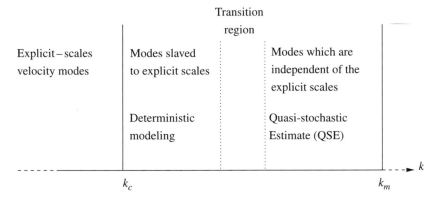

FIG. 11.12 Implications of scale separation.

large scale separation is used to simplify otherwise intractable terms in that equation. We shall return to this matter in Section 11.6.3.

11.6   Application of RG to turbulence at large wave numbers

We begin by restating eqns (11.16) and (11.17) in our current notation, thus:

$$\left(\frac{\partial}{\partial t} + v_0 k^2\right) u_\alpha^+(\mathbf{k}, t) = M_{\alpha\beta\gamma}^+ \int d^3 j \{u_\beta^-(\mathbf{j}, t) u_\gamma^-(\mathbf{k} - \mathbf{j}, t)$$

$$+ 2u_\beta^-(\mathbf{j}, t) u_\gamma^+(\mathbf{k} - \mathbf{j}, t) + u_\beta^+(\mathbf{j}, t) u_\gamma^+(\mathbf{k} - \mathbf{j}, t)\}, \qquad (11.72)$$

and

$$Q^+(k) = \langle u^+(\mathbf{k}, t) u^+(-\mathbf{k}, t)\rangle, \qquad (11.73)$$

for the band-filtered spectral density.

In this section, we again use an abbreviated notation, with all indices and variables contracted into a single subscript, for example, $u_\alpha(\mathbf{k}, t) \to u_k$. We conditionally average the low-wave number filtered NSE as given by (11.60),

$$L_{01} u_k^- = M_k^- u_j^- u_{k-j}^- + M_k^- \langle u_j^+ u_{k-j}^+ \rangle_c, \tag{11.74}$$

for $j, k - j \sim k_{\max}$ (asymptotically), where $L_{01} \equiv \partial/\partial t + \nu_0 k^2$. We note that the cross terms of the form $\langle u^+ u^- \rangle_c = \langle u^+ \rangle_c u^-$ vanish under the asymptotic conditional average and we are left with the term involving $M_k^- \langle u_j^+ u_{k-j}^+ \rangle_c$. It should be emphasized that this term is not related in any simple or obvious way to an ordinary ensemble average, which would be constant in time, for a stationary flow and in fact is still a random variable fluctuating on $0 \le k \le k_1$.

In practice, we evaluate the conditional average by using the high-wave number filtered NSE to solve iteratively for $\langle u_j^+ u_{k-j}^+ \rangle_c$. We do this by writing (11.72) as an equation for u_j^+ (introducing the new dummy variable p), multiplying through by u_{k-j}^+ and conditionally averaging. Then we repeat the process in reverse order, and add the resulting two equations together. This allows us to treat the one tricky point:

$$u_j^+ \frac{\partial}{\partial t} u_{k-j}^+ + u_{k-j}^+ \frac{\partial}{\partial t} u_j^+ = \frac{\partial}{\partial t} \left(u_j^+ u_{k-j}^+ \right). \tag{11.75}$$

All other steps are straightforward and the resulting equation is

$$L_{02} \langle u_j^+ u_{k-j}^+ \rangle_c = 2M_j^+ \left\{ 2u_p^- \langle u_{j-p}^+ u_{k-j}^+ \rangle_c + \langle u_p^+ u_j^+{}_{-p} u_{k-j}^+ \rangle_c \right\}, \tag{11.76}$$

where $L_{02} \equiv \partial/\partial t + \nu_0 j^2 + \nu_0 |\mathbf{k} - \mathbf{j}|^2$. Or, inverting the linear operator, we have

$$\langle u_j^+ u_{k-j}^+ \rangle_c = 2L_{02}^{-1} M_j^+ \left\{ 2u_p^- \langle u_{j-p}^+ u_{k-j}^+ \rangle_c + \langle u_p^+ u_{j-p}^+ u_{k-j}^+ \rangle_c \right\}, \tag{11.77}$$

for the requisite conditional average in (11.74). The first term on the right hand side of eqn (11.77) is the leading order term but we can go on and solve to any order, at least in principle.

Next we use the NSE to solve for the triple u^+ conditional moment on the right hand side of (11.77), obtaining

$$\langle u_p^+ u_{j-p}^+ u_{k-j}^+ \rangle_c = 2L_{03}^{-1} M_p^+ u_q^- \langle u_{p-q}^+ u_{j-p}^+ u_{k-j}^+ \rangle_c$$

$$+ L_{03}^{-1} M_p^+ \langle u_q^+ u_{p-q}^+ u_{j-p}^+ u_{k-j}^+ \rangle_c + 2 \text{ pairs of similar terms.} \tag{11.78}$$

Now solve for the triple u^+ conditional moment on the right hand side of (11.78) in terms of quadruple moments, thus:

$$\langle u_{p-q}^+ u_{j-p}^+ u_{k-j}^+ \rangle_c = 2L_{03}^{-1} M_{p-q}^+ u_i^- \langle u_{p-q-l}^+ u_{j-p}^+ u_{k-j}^+ \rangle_c$$

$$+ L_{03}^{-1} M_{p-q}^+ \langle u_l^+ u_{p-q-l}^+ u_{j-p}^+ u_{k-j}^+ \rangle_c + 2 \text{ pairs of similar terms.} \tag{11.79}$$

As required by spatial homogeneity, in the asymptotic evaluation, all wave vector arguments in each conditional moment must sum to zero. We can see that the moment on the left hand side of eqn (11.79) will be zero unless $\mathbf{k} = \mathbf{q}$, as will the quadruple moment on the right hand side. However, the triple moment on the right hand side of (11.79) will only be nonzero if $\mathbf{k} - \mathbf{q} - \mathbf{l} = 0$, which implies $\mathbf{l} = 0$, and hence

$$u_l^- \to u^-(0) = 0. \tag{11.80}$$

Thus the three triple moments on the right hand side of eqn (11.79) give zero, and hence substituting from that equation for the triple u^+ moment into (11.76) we have

$$L_{02}\langle u_j^+ u_{k-j}^+\rangle_c = 4M_j^+ u_p^-\langle u_{j-p}^+ u_{k-j}^+\rangle_c$$

$$+ 4M_j^+ L_{03}^{-1} M_p^+ L_{03}^{-1} M_{p-q}^+ u_q^- \langle u_l^+ u_{p-q-l}^+ u_{j-p}^+ u_{k-j}^+\rangle_c$$

$$+ 2M_j^+ L_{03}^{-1} M_p^+ \langle u_q^+ u_{p-q}^+ u_{j-p}^+ u_{k-j}^+\rangle_c + \text{similar terms.} \tag{11.81}$$

We can continue to iterate the above approach to obtain the higher-order terms in the (conditional) moment hierarchy. As is the case with the above triple moment, all higher-order *odd* moments of u^+ vanish, and hence again inverting the L_{02} operator in (11.81), we can write the required conditional average as

$$\langle u_j^+ u_{k-j}^+\rangle_c = \int ds\, A(\mathbf{k}, t - s) u^-(\mathbf{k}, s), \tag{11.82}$$

where $A(\mathbf{k}, t - s)$ is an expansion in even-order moments of u^+, the form of which may be inferred (for the first two orders) to be:

$$A(\mathbf{k}, t - s) = 4L_{02}^{-1} M_j^+ \langle u_{j-p}^+ u_{k-j}^+\rangle_c$$

$$+ 4L_{02}^{-1} M_j^+ L_{03}^{-1} M_p^+ L_{03}^{-1} M_{p-q}^+ \langle u_l^+ u_{p-q-l}^+ u_{j-p}^+ u_{k-j}^+\rangle_c + \text{similar terms.} \tag{11.83}$$

Substituting back, eqn (11.74) for the filtered NSE may thus be rewritten as

$$L_{01} u_k^- = M_k^- u_j^- u_{k-j}^- + M_k^- \int ds\, A(\mathbf{k}, t - s) u^-(\mathbf{k}, s). \tag{11.84}$$

11.6.1 Perturbative calculation of the conditional average

We introduce a perturbation expansion based upon the local Reynolds number at $k = k_C$. To do this, we first introduce the dimensionless variables ψ_α, $\mathbf{k}'$ and t' according to

$$u_\alpha(\mathbf{k}, t) = V(k_C)\psi_\alpha(\mathbf{k}', t'); \tag{11.85}$$

$$\mathbf{k}' = \mathbf{k}/k_C; \tag{11.86}$$

$$t' = t/\tau(k_C), \tag{11.87}$$

where $V(k_C)$ is the r.m.s. velocity at k_C and $\tau(k_C)$ is a representative timescale. Having defined these variables, we can then introduce the *local* Reynolds number

$$R(k_C) = \frac{k_C^2 V(k_C)}{\nu}, \tag{11.88}$$

a result which we can show satisfies the condition:

$$R(k_C) \leq \left(\frac{\alpha}{2\pi}\right)^{1/2} \left(\frac{k_C}{k_d}\right)^{-4/3}, \tag{11.89}$$

where α is the Kolmogorov constant. For any realistic choice of value of α this implies that $R(k_C)$ is less than unity for any k_C greater than $0.6k_d$.

11.6.2 Truncation of the moment expansion

Formally we introduce a perturbation series in ψ^+

$$\psi^+ = \psi_0^+ + R(k_C)\psi_1^+ + R^2(k_C)\psi_2^+ + \cdots \tag{11.90}$$

We find, to $\mathcal{O}(R^4)$,

$$L_0' \psi_{k'}^- - R(k_C) M_{k'}^- \psi_{j'}^- \psi_{k'-j'}^-$$

$$+ 4R^2(k_C) M_{k'}^- L_{02}'^{-1} M_{j'}^+ \psi_{p'}^- \langle \psi_{j'-p'}^+ \psi_{k'-j'}^+ \rangle_c$$

$$+ 4R^4(k_C) M_{k'}^- L_{02}'^{-1} M_{j'}^+ L_{03}'^{-1} M_{p'}^+ L_{03}'^{-1} M_{p'-q'}^+$$

$$\times \psi_{q'}^- \langle \psi_{l'}^+ \psi_{p'-q'-l'}^+ \psi_{j'-p'}^+ \psi_{k'-j'}^+ \rangle_c + \text{similar terms.} \tag{11.91}$$

To a first approximation, assuming that k_C is in the region where $R(k_C) < 1$, it is reasonable to neglect terms of order R^4 and greater as being negligible with respect to the order R^2 term. Accordingly, in order to simplify our calculation, we truncate at second order, thus:

$$L_0' \psi_{k'}^- = R(k_C) M_{k'}^- \psi_{j'}^- \psi_{k'-j'}^-$$

$$+ 4R^2(k_C) M_{k'}^- L_{02}'^{-1} M_{j'}^+ \psi_{p'}^- \langle \psi_{j'-p'}^+ \psi_{k'-j'}^+ \rangle_c + \text{similar terms.} \tag{11.92}$$

11.6.3 The RG calculation of the effective viscosity

We rewrite the low-wave number filtered NSE, from (11.84), where we have truncated the expansion in (11.83) at lowest order, as

$$\left(\frac{\partial}{\partial t} + \nu_0 k^2\right) u_k^- = M_k^- u_j^- u_{k-j}^- + \frac{4M_k^- M_j^+ \langle u_{j-p}^+ u_{k-j}^+ \rangle_c}{\nu_0 j^2 + \nu_0 |\mathbf{k} - \mathbf{j}|^2} u_p^-, \tag{11.93}$$

and the denominator in the second term corresponds to L_{02}^{-1}, with a Markovian approximation.

The conditional average may be performed, as a stochastic estimate using our hypothesis,

$$\lim_{\delta u^- \to 0} \langle u^+_{j-p} u^+_{k-j} \rangle_c = \lim_{|j-p|,|k-j| \to k_m} \langle u^+_{j-p} u^+_{k-j} \rangle + \text{higher order terms.} \qquad (11.94)$$

Since the velocity field is homogeneous, isotropic and stationary, we have the result that

$$\langle u^+_{j-p} u^+_{k-j} \rangle = Q^+(|k - j|) D_{k-j} \delta(k - p). \qquad (11.95)$$

In order to evaluate the limit involved in the conditional average we write $\langle u^+_{j-p} u^+_{k-j} \rangle_c$ in terms of a Taylor series expansion of $Q^+(|k - j|)$ about k_m.

Introducing this step, the mode-coupling term may be rewritten, as

$$\frac{4 M^-_k M^+_j \langle u^+_{j-p} u^+_{k-j} \rangle_c}{\nu_0 j^2 + \nu_0 |k - j|^2} u^-_p = -\frac{L(k, j) \left(Q^+(l)\big|_{l=k_m} + (l - k_m) \frac{\partial Q^+(l)}{\partial l}\big|_{l=k_m} \right)}{\nu_0 j^2 + \nu_0 |k - j|^2} u^-_k, \qquad (11.96)$$

where $l = |k - j|$, and $L(k, j) = -2 M^-_k M^+_j D_{k-j}$, as given by eqn (6.12).

Thus we may rewrite the explicit scales-equation as

$$\left(\frac{\partial}{\partial t} + \nu_1(k)k^2 \right) u^-_k = M^-_k u^-_j u^-_{k-j}, \qquad (11.97)$$

where

$$\nu_1(k) = \nu_0 + \delta \nu_0(k), \qquad (11.98)$$

and

$$\delta \nu_0(k) = \frac{1}{k^2} \int d^3 j \, \frac{L(k, j) \left(Q^+(l)\big|_{l=k_m} + (l - k_m) \frac{\partial Q^+(l)}{\partial l}\big|_{l=k_m} \right)}{\nu_0 j^2 + \nu_0 |k - j|^2}. \qquad (11.99)$$

11.6.4 Recursion relations for the effective viscosity

The above procedure may be extended to further wave number shells as follows:

1. Set $u^-_k = u_k$ in the filtered equation, so that we have a new NSE with effective viscosity $\nu_1(k)$ defined on the interval $0 < k < k_1$.

2. Decompose into u^- and u^+ modes at wave number k_2, such that u^+ is now defined in the band $k_2 \leq k \leq k_1$.

   Repeat the above procedure to eliminate modes in the band $k_2 \leq k \leq k_1$.

Define the nth shell in this procedure by

$$k_n = (1 - \eta)^n k_m, \quad 0 \le \eta \le 1, \tag{11.100}$$

then by induction:

$$v_{n+1}(k) = v_n(k) + \delta v_n(k); \tag{11.101}$$

and, similarly,

$$\delta v_n(k) = \frac{1}{k^2} \int d^3 j \, \frac{L(\mathbf{k}, \mathbf{j}) \left(Q^+(l)\big|_{l=k_n} + (l - k_n) \frac{\partial Q^+(l)}{\partial l}\big|_{l=k_n} \right)}{v_n(j)j^2 + v_n(|\mathbf{k} - \mathbf{j}|)|\mathbf{k} - \mathbf{j}|^2}. \tag{11.102}$$

The fixed point of the RG calculation is indicated by the scaled forms of the above equations becoming invariant under successive iterations. To obtain the scaled forms of the equations, assume that the energy spectrum in the band is of the form $E(k) = \alpha \varepsilon^r k^s$ and introduce the scaling transformation $k = k_n k'$. If we then impose the consistency requirement that $v_n(k)$ and $\delta v_n(k)$ scale in the same manner, we obtain the scaled recursion relation

$$\tilde{v}_{n+1}(k') = h^{4/3} \tilde{v}_n(hk') + h^{-4/3} \delta \tilde{v}_n(k'), \tag{11.103}$$

where

$$\delta \tilde{v}_n(k') = \frac{1}{4\pi k'^2} \int d^3 j' \, \frac{L(\mathbf{k}', \mathbf{j}')Q'}{\tilde{v}_n(hj')j'^2 \mid \tilde{v}_n(hl')l'^2}, \tag{11.104}$$

for the wave number bands $0 \le k' \le 1$ and $1 \le j', l' \le h^{-1}$ and

$$Q' = h^{11/3} - \frac{11}{3} h^{14/3}(l' - h^{-1}) + \text{higher order terms}, \tag{11.105}$$

where h is defined by the relation $k_{n+1} = hk_n$, that is $h = 1 - \eta$.

In Fig. 11.13 we plot the result of iterating eqns (11.103) and (11.104) for a fixed value of the scaled wave number. It may be seen that for three widely different starting values for the molecular viscosity the iteration reaches a fixed point in fewer than five or six cycles. As usual, the discrete values have been joined by lines in order to make the overall behavior clearer.

In Fig. 11.14 we show the evolution of the local Reynolds number for three different values of the bandwidth parameter η. Note from eqn (11.91), that the perturbation expansion is effectively in terms of $R^2(k_c)$. Thus the effective expansion parameter at the fixed point is $\sim 0.4^2 = 0.16$. Again there is a reassuring insensitivity to choices of arbitrary parameters and this is particularly encouraging in Fig. 11.15 where we show the predicted Kolmogorov spectral prefactor α as a function of bandwidth at the fixed point. It should be noted that α is the observable quantity, in this theory, as it can be measured, whereas the effective viscosity cannot.

Lastly, in Fig. 11.16 we can see the behavior of the actual renormalized viscosity as the iteration proceeds. On this scale, the molecular viscosity is unity and the renormalized viscosity builds up to about 50 or 60 times the molecular value at the fixed point. This is

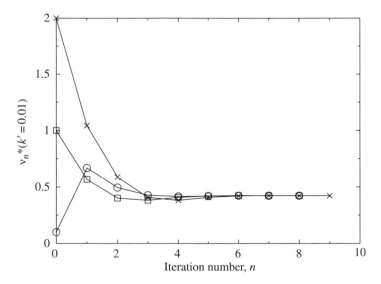

FIG. 11.13 Variation of subgrid viscosity with iteration cycle.

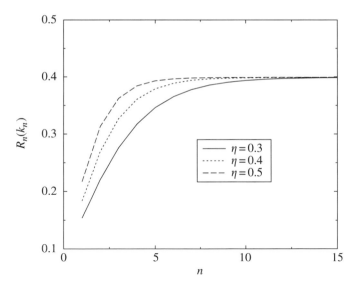

FIG. 11.14 Evolution of local Reynolds number with iteration cycle.

reasonably in agreement with what is known about *mean* turbulent viscosities, while the plateau behavior towards small values of k is in accord with what is known about the effects of scale separation. However, the behavior near the cut-off wave number is known to be incorrect, largely due to our neglect of strong coupling effects in the neighborhood of the cut-off. At this stage, this theory may claim to have some promising features, but it is still an active field of research.

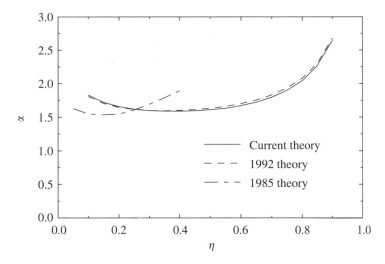

FIG. 11.15 Variation of the Kolmogorov prefactor with bandwidth.

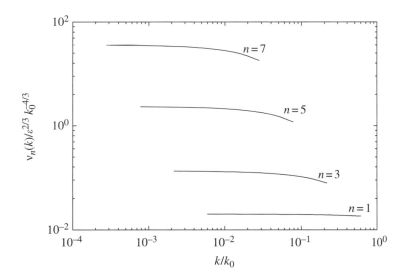

FIG. 11.16 Evolution of unscaled viscosity with iteration cycle.

Further reading

In addition to the references cited in Section 11.3.2, additional discussion of the application of RG to stirred fluid motion at small wave numbers can be found in [20] and [21], while the formal application of field-theoretic methods to turbulence is dicussed in the monograph [1]. Some background discussion of the work of Sections 11.4–11.6 can be found in [20] and [21], but for a more up-to-date treatment the interested reader is referred to [23].

PART IV

APPENDICES

APPENDIX A

STATISTICAL ENSEMBLES

A.1 Statistical specification of the N-body assembly

Formally we consider an assembly of N identical particles in a box. In general, the number N is very large. For instance, for air at STP, N is of the order of 3×10^{19} molecules in one cubic centimeter of gas. Later on we shall have to take into consideration the nature of both the molecules and the box. But, for the present, the general concept will serve our purposes, although, for sake of definiteness, we can simply imagine a macroscopic box with impermeable, insulating walls, holding a gas which may be visualized as consisting of hard spheres to represent molecules.

If we specify the state of our assembly at the macroscopic (i.e. thermodynamic) level, then we usually require only a few numbers, such as N particles in a box of volume V, at temperature T, and with total energy E or pressure P. Such a specification is known as a *macrostate*, and we write it as:

$$macrostate \equiv \{E, V, N, T, P, \ldots\}.$$

On the other hand, we may specify the state of our assembly at the microscopic level. For instance, in the classical assembly we can do this by giving the positions and velocities of all the individual molecules. Evidently this requires of the order of $6N$ numbers. The result is known as the *microstate* of the assembly. In a quantum assembly, one might specify the numbers of particles occupying each of the discrete quantum states and this also would require of the order of $6N$ numbers. However, on both classical and quantum pictures, the microstates are rapidly changing functions of time, even in isolated assemblies.

It is, of course, evident that there will be many ways in which the microscopic variables of an assembly can be arranged. This means that for any one macrostate, there will be many possible microstates. We define the

$$statistical\ weight \equiv \Omega(E, V, N, \ldots)$$

of a particular macrostate $\{E, V, N, \ldots\}$ as the number of microstates corresponding to that particular macrostate.

For the moment, we invoke a very simple quantum mechanical description of the assembly, in which each particle has access to states with energy levels

$$\epsilon_0, \epsilon_1, \epsilon_2, \ldots, \epsilon_s.$$

Then, in this description, a microstate of the assembly is the particular way in which the N particles are divided up among the given energy levels, thus:

- n_0 particles on level ϵ_0
- n_1 particles on level ϵ_1

  $\vdots$

- n_s particles on level ϵ_s

Also, the total energy of the assembly is found by adding up all the energies of the constituent particles, thus:

$$E = \sum_s n_s \epsilon_s, \tag{A.1}$$

such that the constraint:

$$\sum_s n_s = N, \tag{A.2}$$

is satisfied.

This way of expressing the energy of an assembly in terms of the number of single particles on a level is known as the *occupation number representation*. If we know the energy levels of the assembly, then we may simply express the microstate as

$$microstate \equiv \{n_0, n_1, n_2, \ldots, n_s\} \equiv \{n_s\}.$$

A.2   The basic postulates of equilibrium statistical mechanics

We now make two basic postulates about the microscopic description of the assembly.

1. First, we assume that all microstates are equally likely. Then, just as we would reason that the probability of a "head" or "tail" is $p = 1/2$ when we toss a coin, or that the probability of any particular outcome of the throw of a die is $p = 1/6$, so this leads us to the immediate conclusion that the probability of any given microstate occurring is given by

$$p(\{n_s\}) = 1/\Omega,$$

   where Ω is the total number of such equally likely microstates.

2. Second, we assume that the Boltzmann definition of the entropy, in the form

$$S = k \ln \Omega, \tag{A.3}$$

   where k is the Boltzmann constant, may be taken as being equivalent to the usual thermodynamic entropy. In particular, we shall assume that the entropy S, as defined by eqn (A.3), takes a maximum value for an isolated assembly in equilibrium.

These assumptions lead to a consistent and successful relationship between microscopic and macroscopic descriptions of matter. They may therefore be regarded as justifying themselves in practice. However, although they are the key to statistical physics, they are in the end just assumptions. We now consider the way in which we can put them to use.

A.3   Ensemble of assemblies in energy contact

Let us consider an assembly in a heat reservoir, which is held at a constant temperature, and with which it can exchange energy. Then the energy of the assembly will fluctuate randomly with time about a time-averaged value $\overline{E}$ which will correspond to the macroscopic energy of the assembly when at the temperature of the heat reservoir.

Alternatively, we may imagine a "thought experiment" in which we have a large number m of these N-particle assemblies (see Fig. A.1), each free to exchange energy with a heat reservoir. Then, at one instant of time, each assembly will be in a particular state and we can evaluate the mean value of the energy by taking the value for each of the assemblies, adding them all up, and dividing the sum by m to obtain a value $\langle E \rangle$. It should be clearly understood that while all these assemblies are identical, they are also completely independent of each other.

It is usual to refer to the assembly of assemblies as an *ensemble* and hence to call the quantity $\langle E \rangle$ the ensemble average of the energy E. Then the assumption of ergodicity, is equivalent to the assertion:

$$\overline{E} = \langle E \rangle,$$

where $\overline{E}$ is the average value of E with respect to time. It should be noted that the number m of assemblies in the ensemble is quite arbitrary (although, it is a requirement, in principle, that it should be large) and is not necessarily equal to N. In fact it is sometimes convenient to make the two numbers the same, although we shall not do that here.

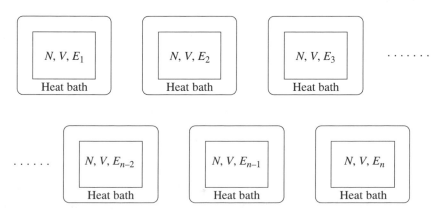

FIG. A.1   An ensemble of assemblies, each in a heat bath and each frozen at an instant of time.

A.4   Entropy of an assembly in an ensemble

Now consider a formal N-particle representation for each assembly. That is, in principle at least, we assume that any microstate is a solution of the N-body Schrödinger equation. We represent the microstate corresponding to a quantum number i by the symbolic notation $| i \rangle$, and associate with it an energy eigenvalue E_i, with the probability of the assembly being in this microstate denoted by p_i. Then, our aim is to obtain the probability distribution p_i. For the case of stationary equilibrium ensembles, we shall do this by maximizing the entropy, so our immediate task is to generalize Boltzmann's definition of the entropy, as given by eqn (A.3).

Generalizing our previous specification of an assembly, we consider the ensemble to be in a state given by:

- m_1 assemblies in state $| 1 \rangle$

- m_2 assemblies in state $| 2 \rangle$

  $\vdots$

- m_i assemblies in state $| i \rangle$

We should note that the sum of all the probabilities must be unity, corresponding to certainty, so that we have the condition

$$\sum_i p_i = 1,$$

with the summation being over all possible values of integer i.

Bearing in mind that each assembly is a macroscopic object and therefore capable of being labeled,[57] we work out the number of ways in which we can distribute the m *distinguishable* assemblies among the available states. To do this, we work out the number of ways of choosing m_1 assemblies from m, thus:

$$\binom{m}{m_1} = \frac{m!}{m_1!\,(m - m_1)!};$$

and the number of ways of choosing m_2 from the remaining $m - m_1$ is:

$$\binom{m - m_1}{m_2} = \frac{(m - m_1)!}{m_2!(m - m_1 - m_2)!};$$

and so on, giving a total number of ways:

$$\binom{m}{m_1}\binom{m - m_1}{m_2}\cdots = \frac{m!(m - m_1)!\ldots}{m_1!m_2!(m - m_1)!\ldots},$$

[57] It is a basic tenet of quantum mechanics that microscopic particles are indistinguishable.

and making the cancellations, we arrive at the statistical weight Ω_m of the ensemble state as:

$$\Omega_m = \frac{m!}{m_1! m_2! \ldots m_i!}. \qquad (A.4)$$

We now invoke the Boltzmann definition of the entropy, and apply it to the number of ways in which the ensemble can be arranged. Denoting the entropy of the ensemble by S_m, we may use eqn (A.3) to write

$$S_m = k \ln \Omega_m = k \ln m! - [k \ln m_1! + k \ln m_2! + \cdots + k \ln m_i!], \qquad (A.5)$$

where we have substituted from (A.4) for the statistical weight Ω_m.

At this stage we resort to Stirling's approximation (which is valid for very large values of m):

$$\ln m! = m \ln m - m,$$

and application of this to eqn (A.5) yields

$$S_m = km \ln m - km - \left[k \sum_i m_i \ln m_i - k \sum_i m_i \right]. \qquad (A.6)$$

Therefore, as $\sum_i m_i = m$, we may write the total entropy of the ensemble as

$$S_m - k \left[m \ln m - \sum_i m_i \ln m_i \right] = -km \sum_i p_i \ln p_i, \qquad (A.7)$$

where we have made the substitution $m_i = p_i m$. However, S_m is the total entropy of the ensemble; that is, of the m assemblies. Thus, as the entropy is, in the language of thermodynamics, an extensive quantity, it follows that the entropy of a single assembly within the ensemble is

$$S = S_m / m = -k \sum_i p_i \ln p_i \equiv -k \langle \ln p \rangle,$$

where the sum is over all possible states $| i \rangle$ of the assembly. The equivalence on the extreme right hand side of the equation is added for completeness, as it is a succinct form of the result which will be useful later. Note that the angle brackets just denote the average value.

It is worth highlighting this result by writing it as:

$$S = -k \sum_i p_i \ln p_i. \qquad (A.8)$$

This is often known as the Gibbs form of the entropy. In this form it is the basis of our ensemble theory.

Thus, to sum up the work of this section, we note that the equivalent of maximizing $\ln \Omega$ for the isolated assembly, is to maximize the entropy given by eqn (A.8) for a single assembly within the ensemble. This allows us to recast the method of the most probable distribution into a much clearer, more general, and more powerful form.

A.5 Principle of maximum entropy

Continuing to use the microscopic representation which we introduced in the preceding section, we may express mean values in terms of the probability distribution, in the usual way. For instance, the mean value of the energy of an assembly may be written as

$$\overline{E} = \langle E \rangle = \sum_{i=1}^{r} p_i E_i, \tag{A.9}$$

where p_i represents the probability of the assembly being in the state $| \, i \rangle$, such that $1 \leq i \leq r$, with energy eigenvalue E_i, and with the probability distribution normalized to one, thus:

$$\sum_{i=1}^{r} p_i = 1.$$

(It should be noted that, strictly speaking, we have now introduced a third kind of average: the expectation value. However, we shall in general treat all the methods of taking averages as being equivalent, and use the overbar or angle brackets as convenient in a given situation.) Then, by a stationary assembly, we mean one in which the mean energy, as given by eqn (A.9), is constant with respect to time. Thus, the assembly (if not isolated) fluctuates between states, with its instantaneous energy E_i varying stepwise with time. That is, E_i fluctuates randomly and rapidly about a constant mean value $\overline{E}$.

We know from thermodynamics that the entropy of an isolated system always increases. In the present case we can think of the isolated system consisting of either the combination "assembly plus reservoir" or (and this is perhaps better) the entire ensemble. It is the latter interpretation which has been taken in the proceeding section when we obtained eqn (A.8). It follows that any change in the entropy must satisfy the general condition

$$dS \geq 0,$$

so that at thermal (and statistical) equilibrium, the equality applies, and our general condition becomes

$$dS = 0, \tag{A.10}$$

corresponding to a maximum value of the entropy. The method of finding the most probable distribution now becomes the method of choosing p_i such that the entropy, as given by eqn (A.8), is a maximum. That is, if we vary the distribution by an amount δp_i from the most probable value, the corresponding change in the entropy must satisfy the equation

$$dS/dp_i = 0. \tag{A.11}$$

Thus, we find the most probable distribution by solving eqn (A.11), subject to any constraints which are applied to the assembly.

A.6 Variational method for the most probable distribution

Formally we now set up our variational procedure. Using eqn (A.8) for the entropy S, we consider the effect on S of a small change in the probability p_i. That is, we let $p_i \to p_i + \mathrm{d}p_i$ such that $S \to S + \mathrm{d}S$ and substitute these two variations into eqn (A.8) in order to calculate $\mathrm{d}S$. Then, invoking eqn (A.11) for the equilibrium condition, we set our expression for $\mathrm{d}S$ to zero to obtain

$$\mathrm{d}S = -k \sum_i \{\ln p_i + 1\}\mathrm{d}p_i = 0. \tag{A.12}$$

This equation must be solved for p_i, subject to the various constraints imposed on the assembly. In addition to the requirement (which applies to all cases) that the distribution must be normalized to unity, we shall assume for generality that the assembly is subject to the additional constraint that two associated mean values $\bar{x}$ and $\bar{y}$, say, are invariants. Evidently x and y can stand for any macroscopic variable such as energy, pressure, or particle number. Thus, we summarize our constraints in general as:

$$\sum_i p_i = 1, \tag{A.13}$$

$$\langle x \rangle = \sum_i p_i x_i = \bar{x}, \tag{A.14}$$

and

$$\langle y \rangle = \sum_i p_i y_i = \bar{y}. \tag{A.15}$$

In order to handle the constraints, we make use of Lagrange's method of undetermined multipliers. We illustrate the general approach by considering the first constraint: the normalization requirement in (A.13). If we vary the right hand side of (A.10), it is obvious that the variation of a constant gives zero, thus:

$$\mathrm{d} \sum_i p_i = 0. \tag{A.16}$$

On the other hand, if we make the variation $p_i \to p_i + \mathrm{d}p_i$ inside the summation, then we have

$$\sum_i (p_i + \mathrm{d}p_i) = 1 \Rightarrow \sum_i \mathrm{d}p_i = 0. \tag{A.17}$$

In other words, if we make changes to the individual probabilities that specific levels will be occupied, then the sum of these changes must add up to zero in order to preserve the normalization of the distribution. Clearly this statement is not affected if we multiply it through by some factor λ_0 and it follows that we are free to subtract

$$\lambda_0 d \sum_i p_i = \lambda_0 \sum_i d p_i = 0, \tag{A.18}$$

from the middle term of eqn (A.13) without effect, where λ_0 is a multiplier which is to be determined.

This procedure goes through in just the same way for our two general constraints as well. It should be borne in mind that varying the distribution of the way assemblies are distributed among the permitted states does not affect the eigenvalues associated with those states. Formally, therefore, we introduce the additional Lagrange multipliers λ_x and λ_y, so that eqn (A.13) can be rewritten in the form

$$\sum_i \{-k(\ln p_i + 1) - \lambda_0 - \lambda_x x_i - \lambda_y y_i\} d p_i = 0. \tag{A.19}$$

That is, we are just subtracting three quantities, each of which is equal to zero, from the middle term of eqn (A.13). Note that we have dropped the term dS for simplicity.

As this relation holds for abitrary states $| i \rangle$, it follows that the expression in curly brackets must vanish, thus:

$$k \ln p_i = -k - \lambda_0 - \lambda_x x_i - \lambda_y y_i.$$

Then, dividing across by the factor k and inverting the logarithm, this further implies that the required distribution must take the form:

$$p_i = \exp(-1 - \lambda_0/k) \exp(-[\lambda_x x_i + \lambda_y y_i]/k). \tag{A.20}$$

The prefactor is now chosen to ensure that the distribution satisfies the normalization requirement (A.14). Thus, summing over the index i and invoking eqn (A.14), we find

$$\exp(-1 - \lambda_0/k) = \frac{1}{\sum_i \exp(-[\lambda_x x_i + \lambda_y y_i]/k)} = \frac{1}{\mathcal{Z}}, \tag{A.21}$$

where $\mathcal{Z}$ is the partition function. Rearranging, we obtain an explicit expression for the partition function as

$$\mathcal{Z} = \sum_i \exp(-[\lambda_x x_i + \lambda_y y_i]/k). \tag{A.22}$$

Clearly this procedure is equivalent to fixing a value for the Lagrange multiplier λ_0. The other two multipliers are to be determined when we decide on a particular ensemble. At this stage, therefore, our general form of the probability of an assembly being in state $| i \rangle$ is

$$p_i = \frac{\exp(-[\lambda_x x_i + \lambda_y y_i]/k)}{\mathcal{Z}}. \tag{A.23}$$

It can be readily verified that the assembly invariants are related to their corresponding Lagrange multipliers by

$$\langle x \rangle = -k\partial \ln \mathcal{Z}/\partial \lambda_x, \tag{A.24}$$

$$\langle y \rangle = -k\partial \ln \mathcal{Z}/\partial \lambda_y. \tag{A.25}$$

In Appendix B, we apply these ideas to the so-called canonical ensemble, where the mean energy is fixed and this is the constraint.

APPENDIX B

FROM STATISTICAL MECHANICS TO THERMODYNAMICS

B.1 The canonical ensemble

We now apply the results of Appendix A to an assembly in which the mean energy is fixed but the instantaneous energy can fluctuate. Then, by considering a simple thermodynamic process and comparing the microscopic and macroscopic formulations, we show that k can be identified as the Boltzmann constant. With this step, the general equivalence of microscopic and macroscopic formulations is established.

Now, the particle number is the same for every assembly in the ensemble and is therefore constant with respect to the variational process. So, it is worth showing that a constraint of this type is essentially trivial. Suppose, in general, that y is any macroscopic variable which does not depend on the state of the assembly. Then we have for its expectation value,

$$\langle y \rangle = \sum_i y p_i = y \sum_i p_i,$$

and when we make the variation in p_i this yields the condition

$$y \sum_i \mathrm{d}p_i = 0.$$

It may be readily verified, by rederiving eqn (A.22) for this case, that λ_y can be absorbed into λ_0 when we set the normalization. In effect this means, that when y is independent of the state of the assembly, we may take $\lambda_y = 0$.

In the canonical ensemble, the only nontrivial constraint is on the energy. Accordingly, we put $x = E$ and $\lambda_y = 0$ in eqns (A.23) and (A.22) to obtain

$$p_i = \frac{\exp(-\lambda_E E_i / k)}{\mathcal{Z}}, \tag{B.1}$$

with partition function $\mathcal{Z}$ given by

$$\mathcal{Z} = \sum_i \exp[-\lambda_E E_i / k]. \tag{B.2}$$

Also, from eqn (A.24), with the replacement $x = E$, we have the mean energy of the assembly as

$$\overline{E} = -k \partial \ln \mathcal{Z} / \partial \lambda_E. \tag{B.3}$$

FIG. B.1 Compression of a gas by a piston in a cylinder.

B.1.1   Identification of the Lagrange multiplier

As an example of a simple thermodynamical process, let us consider the compression of a perfect gas by means of a piston sliding in a cylinder. We can describe the relationship between the macroscopic variables during such a process by invoking the combined first and second laws of thermodynamics, thus:

$$d\overline{E} = T\,dS - P\,dV. \tag{B.4}$$

It should be noted (see Fig. B.1) that for a compression, the volume change is negative, which makes the pressure work term positive, indicating that work is done on the gas by the movement of the piston.

Now let us use our statistical approach to derive the equivalent law from microscopic considerations. Equation (A.9) gives us our microscopic definition of the total energy of an assembly. From quantum mechanics, we know that changing the volume of the "box" must change the energy levels and also the probability of the occupation of any level, such that the change in "box size" $V \rightarrow V + dV$ implies changes in energy levels $E_i \rightarrow E_i + dE_i$ and changes in the probability of occupying a level $p_i \rightarrow p_i + dp_i$. It follows therefore from (A.9) that the change dV must give rise to a change in the mean energy, and that this is

$$d\overline{E} = \sum_i E_i\,dp_i + \sum_i p_i\,dE_i. \tag{B.5}$$

Evidently we wish to change this equation into a form in which it can be usefully compared with eqn (B.4). The second term on the right hand side can be put into the requisite form immediately, using the normal rules for partial differentiation, thus:

$$d\overline{E} = \sum_i E_i\,dp_i + \sum_i p_i \frac{\partial E_i}{\partial V}\,dV. \tag{B.6}$$

Obviously the second term now gives us a microscopic expression for the thermodynamic pressure, but we shall defer the formal comparison until we have dealt with the first term, which of course we wish to relate to $T\,dS$. We do this in a less direct way, by deriving a microscopic expression for the change in entropy dS.

Intuitively, we associate the change in entropy through eqn (A.8) with the change in the probability distribution, and this may be expressed mathematically as[58]

$$dS = \sum_i \frac{\partial S}{\partial p_i} dp_i = -k \sum_i \ln p_i dp_i,$$ (B.7)

where we have invoked eqn (A.8) for S and the normalization condition in the form $\sum_i dp_i = 0$. By substituting from (B.1) for p_i, and again using the condition $\sum_i dp_i = 0$, we may further write our expression for the change in the entropy as

$$dS = \lambda_E \sum_i E_i dp_i,$$ (B.8)

which, with a little rearrangement, allows us to rewrite the first term on the right hand side of (B.5), which then becomes

$$d\bar{E} = \frac{dS}{\lambda_E} + \left(\sum_i p_i \frac{\partial E_i}{\partial V}\right) dV.$$ (B.9)

Comparison with the thermodynamic expression for the change in mean energy, as given by eqn (B.4), then yields the Lagrange multiplier as

$$\lambda_E = 1/T,$$ (B.10)

along with an expression for the thermodynamic pressure P in terms of the microscopic description, thus

$$P = -\sum_i p_i \partial E_i/\partial V.$$ (B.11)

The latter equation can be used to introduce the instantaneous pressure P_i, such that the mean pressure takes the form

$$P = \sum_i p_i P_i,$$ (B.12)

from which it follows that the instantaneous pressure is given by

$$P_i = -\partial E_i/\partial V.$$ (B.13)

B.1.2 General thermodynamic processes

The above procedure can be generalized to any macroscopic process in which work is done such that the mean energy of the assembly remains constant. For instance, a variation in the magnetic field acting on a ferromagnet, will do work on the magnet and in the process

[58] Hint: rewrite eqn (A.8) for S in terms of a new dummy variable m, (say) and differentiate with respect to p_i.

increase its internal energy. Accordingly, we may extend the above analysis to other systems by writing the combined first and second laws as

$$d\overline{E} = TdS + \sum_{\alpha} H_{\alpha}dh_{\alpha}, \tag{B.14}$$

where H_{α} is any thermodynamic force (e.g. pressure exerted on a gas, or the magnetic field applied to a specimen of magnetic material) and h_{α} is the corresponding displacement, such as volume of the gas or magnetization of the specimen.

For example, if we take $H_1 = P$, the gas pressure, and $h_1 = V$, the system volume (but $dh_1 = -dV$), and assume that no other thermodynamic forces act on the system, then we recover eqn (B.4). Evidently the analysis which led to eqn (B.5), for the macroscopic pressure, can be used again to lead from eqn (B.8) to the more general result

$$H_{\alpha} = \sum_{i} p_i \partial E_i / \partial h_{\alpha}. \tag{B.15}$$

of which eqn (B.5) is a special case.

B.1.3   Equilibrium distribution and the bridge equation

With the identification of the Lagrange multiplier as the inverse absolute temperature, we may now write eqn (B.1) for the equilibrium probability distribution of the canonical ensemble in the explicit form

$$p_i = \frac{\exp[-E_i/kT]}{\mathcal{Z}}, \tag{B.16}$$

with the partition function $\mathcal{Z}$ given by

$$\mathcal{Z} = \sum_{i} \exp[-E_i/kT]. \tag{B.17}$$

From eqn (B.3), we may write an equally explicit form for the mean energy of the assembly as

$$\overline{E} = kT^2 \partial \ln \mathcal{Z}/\partial T. \tag{B.18}$$

In the language of quantum mechanics, this is the equilibrium distribution function for the canonical ensemble *in the energy representation*, as our quantum description of an assembly is based on the energy eigenvalues available to it. We may also introduce other thermodynamic potentials in addition to the total energy $\overline{E}$, by substituting the above form for p_i as given by eqn (B.16) into eqn (A.8) to obtain an expression for the entropy in terms of the partition function and the mean energy, thus:

$$S = k \ln \mathcal{Z} + \overline{E}/T. \tag{B.19}$$

Or, introducing the Helmholtz free energy F by means of the usual thermodynamic relation $F = \overline{E} - TS$, we may rewrite the above equation as

$$F = -kT \ln \mathcal{Z}. \tag{B.20}$$

This latter result is often referred to as the *bridge equation*, as it provides a bridge between the microscopic and macroscopic descriptions of an assembly. The basic procedure of statistical physics is essentially to obtain an expression for the partition function from purely microscopic considerations, and then to use the bridge equation to obtain the thermodynamic free energy.

We finish off the work of this section by noting the useful contraction

$$\beta = \frac{1}{kT}. \tag{B.21}$$

We shall use this abbreviation from time to time, when it is convenient to do so.

B.2 Overview and summary

In this section we collect together the results which we will need for applications. We begin with the general results for an ensemble subject to constraints, as derived in Appendix A. The general form of the probability of an assembly being in state $| i \rangle$ is

$$p_i = \frac{\exp(-[\lambda_x x_i + \lambda_y y_i]/k)}{\mathcal{Z}},$$

with the partition function given by

$$\mathcal{Z} = \sum_i \exp(-[\lambda_x x_i + \lambda_y y_i]/k).$$

The assembly invariants are related to their corresponding Lagrange multipliers by

$$\langle x \rangle = -k \partial \ln \mathcal{Z}/\partial \lambda_x,$$
$$\langle y \rangle = -k \partial \ln \mathcal{Z}/\partial \lambda_y.$$

B.2.1 The canonical ensemble

The equilibrium probability distribution of the canonical ensemble in the energy representation is

$$p_i = \frac{\exp[-E_i/kT]}{\mathcal{Z}},$$

with the partition function $\mathcal{Z}$ given by

$$\mathcal{Z} = \sum_i \exp[-E_i/kT].$$

The mean energy of the assembly is

$$\overline{E} = kT^2 \partial \ln \mathcal{Z}/\partial T,$$

and the bridge equation for the Helmholtz free energy is

$$F = -kT \ln \mathcal{Z}.$$

APPENDIX C

EXACT SOLUTIONS IN ONE AND TWO DIMENSIONS

C.1 The one-dimensional Ising model

We assume, to begin with, that the interaction strength is spatially nonuniform (i.e varies from one lattice site to another) and that there is no external magnetic field, so that eqn (2.7) for the Ising Hamiltonian may be written as:

$$H = - \sum_{i=1}^{N-1} J_i S_i S_{i+1},$$

where $S_i = \pm 1$ and J_i is the interaction energy between the spins on sites i and $i + 1$. Note that the summation adds up the contributions from the nearest-neighbor pairs and that the upper limit $i = N - 1$ correctly completes the summation with the term $S_{N-1} S_N$. Next we make the convenient substitution

$$K_i \equiv J_i / kT,$$

where K_i is called the *coupling constant*.

For a linear chain of N spins, as illustrated in Fig. 2.1 , the partition function may be written as:

$$Z_N \equiv Z_N(K_1, K_2, \ldots, K_{N-1}) = \sum_{s_1=-1}^{1} \sum_{s_2=-1}^{1} \cdots \sum_{s_N=-1}^{1} \exp \left\{ \sum_{i=1}^{N-1} K_i S_i S_{i+1} \right\}.$$

Now let us consider the effect of adding one extra spin to the chain. From the preceding equation, we have

$$Z_{N+1} = \sum_{s_1=-1}^{1} \sum_{s_2=-1}^{1} \cdots \sum_{s_N=-1}^{1} \exp \left\{ \sum_{i=1}^{N-1} K_i S_i S_{i+1} \right\} \sum_{s_{N+1}=-1}^{1} \exp\{K_N S_N S_{N+1}\}, \quad \text{(C.1)}$$

and if we carry out the last summation, this becomes

$$\sum_{s_{N+1}=-1}^{1} \exp\{K_N S_N S_{N+1}\} = \exp\{-K_N S_N\} + \exp\{K_N S_N\} = 2 \cosh(K_N S_N). \quad \text{(C.2)}$$

Now, S_N only takes values ± 1 and therefore this expression is independent of spin orientation, as cosh is an even function of its argument. Hence, putting $S_N = 1$, we have

$$Z_{N+1} = 2Z_N \cosh K_N. \tag{C.3}$$

In other words, the two partition functions are connected by a recursion relation, which we may iterate as follows: we set $N = 1$ and calculate the partition function corresponding to $N = 2$, thus:

$$Z_2 = 2Z_1 \cosh K_1; \tag{C.4}$$

and setting $N = 2$ we obtain the partition function corresponding to $N = 3$, thus:

$$Z_3 = 2Z_2 \cosh K_2 = 2^2 Z_1 \cosh K_1 \cosh K_2; \tag{C.5}$$

and so on. Then, by a process of induction, we obtain the general case

$$Z_{N+1} = Z_1 2^N (\cosh K_1 \cosh K_2 \cdots \cosh K_N). \tag{C.6}$$

Thus, in order to work out the partition function for a linear chain of $N + 1$ spins, we only need to know single-spin partition function Z_1.

Now $Z_1 = 2$ because we are dealing with a one spin system, thus there are no interactions and so:

$$\text{the sum over states} = \text{number of states} = 2.$$

Then substituting $Z_1 = 2$ into equation (C.6) yields

$$Z_N = 2^N \prod_{i=1}^{N-1} \cosh K_i. \tag{C.7}$$

We may specialize this result to the more usual uniform case, in which $K_i = K \ \forall i$, hence the final form of the partition function is

$$Z_N = 2^N \cosh^{N-1} K. \tag{C.8}$$

We conclude this theory by obtaining an expression for the pair correlation of spins. Referring back to Section 7.1.2, we define the pair correlation as

$$G_k(r) = \langle S_k S_{k+r} \rangle,$$

where r is measured in units of lattice constant. This may be evaluated using the normal expectation value for the canonical ensemble:

$$\langle S_k S_{k+r} \rangle = Z_N^{-1} \sum_S S_k S_{k+r} \exp \left\{ \sum K_i S_i S_{i+1} \right\}. \tag{C.9}$$

We can easily show that this takes the form

$$G_k(r) = \prod_{i=1}^{r} \tanh K_{k+i-1}, \tag{C.10}$$

and, again specializing to the case of uniform interaction strength,

$$G_k(r) = \langle S_k S_{k+r} \rangle = \tanh^r K. \tag{C.11}$$

C.2   Bond percolation in $d = 2$

We consider bond percolation in two dimensions on a square lattice, as discussed in Section 2.2.1. We wish to demonstrate the exact result for the critical probability, $p_c = 0.5$. To do this, we work with dual lattices.

Our first lattice is as illustrated in Fig. 2.4 and has an associated probability p for the probability of any two sites being joined by a bond. We call this the p-lattice.

Now construct a second lattice by: (1) placing sites at the center of each unit cell in the (original) p-lattice; (2) putting bonds connecting any sites which are not separated by bonds on the p-lattice.

Then we can say that the new lattice is of the same shape and size as the first lattice, but has bonds with probability

$$q = 1 - p. \tag{C.12}$$

We call this new lattice the q-lattice.

Now if the old lattice has $p > p_c$, then by definition of p_c there must be at least one continuous path (or a cluster of infinite size) crossing the lattice. If this is the case, then there cannot be such a path on the q-lattice as it would have to cross the path on the p-lattice somewhere and this is forbidden by the rules for constructing the q-lattice.

Accordingly, we deduce that:

$$\text{if } p > p_c \quad \text{then} \quad q < q_c;$$

and conversely:

$$\text{if } p < p_c \quad \text{then} \quad q > q_c.$$

Thus we conclude that when the p-lattice is at the critical point, so also is the q-lattice, and eqn (C.12) implies that

$$q_c = 1 - p_c. \tag{C.13}$$

However, as the two systems are identical, we also must have

$$q_c = p_c, \tag{C.14}$$

and reconciling these two statements leads at once to the conclusion:

$$p_c = \tfrac{1}{2}. \tag{C.15}$$

APPENDIX D

QUANTUM TREATMENT OF THE HAMILTONIAN
N-BODY ASSEMBLY

In principle, quantum mechanics gives a rigorous theory for any N-body system and in practice this is the case for microscopic systems such as many-electron atoms, where one can calculate the system wavefunction with appropriate symmetry. However, with macroscopic systems, the difficulty of describing $6N$ degrees of freedom is not evaded merely by moving from a classical to quantum description: we still have to specify an enormous number of initial conditions in order to solve the Schrödinger equation for $\sim 10^{23}$ bodies.

Also, there is the fundamental problem that it is not strictly possible to describe a macroscopic assembly in terms of a quantum mechanical stationary state. There are two reasons for this:

1. It is a consequence of the extremely large number of energy levels in a macroscopic assembly that any interaction with the outside world—however small—will be large compared with the difference between these levels. Indeed, normally the energy levels for a macroscopic system will effectively form a continuum.

2. Suppose the assembly—due to interaction with the outside world—changes energy by an amount ΔE. Then, by the uncertainty principle, the time needed to bring the macroscopic assembly back into a stationary state would be very long. The uncertainty principle gives us $\Delta t \sim \hbar / \Delta E$. For stationarity we have $\Delta E = 0$, and so $\Delta t \to \infty$. That is, on the basis of a quantum mechanical picture, the assembly will take an infinite time to return to equilibrium.

For these reasons, it is difficult to describe a macroscopic assembly in terms of the wavefunction of a pure state, due to the inherent difficulty of possessing sufficient information or of representing such information to a sufficient degree of accuracy. Instead, we resort to an incomplete description called the *density matrix* which gives mean values which are sufficiently accurate for practical purposes. However, it is perhaps worth remarking that there is no difference in principle between what we are doing now and our procedure in the classical description. In a classical treatment one uses an ensemble average to smooth out the awkward delta function structure which corresponds to a single-particle description and obtain the density distribution which corresponds to a continuum. Here we shall use an ensemble average to smooth out the instabilities in the quantum mechanical description in order to obtain the density matrix.

D.1   The density matrix ρ_{mn}

We begin by neglecting all interactions with the outside world. Then the *stationary states* of the assembly are described by the complete orthonormal set $\{\phi_n(\mathbf{q})\}$, where

$$\{\mathbf{q}\} \equiv \text{all space coordinates};$$

$$\{n\} \equiv \text{all the quantum numbers of stationary states};$$

$$E_n \equiv \text{the associated energy eigenvalues}.$$

The $\{\phi_n(\mathbf{q})\}$ are just the solutions of the Schrödinger equations for the assembly, on the assumption that all outside perturbations can be neglected.

Next, allow the system to interact with the outside world and assume the state of the assembly at some time t to be completely described by a wavefunction ψ. This may be expanded as:

$$\psi = \sum_n c_n \phi_n(\mathbf{q}) \tag{D.1}$$

where, in general, the coefficients $\{c_n\}$ depend on time.

Now, according to quantum mechanics, any variable A associated with this state has the expectation value:

$$\langle A \rangle_\psi = \int \psi^* A\, \psi \mathrm{d}\mathbf{q}, \tag{D.2}$$

where the integral is over all the relevant coordinates of the system. We should note that this is just the usual quantum mechanical average for any particular state of the assembly. The change to the incomplete description must be made by some kind of averaging over the various ψ states of the assembly. We shall do this in two steps.

First, substitute from (D.1) for ψ^* and ψ into (D.2) to obtain

$$\langle A \rangle_\psi = \sum_{n,m} c_n^* c_m \int \phi_n^*(\mathbf{q}) A \phi_m(\mathbf{q}) \mathrm{d}\mathbf{q}.$$

Second, average over fluctuations in ψ:

$$\langle\langle A \rangle_\psi\rangle = \sum_{n,m} \langle c_n^* c_m \int \phi_n^*(\mathbf{q}) A \phi_m(\mathbf{q}) \mathrm{d}\mathbf{q}\rangle. \tag{D.3}$$

Now, the left hand side of this expression may be simplified by a re-naming: we are interested in the statistical average of some observable A, and the fact that this average is made up of two different averages is of no practical importance and so we put

$$\langle\langle A\rangle_\psi\rangle \equiv \langle A\rangle.$$

On the right hand side, we note that the stationary states are not affected by the ensemble average, as it just acts on the $c_n^* c_n$. Accordingly, we may write eqn (D.3) as

$$\langle A\rangle = \sum_{n,m}\langle c_n^* c_m\rangle \int \phi_n^*(\mathbf{q}) A \phi_m(\mathbf{q})\mathrm{d}\mathbf{q}. \tag{D.4}$$

The next stage is to define matrix elements of A with respect to the basis $\{\phi_n\}$, thus:

$$A_{nm} = \int \phi_n^* A \phi_m \mathrm{d}\mathbf{q}, \tag{D.5}$$

and the density matrix ρ_{mn} as

$$\rho_{mn} = \langle c_n^* c_m\rangle. \tag{D.6}$$

Note: it is essential to pay attention to the fact that the order of indices is not the same on both sides of the above expression.

Then the expectation value of A may be written

$$\langle A\rangle = \sum_{n,m} \rho_{mn} A_{nm} = \sum_m \left(\sum_n \rho_{mn} A_{nm}\right) = \sum_m (\rho A)_m = \mathrm{tr}(\rho A),$$

where tr is short for "trace" and is the usual expression for the *sum* of the diagonal elements of a matrix. Or, highlighting our final result,

$$\langle A\rangle = \mathrm{tr}(\rho A). \tag{D.7}$$

It should be noted that the trace of an operator is independent of the representation used. Hence this result is very general and independent of our choice of quantum mechanical representation.

D.2   Properties of the density matrix

We can interpret the *density matrix* as being the matrix representation (in the usual quantum mechanical sense) of a density operator ρ, or:

$$\langle c_m^* c_n\rangle \equiv \rho_{nm} = \int \phi_n^* \rho \phi_m \mathrm{d}\mathbf{q}. \tag{D.8}$$

Evidently the matrix elements with respect to any other representation can be found by the appropriate transformations. As is usual in quantum mechanics, the diagonal elements give the probability that the assembly is in its *nth* eigenstate, thus:

$$\rho_{nn} = \langle c_n^* c_n \rangle = \langle |c_n^2| \rangle = p_n.$$

(D.9)

It then follows, from the normalization of the eigenfunctions, that

$$\mathrm{tr}\rho = \sum_n p_n = 1,$$

(D.10)

corresponding to the standard condition

$$\sum_n |c_n^2| = 1.$$

(D.11)

Now consider the density matrix in the *energy representation*, where the ψ_n are the eigenstates of the Hamiltonian H. The analog of (D.8) is

$$\rho_{nm} = \int \psi_n^* \rho \psi_m \mathrm{d}\mathbf{q},$$

(D.12)

where the Schrödinger equation for ψ_n gives

$$\rho_{nm}(t) = \rho_{nm}(0)e^{i(E_n - E_m)t/\hbar}.$$

(D.13)

Hence, only the *diagonal* elements of the density matrix are *stationary in time*. For equilibrium, the expectation values of observables—as given by (D.2)—must be independent of time. Thus, an assembly in equilibrium must be described by a density matrix which is diagonal in the energy representation.

At equilibrium, we therefore have the condition,

$$\rho_{nm} = 0, \quad \text{if } n \neq m.$$

(D.14)

D.3 Density operator for the canonical ensemble

Our earlier treatment of the canonical ensemble in Appendix B was in the energy representation. If we take the density matrix also in the energy representation, then we can equate the probability associated with the diagonal elements, as given by (D.9), with the result given by (B.16). That is,

$$\rho_{nn} = p_n = \frac{e^{-\beta E_n}}{Z}.$$

(D.15)

From (D.12) and (D.15), ρ must satisfy

$$Z^{-1}e^{-\beta E_n}\delta_{nm} = \int \psi_n^* \rho \psi_m \mathrm{d}\mathbf{q},$$

(D.16)

and so

$$\rho = \mathcal{Z}^{-1} e^{-\beta H}. \tag{D.17}$$

From eqn (D.10) we have

$$\text{tr}\,\rho = \mathcal{Z}^{-1} \text{tr} e^{-\beta H} = 1, \tag{D.18}$$

thus

$$\mathcal{Z} = \text{tr} e^{-\beta H}. \tag{D.19}$$

As an aside, we should note that, in going from (D.16) to (D.17), we expand the exponential as

$$e^{-\beta H} = 1 - \beta H + \frac{\beta^2 H^2}{2!} - \cdots,$$

where H is the Hamiltonian and in this context should be interpreted as an operator, with eigenvalues E_n. As the eigenfunctions ψ_n form a complete orthonormal set, they have the usual properties:

$$\int \psi_n^* \psi_m \mathrm{d}\mathbf{q} = \delta_{nm},$$

$$\beta \int \psi_n^* H \psi_m \mathrm{d}\mathbf{q} = \beta E_m \delta_{nm},$$

$$\beta^2 \int \psi_n^* H^2 \psi_m \mathrm{d}\mathbf{q} = \beta^2 \int \psi_m^* H E_m \psi_m \mathrm{d}\mathbf{q} = \beta^2 E_m^2 \delta_{nm},$$

and so on.

Then eqns (D.7), (D.17), and (D.19) can all be combined, to give

$$\langle A \rangle = \frac{\text{tr}(A e^{-\beta H})}{\text{tr}(e^{-\beta H})}. \tag{D.20}$$

This is the properly normalized version of the general result given in eqn (D.7) for the canonical ensemble and is one of the key equations of statistical mechanics.

APPENDIX E

GENERALIZATION OF THE BOGOLIUBOV VARIATIONAL METHOD TO A SPATIALLY VARYING MAGNETIC FIELD

Here we repeat the calculation of Section 7.5.4 with the generalization that the external magnetic field is no longer a constant but varies from one lattice site to the next and is denoted by B_i, for the ith lattice site.

It follows that the effective field $B_E \to B_E^{(i)}$ and the molecular field $B' \to B_i'$ also. Some of the key relationships in Section 7.5 also generalize straightforwardly. Equation (7.66) becomes:

$$F_0 = -\beta^{-1} \sum_i \ln \left[2 \cosh \left(\beta B_E^{(i)} \right) \right], \tag{E.1}$$

and (7.73) becomes

$$\langle S_i \rangle_0 = \tanh \left(\beta B_E^{(i)} \right). \tag{E.2}$$

Then, from (E.1) and (E.2), we may easily obtain:

$$\frac{\partial F_0}{\partial B_E^{(i)}} = \sum_i \langle S_i \rangle_0, \tag{E.3}$$

which is the required generalization of eqn (7.76).

Next we take eqn (7.68) with the equality and generalize to the nonuniform case, thus:

$$F = F_0 - \sum_{i,j} J_{ij} \langle S_i S_j \rangle_0 + \sum_i B_i' \langle S_i \rangle_0, \tag{E.4}$$

and from (7.63), with an obvious generalization to the inhomogeneous case,

$$F = F_0 - \sum_{i,j} J_{ij} \langle S_i S_j \rangle_0 + \sum_i \left(B_E^{(i)} - B_i \right) \langle S_i \rangle_0. \tag{E.5}$$

As before, on the unperturbed model, we treat the spins as independent. Reminding ourselves from (7.61) of the properties of the double sum, we may write:

$$\sum_{i,j} J_{ij} \langle S_i S_j \rangle = \frac{J}{2} \sum_i \sum_{\langle j \rangle} \langle S_i \rangle_0 \langle S_j \rangle_0, \tag{E.6}$$

where $\langle j \rangle$ denotes "sum over the nearest neighbors of each i."

Now we carry out the variation, differentiating F as given by (E.5) and (E.6), with respect to $B_E^{(i)}$, thus:

$$\frac{\partial F}{\partial B_E^{(i)}} = \frac{\partial F_0}{\partial B_E^{(i)}} - \frac{J}{2} \sum_i \sum_{\langle j \rangle} \frac{\partial \langle S_i \rangle_0}{\partial B_E^{(i)}} \langle S_j \rangle_0 + \frac{J}{2} \sum_i \sum_{\langle j \rangle} \langle S_i \rangle_0 \frac{\partial \langle S_j \rangle_0}{\partial B_E^{(i)}}$$
$$+ \left(B_E^{(i)} - B_i \right) \frac{\partial \langle S_i \rangle_0}{\partial B_E^{(i)}} + \sum_i \langle S_i \rangle_0. \tag{E.7}$$

Two points should now be noted:

1. From (E.3), we see that the first term on the right hand side cancels the last term, just as in the homogeneous case.

2. The second term involving the double sum vanishes because $\partial \langle S_j \rangle_0 / \partial B_E^{(i)} = 0$ for $j \neq i$, and j is *never* equal to i.

Then, setting $\partial F / \partial B_E^{(i)} = 0$, and equating coefficients of $\partial \langle S_i \rangle_0 / \partial B_E^{(i)}$, we obtain the condition for an extremum as:

$$B_E^{(i)} - B_i = J \sum_{\langle j \rangle} \langle S_j \rangle_0, \tag{E.8}$$

and this is the inhomogeneous version of (7.78). We may further write this condition in the useful form:

$$B_E^{(i)} - B_i = J \sum_{\langle j \rangle} \tanh \left(\beta B_E^{(j)} \right), \tag{E.9}$$

where we have substituted for $\langle S_i \rangle_0$ from (E.2).

We may conclude by writing down the optimal form of the free energy. From (E.5) and (E.6), along with (E.1) and (E.2), this is:

$$F = -\frac{1}{\beta} \sum_i \ln \left[2 \cosh \left(\beta B_E^{(i)} \right) \right] - \frac{J}{2} \sum_i \sum_{\langle j \rangle} \tanh \left(\beta B_E^{(i)} \right) \tanh \left(\beta B_E^{(j)} \right)$$
$$+ \sum_i \left(B_E^{(i)} - B_i \right) \tanh \left(\beta B_E^{(i)} \right). \tag{E.10}$$

Then substituting from (E.9) into the middle term we have

$$F = -\frac{1}{\beta} \sum_i \ln \left[2 \cosh \left(\beta B_E^{(i)} \right) \right] + \frac{1}{2} \sum_i \left(B_E^{(i)} - B_i \right) \tanh \left(\beta B_E^{(i)} \right). \tag{E.11}$$

Note that this result is *not* a simple generalization of eqn (7.82).

REFERENCES

[1]   L.Ts. Adzhemyan, N.V. Antonov, and A.N. Vasiliev. *The Field Theoretic Renormalization Group in Fully Developed Turbulence*. Gordon and Breach Science Publishers, 1999.

[2]   Daniel J. Amit. *Field Theory, the Renormalization Group, and Critical Phenomena*. World Scientific, 1984.

[3]   A-L. Barabasi and H.E. Stanley. *Fractal Concepts in Surface Growth*. Cambridge University Press, 1995.

[4]   Michel Le Bellac. *Quantum and Statistical Field Theory*. Oxford University Press, 1991.

[5]   J.J. Binney, N.J. Dowrick, A.J. Fisher, and M.E.J. Newman. *The Theory of Critical Phenomena*. Oxford University Press, 1993.

[6]   Herbert B. Callen. *Thermodynamics and an Introduction to Thermostatistics*. John Wiley and Sons, 1985.

[7]   John Cardy. *Scaling and Renormalization in Statistical Physics*. Cambridge University Press, 2000.

[8]   Francesca Colaiori and M.A. Moore. Upper Critical Dimension, Dynamic Exponent, and Scaling Functions in the Mode-Coupling Theory for the Kardar–Parisi–Zhang Equation. *Physical Review Letters*, 86(18): 3946–49, 2001.

[9]   Milton Van Dyke. *Perturbation Methods in Fluid Mechanics*. Academic Press, 1964.

[10] D. Forster, D.R. Nelson, and M.J. Stephen. Large-distance and long-time properties of a randomly stirred fluid. *Physical Review A*, 16(2): 732–49, 1977.

[11] T. Halpin-Healy and Y-C. Zhang. Kinetic roughening phenomena, stochastic growth, directed polymers and all that. *Physics Reports*, 254: 215, 1995.

[12] Kerson Huang. *Statistical Mechanics*. John Wiley and Sons, 1963.

[13] Michio Kaku. *Quantum Field Theory*. Oxford University Press, 1993.

[14] D.C. Leslie. *Developments in the Theory of Turbulence*. Clarendon Press, Oxford, 1973.

[15] S.K. Ma. *Modern Theory of Critical Phenomena*. Benjamin, Philadelphia, 1976.

[16] F. Mandl. *Statistical Physics*. John Wiley and Sons, 1988.

[17] R.D. Mattuck. *A Guide to Feynman Diagrams in the Many-body Problem*. McGraw-Hill, 1976.

[18] Joseph E. Mayer. The Statistical Mechanics of Condensing Systems. 1. *Journal of Chemical Physics*, 5: 67–73, 1937.

[19] Joseph E. Mayer. The Theory of Ionic Solutions. *The Journal of Chemical Physics*, 18(11): 1426–36, 1950.

[20] W.D. McComb. *The Physics of Fluid Turbulence*. Oxford University Press, 1990.

[21] W.D. McComb. Theory of turbulence. *Reports on Progress in Physics*, 58: 1117–206, 1995.

[22] W.D. McComb. *Dynamics and Relativity*. Oxford University Press, 1999.

[23] W.D. McComb and C. Johnston. Conditional mode elimination and scale-invariant dissipation in isotropic turbulence. *Physica A*, 292: 346–82, 2001.

[24] W.D. McComb and A.P. Quinn. Two-point, two-time closures applied to forced isotropic turbulence. *Physica A*, 317: 487, 2003.

[25] Ali Nayfeh. *Pure and Applied Mathematics*. John Wiley and Sons, Inc., New York, 1973.

[26] L.E. Reichl. *A Modern Course in Statistical Physics*. John Wiley and Sons, Inc., New York, 1998.

[27] Edwin E. Salpeter. On Mayer's Theory of Cluster Expansions. *Annals of Physics*, 5: 183–223, 1958.

[28] James T. Sandefur. *Discrete Dynamical Systems: Theory and Applications*. Oxford University Press, Oxford, 1990.

[29] H. Eugene Stanley. *Introduction to Phase Transitions and Critical Phenomena*. Oxford University Press, Oxford, 1971.

[30] Dietrich Stauffer and Amnon Aharony. *Introduction to Percolation Theory*. Taylor and Francis, London, 1994.

[31] E.T. Whittaker and G.N. Watson. *A Course of Modern Analysis*. Cambridge University Press, Cambridge, 1965.

[32] J.M. Yeomans. *Statistical Mechanics of Phase Transitions*. Oxford University Press, Oxford, 1993.

[33] Jean Zinn-Justin. *Quantum Field Theory and Critical Phenomena*. Oxford University Press, Oxford, 1999.

INDEX

4-space, 35
D-dimensional spins, 151
N-body assembly, 295
XY model, 152
ϵ-expansion, 203, 228
   Borel-summed, 258
d-dimensional lattice, 151
k-space RG, 41
n-particle bare propagator, 234
n-particle exact propagator, 235
n-point connected correlation, 161
n-point correlation, 213
ad hoc effective viscosity, 279

amputated diagram, 247
anomalous dimension, 204, 206, 217
   Gaussian model, 216
anomalous dimension of ϕ, 236
anomalous dimension of m, 236
assembly
   macrostate, 39, 49
   mean energy, 304, 309
   microstate, 39, 49
   N-particle, 49
assembly in a heat reservoir, 297
assembly invariants, 308
assembly of assemblies, 297
assembly of N identical particles, 295
asymptotic convergence, 31
asymptotic freedom, 260
asymptotic freedom as $k \to 0$, 270
asymptotic freedom at large-scale separation, 284
asymptotically convergent, 258
attractive fixed point, 48, 190

backscatter, 279
band-filtered spectral density, 284
bare n-point propagator, 238
bare charge, 11
bare coupling constant, 235
bare mass, 230
bare mass m_0, 235
bare propagator, 214, 230, 241
bare quantity, 20
beta-function, 223
   ϕ^4 theory in dimension, 236
binomial expansion, 87
block spins, 40, 183

block variables, 40, 159
blocking the lattice, 147
Bogoliubov variational principle, 162
Bogoliubov inequality
   proof, 163
Bogoliubov mean-field theory, 203
Bogoliubov theorem, 162, 163
book-keeping parameter, 33, 96
Borel summable, 258
bridge equation, 13, 50, 54, 58, 159, 163, 183, 261,
   308, 309
Brownian motion, 110
Burgers equation, 274

Callan–Symanzik beta-function, 223, 236
Callan–Symanzik equation, 236
Callan–Symanzik equations, 234
canonical (scaling) dimension, 217
canonical dimension, 205
canonical ensemble, 304, 311
   equilibrium distribution, 307
chaotic nature of turbulence, 280
charge renormalization, 78
classical assembly, 295
classical stochastic fields, 110
cluster integrals, 99
coarse-graining, 40, 158
coarse-graining operation, 188
coarse-graining transformation, 199
collective field, 164
combinational divergence, 240
compensated spectrum, 142
completing the square, 28
computer experiment, 132
conditional average, 260, 282, 285, 288
   as an approximation, 280
   at large wave numbers, 280
   need for, 264
   perturbative calculation, 286
   quasi-stochastic estimate, 283
conditional moment hierarchy, 286
conditionally sampled ensemble, 281
configurational integral, 95
configurational energy, 54
configurational integral, 98
configurational partition function, 95
connected correlation, 247
connected correlation function in Fourier space, 174
connected correlations, 149, 161, 246